# Polymer Thermodynamics

## Blends, Copolymers and Reversible Polymerization

# Polymer Thermodynamics

## Blends, Copolymers and Reversible Polymerization

Kal Renganathan Sharma Ph.D. PE

**CRC Press**
Taylor & Francis Group
Boca Raton   London   New York

CRC Press is an imprint of the
Taylor & Francis Group, an **informa** business

CRC Press
Taylor & Francis Group
6000 Broken Sound Parkway NW, Suite 300
Boca Raton, FL 33487-2742

First issued in paperback 2017

© 2012 by Taylor & Francis Group, LLC
CRC Press is an imprint of Taylor & Francis Group, an Informa business

No claim to original U.S. Government works

ISBN 13: 978-1-138-11349-7 (pbk)
ISBN 13: 978-1-4398-2639-3 (hbk)

---

**Library of Congress Cataloging-in-Publication Data**

---

Sharma, Kal Renganathan.
    Polymer thermodynamics : blends, copolymers and reversible polymerization / Kal Renganathan Sharma.
        p. cm.
    Includes bibliographical references and index.
    ISBN 978-1-4398-2639-3 (alk. paper)
    1. Polymers. 2. Copolymers. 3. Thermodynamics. I. Title.

QC173.4.P65S53 2011
668.9--dc22                                                                  2010012368

---

**Visit the Taylor & Francis Web site at**
**http://www.taylorandfrancis.com**

**and the CRC Press Web site at**
**http://www.crcpress.com**

*Dedicated to my son, R. Hari Subrahmanyan Sharma (alias Ramkishan, born August 13, 2001), with unconditional love.*

# Contents

# Foreword

Predictions indicate that over one million scientists, engineers, technologists, and associated professional people will work in the nanoscience- and nanotechnology-related industries over the next 10–15 years, and the market for nanotechnology products will reach some $3 trillion by the year 2015. Engineers trained in nanotechnology will attract higher salaries compared with even computer hardware and chemical engineers. Indeed, nanoscale science and technology promise major advances in almost every area of our socio-economic environment.

We can, for instance, expect

The development of hybrid electric vehicles and highly energy-efficient transport systems that can lower the cost of commuter traffic

New photovoltaics that can collect solar power efficiently and cost effectively in deserts regions

The creation of materials with much higher thermal conductivity than copper and aluminum

The synthesis of supremely tough nanocomposite materials

A paradigm shift in the effectiveness of our medical strategies

Major advances in the power of nanomagnetic materials

The development of molecular computers—and much more

The microprocessor and personal computer revolution was largely a consequence of our success in miniaturization, and up until now, computing speed has doubled every 18 months as more and more transistor elements have been packed on a silicon chip. Genetic technology is now showing analogous advances, as the efficiency with which we are able to handle DNA increases dramatically. At the nanoscale, quantum mechanical effects occur that, if they can be harnessed effectively, promise novel and powerful new applications which are quite different from those we are familiar with at the macroscopic level.

The all-carbon hollow cage molecules, the fullerenes and their elongated cousins, the carbon nanotubes (CNT) are stable allotropes that, in addition to graphene, graphite, and diamonds, show fascinating promise as basic materials for novel nanoscale applications. The morphology of materials is a fascinating field and structure-related properties are of key interest in product development and process engineering, resulting in materials with advanced performance in sustainable, environmentally friendly applications.

If all these exciting advances are to be realized, then the next cohort of young scientists, engineers, and technologists must have a sound education in nanoscale science and technology, and this education needs to be integrated into undergraduate curricula in chemical engineering. This text by Dr. Kal Renganathan Sharma, PE, is a welcome and highly effective response to this challenge that must be met if we are to develop the sustainable technologies we shall certainly need to survive into the next century.

**Harold Kroto**
*Department of Chemistry and Biochemistry*
*Florida State University*
*Tallahassee, Florida*

# Preface

This book is a natural outgrowth from the author's industrial research and development in polymer systems with Monsanto Plastics Technology, Indian Orchard, Massachusetts and teaching guest lectures on applied polymer thermodynamics in the advanced chemical engineering thermodynamics at his alma mater West Virginia University, Morgantown, West Virginia. Based on his first year's performance he became a research fellow at Monsanto Co. in 1991. Dr. Sharma has authored 11 books, 526 conference papers, 21 journal articles and 1128 other presentations. He has instructed over 2554 students in 89 semester-long courses. Among other subjects he teaches thermodynamics-I and materials science, introduction to polymer engineering, at Prairie View A & M University, Prairie View, Texas. At PVAMU he serves as Adjunct Professor in chemical engineering department.

According to the Nobel laureate P. Flory finding thermodynamically miscible polymers is difficult to do unless they have specific interactions with each other. Krause found *282* chemically dissimilar polymer pairs that appeared to be miscible; 75% of these were miscible because of specific interactions; 15% of the miscible pairs were due to clever choices of copolymer composition. Since Flory's work, a number of miscible polymer blends have been found. A patent search today on polymer blends returns 63,105 entries. In this work, compatible commercial systems and thermodynamically miscible systems are distinguished from each other.

This book describes in detail the thermodynamic basis for miscibility and the mathematical models that can be used to predict the compositional window of miscibility and construct temperature-phase volume-phase diagrams. The binary interaction model, the solubility parameter approach, and entropic difference models are discussed in detail. The construction of phase envelopes including circular shape, hyperbola, hourglass type, LCST, UCST, merged LCST/UCST using the equation of state theories and thermodynamic models are discussed in detail. Miscible and compatibilized blends are distinguished from each other. The patent literature has a tsunami of patents that preach polymer blends for practical applications. Current textbooks on thermodynamics do not discuss the applications of free energy, enthalpy, and entropy to predict phase behavior of binary polymer blends. Compositional windows of miscibility for PMMA/SAN, PMMA/AMS–AN, TMPC/SAN, TMPC/AMS–AN, PCL/SAN, PVC/SAN, PVC/AMS–AN, PPO/SAN, and PPO/AMS–AN using the binary interaction model are developed and shown as readily usable charts at different volume fractions of the blend and different copolymer compositions.

Nine different equations-of-state, EOS theories are described including Flory Orwoll & Vrij (FOV) Prigogine Square Well cell model, and the Sanchez & Lacombe free volume theory. When the mathematical complexity of the EOS theories increases it is prudent to watch for spurious results such as negative pressure and negative volume expansivity. Although mathematically correct these have little physical meaning in polymer science. The large molecule effects are explicitly accounted for by the lattice fluid EOS theories. The current textbooks on thermodynamics discuss small molecules in detail but say little about polymer thermodynamics. Enough information about 16 polymers is tabulated that can be used by practioners to design product.

The thermodynamic basis to explain miscibility in polymer blends is an exothermic heat of mixing as entropic contributions are small for such systems. Intramolecular repulsions may be an important factor in realizing exothermic heat of mixing. The application of binary interaction model to predict a compositional window of miscibility for copolymer/homopolymer blends, terpolymer blends with common monomers, copolymer blends with common monomers, is illustrated. A 6 $\chi$ parameter expression for free energy of mixing for two copolymers with four monomers is described. The spinodal is derived from stability and miscibility considerations. The physical meaning behind concave and convex curvature of phase envelopes is described.

Stronger Keesom forces lead to hydrogen bonding between two different polymers leading to miscible polymer–polymer systems. The Hildebrandt solubility parameter can be related to the Flory–Huggins interaction parameter by use of regular solution theory. The Hildebrandt solubility parameter can be calculated from one of the nine EOS theories discussed. Hansen fractional solubility parameter values that split the solubility parameter into dispersive, polar and hydrogen bonding components are presented. FTIR spectroscopy can be used to measure hydrogen bonding in polymer blends. Hydrogen-bonded polymer blends may exhibit LCSTs and closed-loop two-phase regions in the phase diagram of binary blend. A phase computer program was developed for simulating compositional window of miscibility.

Phase diagrams of polymer–polymer blend systems and polymer–solvent mixtures can be constructed in the form of temperature as a function of composition. Two important factors, i.e., miscibility and stability, are considered and six different phase diagram types are recognized. The role of polymer architecture in phase behavior is touched upon.

A mathematical framework is provided to account for the multiple glass transition temperatures found in partially miscible polymer blends. The change in entropy of mixing at the glass transition temperature of the blend can be used to account for the observations. A quadratic expression for the mixed glass transition temperature is developed from the analysis. The chain sequence distribution information can be used in estimating the entropy of the copolymer.

Polymer nanocomposites pertain to the synthesis, characterization, and applications of polymer materials with at least one or more dimensions less than 100 nm. According to the Rayleigh criterion the maximum resolution or minimum detectable size achievable was half the wave length of light, i.e., 200 nm. Thermodynamic equilibrium and stability are important considerations in the formation of nanocomposites. LCST and UCST behavior can be seen in polymeric systems. The four thermodynamically stable forms of carbon are diamond, graphite, $C_{60}$ Buckminister fullerenes and CNT carbon nanotubes. Immiscible nanocomposites, intercalated nanocomposites and exfoliated nanocomposites offer different morphologies. There are some interesting biomedical applications of nanocomposites. Electrospinning for producing a bioresorbable nanofiber scaffold is an active pursuit of scientists for tissue engineering applications. Hydroxyapetite-based polymer nanocomposites are used in bone repair and implants. Some commercial applications for polymer nanocomposites include, for example, Toyota's nylon/clay nanocomposite timing belt cover.

Polymer alloy is a commercial polymer blend with improvement in property balance with the use of compatibilizer(s). PVC/SAN blends were found to be miscible and not compatible at certain AN compositions. Miscible and compatible PVC/SAN blends with better properties can be prepared with a different AN composition. PC and LDPE can be blended with each other with EPDM as the compatibilizer. LDPP and PC can be blended together and ABS used as the modifier for the alloy. Polymer alloys using natural polymers can be prepared. Lignin and protein can form interesting blends.

The Flory–Huggins interaction parameter, $\chi$, was related to the binary diffusion coefficient. Fick's laws and generalized Fick's laws of diffusion and their implications during transient events are discussed in detail. The copolymerization composition equations are derived for copolymers and terpolymers in CSTR or at small conversions. For the general case with n monomers the rate equation in the vector form after applying the QSSA, quasi-steady state approximation for free radicals re derived in the vector form. When the eigenvalues of the rate equation become imaginary the monomer concentration can be expected to undergo subcritical damped oscillations as a function of time.

The dyad probabilities for copolymer were calculated as a function of reactivity ratios and monomer composition. A first-order Markov model was developed to predict the chain sequence distribution of SAN and AMS–AN copolymers. The six triad concentrations for SAN copolymer were calculated. Nin dyad and 27 triads for random terpolymers were calculated and tabulated in Tables 11.3, 11.5–11.7.

A Markov model of the third order was developed to repeat DNA sequence in *Homo sapiens* with 660 bases; 256 possible triad probabilities were calculated. A dynamic programming method can be used to align two sequences. Global and local alignment can be sought. PAM and BLOSUM matrices with affine gap penalty sequences were discussed. The time taken using dynamic programming methods is $O(n^2)$. Greedy algorithms can be used to obtain alignment in $O(en)$ for certain sequences with fewer errors. i.e., $e \ll n$. An X drop alignment for global alignment was touched upon.

Thermodynamics of reversible homopolymerization and copolymerization are revisited. The equilibrium rate constant is expressed as a function of free energy. Methods to calculate the ceiling and floor temperature are outlined.

The Clapeyron equation is derived for polymerization reactions from the combined first and second laws of thermodynamics. It relates the rate of change of equilibrium pressure and equilibrium temperature with the enthalpy change of reaction, volume change of reaction and temperature of the reaction. The effect of pressure on the ceiling temperature was illustrated using a worked example. Thermal polymerization kinetics that obey the Dields–Alder mechanism were studied. Expressions for the rate of propagation and molecular weight were derived. The rate of propagation varies as cube of monomer concentration. Subcritical damped oscillations can be expected under certain conditions. Reaction in circle representation of free radical reactions was used to analyze the stability of reactions. Oscillations in concentration were discussed for system of three and four reactions in the circle and the results generalized to n reactions in circle.

The ceiling temperature constraint in the homopolymerization of alphamethyl styrene (AMS) can be circumvented by copolymerization with acrylonitrile (AN) to prepare multicomponent random microstructures that offer higher heat resistance than SAN. The feasibility of a thermal initiation of free radical chain polymerization is evaluated by an experimental study of the terpolymerization kinetics of AMS–AN–Sty. Process considerations such as polyrates, molecular weight of polymer formed, sensitivity of molecular weight, molecular weight distribution, and kinetics to temperature were measured.

A mathematical model for the heat of copolymerization as a function of chain sequence distribution for random copolymers was developed. Four different types of enthalpy of copolymerization, $\Delta H_p$ vs. $X_1$, conversion can be expected when $\Delta H_{11}$ and $\Delta H_{12}$ are (i) both negative, (ii) one negative and the other positive, (iii) one zero and the other positive, and (iv) both positive.

This is a three-part book with the first part devoted to polymer blends, the second to copolymers and glass transition temperature and to reversible polymerization. Separate chapters are devoted to blends: Chapter 1, Introduction to Polymer Blends; Chapter 2, Equations of State Theories for polymers; Chapter 3, Binary Interaction Model; Chapter 4, Keesome Forces and Group Solubility Parameter Approach; Chapter 5, Phase Behavior; Chapter 6, Partially Miscible Blends. The second group of chapters discusses copolymers: Chapter 7, Polymer Nanocomposites; Chapter 8, Polymer Alloys; Chapter 9, Binary Diffusion in Polymer Blends; Chapter 10, Copolymer Composition; Chapter 11, Sequence Distribution of Copolymers; Chapter 12, Reversible Polymerization.

Appendices are provided that include Maxwell's relations, the five laws of thermodynamics, and group contributions to calculate solubility parameters. Experimental methods to measure the binary interaction parameters, solubility parameters and heat capacity changes at glass transition are discussed.

# About the Author

**Kal Renganathan Sharma**, Ph.D., PE, received his three degrees in chemical engineering: a B.Tech. from the Indian Institute of Technology, Chennai, India, in 1985, and his M.S and Ph.D. degrees from West Virginia University, Morgantown, in 1987 and 1990, respectively. He currently serves as adjunct professor with the Department of Chemical Engineering in the Roy G. Perry College of Engineering, Prairie View A&M University, Prairie View, Texas. He is the author of ten books, 19 journal articles, 503 conference papers, and 108 other presentations. He has instructed over 2450 students in 83 semester-long courses. He has held a number of high level positions in engineering academia in India and the United States, and is listed in *Who's Who in America*. He has been commended by the International Biographical Center at Cambridge, UK.

At the time of this writing, Dr. Sharma's earlier book *Bioinformatics: Sequence Alignment and Markov Models* (McGraw Hill Professional, New York, 2009) had been cataloged in over 440 libraries including the Library of Congress, British Library, Cambridge University, MIT, University of California–Berkeley, and Stanford University. He was inducted as fellow of the Indian Chemical Society, Kolkata, and is a member of the New York Academy of Sciences, Phi Kappa Phi, and Sigma Xi. He has won cash awards from Monsanto Plastics Technology, Indian Orchard, Massachusetts and SASTRA University, Thanjavur, India.

# 1 Introduction to Polymer Blends

## LEARNING OBJECTIVES

- Compatible and miscible blends and alloys
- Morphology
- History of polymer blends
- Flory's statement on the difficulty of finding miscible blends
- Miscible region
- Free energy, spinodal, and binodal
- Partially miscible blends—$\Delta\Delta S_m$
- Natural polymers

The emphasis in research and technological development of new products in the polymer industry has changed from synthesis of novel monomers to blending polymers in order to obtain desirable performance properties at a dial-in cost. Polymer blends offer improved performance/cost ratios and flexibility to tailor-make the product to suit customer needs. Polymer blends are intimate mixtures of two or more polymers. The individual components may be melt-mixed, solution blended, co-precipitated, controlled devolatilized, or coagulated before final processing. Finished articles are converted from blends by extrusion or molding processing operations.

Polymer blends are a way to obtain properties in thermoplastic materials not easily achievable in a single polymer. All important performance properties can be improved by blend systems. Notable among the properties are flow, mechanical strength, thermal stability, and cost. There is, therefore, interest in describing theories that can be used to predict polymer blend system properties, given the constituent polymer properties. Often, the structure of the constituent polymer plays a vital role in the performance properties of the polymer blend system.

On a microscopic scale, the blends may be homogeneous or heterogeneous but should not have any inhomogeneity on a macroscopic scale. Some polymer blends are as follows:

*Compatible blend*—a polymer blend with improvement in property balance compared with the blend components.

*Incompatible blend*—a polymer blend with a lack of improvement in property balance compared with the blend components.

*Polymer alloys*—commercial polymer blends with improvements in property balance through the use of *compatibilizers*. They exhibit an interface.

1

*Miscible blend*—a polymer blend that is homogenous at a microscopic scale (i.e., achieving a state of equilibrium at a molecular level).

*Immiscible blend*—a polymer blend that is at a phase-separated state of mixing at a molecular level with the composition of the separated phases pure or identical to the pure components prior to blending.

*Partially miscible blend*—a polymer blend that is at a phase-separated state of mixing at a molecular level with the composition of the separated phases impure or not identical to the pure components prior to blending.

*Interpenetrating polymer network* (IPN)—a combination of two polymers in network form, at least one of which is synthesized or cross-linked in the immediate presence of the other.

The *morphologies* of the miscible, immiscible, and partially miscible polymer blends are distinct from each other. In an immiscible blend, two phases are present: the discrete phase (domain), which is lower in concentration, and the continuous phase, which is higher in concentration. The miscible polymer blends exhibit single-phase morphology. Partially miscible polymer blends may form completely miscible blends at a different composition. The two phases may not have a well-defined boundary. Each component of the blend penetrates the other phase at a molecular level. The molecular mixing that occurs at the interface of a partially miscible two-phase blend can stabilize the domains and improve the interfacial adhesion. A compatible blend that has commercial possibilities may be immiscible, and a miscible polymer blend may lack commercial applications due to other factors such as cost, source of raw materials, safety, and environmental issues such as recyclability.

## 1.1   HISTORY OF POLYMER BLENDS

By 1992, the worldwide production of plastics was twice that of steel. It is expected that manufacture of polymers will grow by a factor of 10 prior to market saturation. This all started with natural polymers, one of which, shellac, was used as a hot varnish in India in 2000 BC. Christopher Columbus brought along another product, rubber, on his second voyage. Gutta percha was introduced in the second half of the seventeenth century. The first synthetic polymer made was nitrocellulose in 1833. Then in 1851, N. Goodyear patented ebonite, a thermosetting material created by heating natural rubber with sulfur.

The first polymer blend patent appeared in 1846 in Birmingham, Alabama. A. Parkes blended gutta percha with natural rubber. Soon after the invention of nitrocellulose, blends of nitrocellulose with natural rubber were obtained. The first blend of two synthetic polymers came about in 1928 when polyvinyl chloride was mixed with polyvinyl acetate. By 1993, the polymer blend patent literature reached about 3000 patents a year [1].

Blending and mixing are important unit operations in the polymer blend industry. Hancock patented the first internal mixer in 1923. Freyburger in 1876 improved the device to a more efficient counter-rotating twin-shaft machine. The Banbury mixer is a modification of this patent. The Farrel Co. introduced two-roll mills, then the Ram extruder surfaced in 1797. Screw machines that were capable of manufacturing macaroni originated in the 18th century. *Screw extruders* came about in the late nineteenth century in Great Britain. Electric extruders were developed in the early part of the twentieth century. The World War II era saw the development of myriad polymer processing devices such as crosshead dies, co-extruders, filament extruders, film blowers, breaker plates, and screen packs. Venting and two-stage screws in extruders were developed in the 1950s. It would be 30 years before data acquisition and the computer control of machines began in the 1980s.

Baker [2] patented a twin-screw co-rotating device now used for extrusion and other polymer processing operations. Counter-rotating machines came about in 1895 [3].

Polymer blends as products have been the result of billions of dollars of research and technology development expenditures. By 1982, the per annum worldwide sales of polyphenylene/polystyrene exceeded US $1 billion. Polyvinyl chloride/acrylonitrile butadiene styrene (ABS) blends have captured the markets worldwide. Compatibilized nylon/ABS blends appeared on the market and were sold under the name Triax 1000 by Monsanto with step-change improvement in product performance properties.

Blends often offer better processability by reduction in viscosity. Polymer blends are also inherently recyclable.

Lyondell Basell Polymers sells compounded polyolefins. The base polyolefin is blended with impact modifiers, fillers and reinforcements, pigments, and additives. Customer applications include automotive components, electrical appliances, wire and cable, pipe and sheet, building and construction, and furniture. The properties of the product include a broad range of stiffness from very soft to very rigid, excellent aesthetic characteristics and colorability, good processability, high stiffness/impact balance, high mar and scratch resistance, low carbon emissions achievable, and excellent recyclability.

Polypropylene was blended with ethylene propylene diene elastomer (EPDM) to form dynamically vulcanized, tough composition-dependent low-temperature modulus and impact strength by companies such as Monsanto, Novacor, Mitsui Petrochem. Polyvinyl chloride (PVC) and poly(methyl methacrylate) (PVC/PMMA) blends were prepared to make products that are thermoformable, are flexible sheets with high-impact strength, have excellent outdoor performance, good flame, and are chemical and solvent-resistant such as those by Rohm and Haas, Nippon, Mitsubishi Rayon. B. F. Goodrich, Polysar, Showa Denka prepared PVC/nitrile rubber (NBR) blends with nitrile butadiene rubber to offer good processability; fast calendaring and extrusion; and impact and tear strength; oil, fuel, chemical, abrasion, weathering and ozone resistance; antistatic, flame; and moisture resistance. ABS/PVC blends were commercialized by General Electric and offer good processability, high impact strength, flame retardancy. Polyamide/polypropylene blends were introduced by DSM, Atochem, and Mitsubishi and offer low water absorption, dimensional stability, low density, low liquid and vapor permeability, moderate impact strength, and good resistance to alcohols and glycols.

Compatibilized polyamide/ABS blends were sold by Monsanto, BASF, Sumitomo Dow, and GE with high heat and chemical resistance, good flow, and low-temperature impact strength, moisture sensitivity and cost. Polycarbonate, PC/ABS blends were commercialized by GE Plastics, Bayer, Dow, and Monsanto. These products have good processability, toughness, high temperature and delamination resistance, and low temperature impact strength. PC/polystyrene (PS) blends, with good weatherability and thermal stability, were offered by Mitsubishi Chemical. Monsanto and BASF have prepared polybutylene terephthalate (PBT)/ABS blends with high gloss, stiffness, mechanical strength, high heat resistance, and chemical resistance.

PBT/PC blends were commercialized by Dow, Mobay, Bayer, BASF, and GE Plastics with high modulus, heat resistance, tensile and impact strength, solvent resistance, and UV stability. Mobay and Dow commercialized polyethylene terephthalate (PET)/PC blends with solvent resistance, low-temperature impact, and tensile strength. PBT/PET blends with good surface appearance were introduced by GE Plastics.

Polyaramide and Par/PET blends were sold by Amoco with high heat resistance and impact strength. Celanese introduced the polyoxymethylene (POM)/PBT and POM/thermoplastic urethane (TPU) blends. Polyphenylene ether polyamide blends (PPE)/PA blends were prepared by GE Plastics with good solvent and heat resistance.

Some of the important events in the history of polymer blend technology development are in Table 1.1.

## 1.2   FLORY–HUGGIN'S SOLUTION THEORY—AND BEYOND

Paul J. Flory, Nobel laureate and a pioneer in the field of polymer chemistry, wrote:

"...the critical value of the interaction energy is so small for any pair of polymers of high molecular weight that it is permissible to state as a principle of broad generality

**TABLE 1.1**
**Chronology of Polymer Blend Technology**
**Development 1846–1990**

| | |
|---|---|
| 1846 | Nitrile rubber/gutta purcha |
| 1911 | Solution-grafted block copolymer of styrene butadiene |
| 1925 | Solution polymerization of high impact polystyrene (HIPS) |
| 1936 | PVC/PMMA blend |
| 1941 | Neoprene |
| 1946 | ABS (mechanical) |
| 1950 | ABS (grafted) |
| 1951 | PVC/ABS blend |
| 1959 | PA/PVC |
| 1963 | PA/PMMA |
| 1970 | PPE/HIPS |
| 1975 | PA/EPDM |
| 1980 | PMMA/styrene–acrylonitrile |
| 1985 | PC/PVCF |
| 1990 | Polyphenylene sulfide/PBT |

that two high polymers are mutually compatible with one another only if their free energy of interaction is favorable, i.e., negative. Since the mixing of a pair of polymers like the mixing of simple liquids, in the great majority of cases is endothermic, incompatibility of chemically dissimilar polymers is observed to be the rule and compatibility is the exception. The principal exceptions occur among pairs possessing polar substituents which interact favorably with one another" [4].

$$\Delta G_m = RT \left( n_1 \ln \phi_1 + n_2 \ln \phi_2 + n_1 \phi_2 \chi_{12} \right) \tag{1.1}$$

So, according to Flory, finding thermodynamically miscible polymers is difficult to do unless they have specific interactions with each other due to chemical interactions. Very few miscible polymer blends of nonpolar or weakly polar homopolymers were identified during a 30-year-period following the publication of the book by Flory [4]. During the 1980s, however, a steadily increasing number of miscible systems have been reported. Some compatible polymer blends form a hydrogen bond between them.

Krause [5] reviewed the miscible polymer pairs reported in the literature. She found that there were 282 chemically dissimilar polymer pairs that appeared to be miscible in the amorphous state at room temperature. She found that 75% of these were miscible on account of specific interactions such as hydrogen bonding; 15% of the miscible pairs were due to clever choice of copolymer composition. The compositional window over which copolymers are miscible with each other can be calculated from the Flory–Huggins theory.

In later chapters, we shall see that the free energy of the polymer blend can be written as a sum of enthalpic and entropic contributions. Although the entropic contribution for high macromolecules would be low, the enthalpic contribution can be negative, leading to miscible regions. The intramolecular repulsion as a contributor in a negative enthalpy of mixing will be discussed in detail under the binary interaction model. The solubility parameter approach can also be used to calculate an interaction parameter. Matching solubility parameter values predict miscibility. Equation of state theories can be used to construct the phase diagrams for systems that do exhibit miscibility under some conditions. The lower consolute solution temperature (LCST) and upper consolute solution temperature (LCST and UCST) denote the phase boundaries of miscible and immiscible regions.

Equation (1.1) is rewritten as

$$\Delta G_m = \Delta H_m - T\Delta S_m, \tag{1.2}$$

where $G$ is the Gibbs free energy, $H$ the enthalpy, and $S$ the entropy of the system.

The free energy, $G$, of a system is the amount of energy that can be converted to work at constant temperature and pressure. It is named after the thermodynamicist Gibbs. Helmholtz free energy, $H$, of a system, is the amount of energy that can be converted to work at constant temperature. Enthalpy was first introduced by Clapeyron and Clausius in 1827 and represented the useful work done by a system. Entropy of a system, S, represents the unavailability of the system energy to do work.

It is a measure of randomness of the molecules in the system, and is central to the quantitative description of the second law of thermodynamics. Internal energy, U, is the sum of the kinetic energy, potential energy, and vibrational energy of all the molecules in the system.

The binary interaction model would be later shown to estimate an exothermic enthalpy of mixing for miscible blends. For stability at constant temperature and pressure:

$$\frac{\partial^2 G}{\partial \phi^2} \geq 0. \tag{1.3}$$

The *spinodal curve* can be constructed from the equation:

$$\frac{\partial^2 G}{\partial \phi^2} = 0. \tag{1.4}$$

The *binodal curve* can be constructed from the equation:

$$\frac{\partial^2 G}{\partial \phi^2} > 0. \tag{1.5}$$

The region between the binodal and spinodal curve is the metastable region. The phase separated and miscible regions are demarcated by the binodal curves (Figure 1.1).

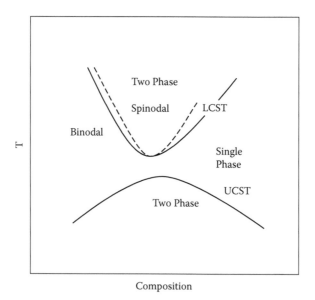

Composition

**FIGURE 1.1**   Polymer polymer phase diagram with LCST, UCST, and spinodal curve.

## 1.3  MISCIBLE POLYMER BLENDS

Miscibility in polymer blends has been studied by both theoreticians and experimentalists. The number of polymer blend systems that have been found to be thermodynamically miscible has increased in the past 20 years. Systems have also been found to exhibit the upper or lower critical solution temperatures. So complete miscibility is found only in limited temperature and composition ranges. A large number of polymer pairs form two-phase blends. This is consistent with the small entropy of mixing that can be expected of high polymers. These blends are characterized by opacity, distinct glass transition temperatures, and deteriorated mechanical properties. Some two-phase blends have been made into composites with improved mechanical properties. Often, incompatibility is the general rule, and miscibility or even partial miscibility is the exception.

A commonly used method for establishing miscibility in polymer–polymer blends is through measurement of the glass transition temperature of the blend, compared with the glass transition temperature of the constituent values. A miscible polymer blend is expected to exhibit a single glass transition temperature. Partial miscibility is indicated by two glass transitions different from the homopolymer constituent values. In cases where specific interactions occur in the blend system, the glass transition may go through a maximum with composition.

Polycarbonate (PC) and aliphatic polyesters were found to form a miscible polymer blend. Kambour et al. [6] reported that copolymer of styrene and bromostyrene formed a miscible pair with copolymer of xylenyl and bromoxylenyl ether.

Amoco [7] reported several polyaryl ether-sulfones to be miscible with each other. The miscible blend comprises a 1,4-arylene unit separated by ether oxygen and another resin 1,4 arylene separated by an $SO_2$ radical. The miscible blends showed a single glass transition temperature in between the constituent values. The blend was transparent. These can be used for printing wiring board structures, electrical connectors, and other fabricated articles that require high heat and chemical resistance and good dimensional and hydrolytic stability.

Differential scanning calorimetry (DSC) experiments indicated that atactic polystyrene and polyvinyl methyl ether (PVME) form miscible blends [8,9]. Syndiotactic and isotactic polystyrene when blended with PVME, phase separate at all temperatures above the glass transition temperature of PVME. Only weak van der Waals interactions between the phenyl rings in polystyrene with the methoxy group of PVME were detected using 2-dimensional nuclear magnetic resonance (NMR) spectroscopy.

Goh, Paul, and Barlow [10] found that an alpha-methylstyrene–acrylonitrile (AMS–AN) copolymer at 50 mole fraction AN formed miscible polymer blends with poly(methyl methacrylate) (PMMA) and with poly(ethyl methacrylate) (PEMA). AMS–AN copolymer did not form miscible blends with polyacrylates or polyvinyl acetate. The miscible blends were found to exhibit lower critical solution temperature (LCST) behavior.

PS forms a miscible blend with polyxylenyl ether and with tetramethyl bisphenol polycarbonate. Styrene acrylonitrile (SAN) copolymers can form miscible polymer blends with PMMA over some compositional window of AN and with polyvinyl chloride (PVC). PS and poly(cyclohexyl methacrylate) were reported to be miscible. Spherulite crystals of ε-PCL (epsilon polycaprolactone) were found to be miscible with PS. Polyphenylene oxide/polystyrene (PPO/PS) was found to mix at a molecular level.

Other examples of miscible polymer blends are

PS/poly-o-chlorostyrene
PMMA/PC
PS/poly-co-4-bromostyrene
Polyvinyl fluoride/PMMA or PEMA or polymethyl acrylate (PMA) or poly-
    ethylene acrylate (PEA)
PMMA/PEO, polyethylene oxide
PC/lithium salt of sulfonated polystyrene
PS/polyphenylmethyl siloxane
Chlorinated PE/chlorinated polybutadiene
PS/carboxylated PPO
PPO/poly(alpha-methylstyrene)
PS/styrene/co-bromostyrene

Specific interactions between chain segments can be of four different kinds [11]:

Hydrogen bonding
Acid/base interactions
Dipole/dipole interactions
Ion/dipole interactions

Examples of reported miscible polymer blends in the literature are

PEO/PMMA or polyvinyl-pyrrolidone (PVP)
PHE/PEO
PVC/PBT
Polyvinyl phenol/PMMA
PBT/PVC
PCL/PVC
PAA/PEO
PVP/PVPh

## 1.4   PARTIALLY MISCIBLE POLYMER BLENDS

Several polymer blends that are commercial products in the industry are partially miscible. Partially miscible polymer blends are those that exhibit some shift from their pure component glass transition temperatures. Thus, a binary miscible polymer blend will exhibit one glass transition temperature and a partially miscible polymer blend will exhibit two distinct glass transition temperatures different from their pure component glass transition temperatures. Some experimental systems that have been reported as partially miscible polymer blends are PET and poly(hydroxy butyrate) (PHB) and PC/SAN. Later on, in a separate chapter, a mathematical framework to calculate these multiple glass transition temperatures will be provided. This is achieved by a revisit to the Couchman equation and includes the differences in entropy of mixing in the glassy and rubbery states of the blends. The two polymer blend $T_g$s that are only partially miscible can be calculated from a quadratic equation (Sharma [12–22]):

$$AT_g^2 + BT_g + C = 0.$$

(1.6)

$A$, $B$, and $C$ can be calculated from the change in heat capacity of the polymer constituent before and after glass transition, change in heat capacity of the polymer blend before and after glass transition, change in entropy, change of mixing before and after the glass transition and the glass transition temperatures of the pure component polymers. When the change in entropy of mixing of the blend during glass transition is zero, the system reverts to a miscible polymer blend the glass transition temperature of the blend is found to be the geometric mean of the pure component glass transition temperatures. When the change in entropy of mixing of the polymer blend during glass transition is much larger, compared with the change in heat capacity of the polymer blend during glass transition, the system reverts to immiscible blends. The sum of the roots of the quadratic equation becomes the sum of the pure component glass transition temperatures, and the product of the glass transition temperatures is the product of the two pure component glass transition temperatures. The systems that are partially miscible will have change in entropy of mixing somewhere in between the miscible and immiscible blends.

## 1.5   NATURAL POLYMERS

Proteins are large bioorganic compounds that are polymeric in nature. They are made of amino acids arranged in a linear chain and joined together between the carboxyl of one amino acid and the amine nitrogen of another by a bond that is called a *peptide bond*. The name *protein* comes from the Greek word *prota*, meaning "of primary importance." These were first described and named by Berzelius in 1838. However, their central role in living organisms was not fully appreciated until 1926, when Sumner showed that the enzyme *urease* was a protein. The first protein structures to be solved included insulin and myoglobin; the former by Sir Frederick Sanger [23] who won a 1958 Nobel Prize for this work, and the latter by Perutz and Kendrew in 1958. Both proteins' three-dimensional structures were among the first determined by x-ray diffraction analysis; the myoglobin structure won the Nobel Prize in chemistry for its discoverers.

The sequence of amino acids in a protein is defined by genes and encoded in the genetic code. Although this genetic code specifies the 20 different amino acids, the residues in a protein are often chemically altered in post-translational modification, either before the protein can function in the cell or as part of control mechanisms. Proteins associate to form complexes that are stable. They can work in concert in order to achieve a particular function. They participate in every function of the cell. Many proteins are enzymes that catalyze biochemical reactions. They are vital to metabolism. The cell shape is maintained by a system of scaffolding. Proteins in the cytoskeleton form the system of scaffolding. They are also important in cell signaling, immune responses, cell adhesion, and cell cycle. Protein is also a necessary component in our diet, since animals cannot synthesize all the amino acids and must obtain essential amino acids from food. Through the process of digestion, animals break down ingested protein into free amino acids that can be used for protein synthesis.

During the formation of the polypeptide polymeric chain, one water molecule is lost per amino acid. This is why the constituents of the proteins are called *amino acid residues*. There are four different types of protein structures recognized in the field. These are described in the subsequent text.

### 1.5.1 PRIMARY STRUCTURE

The primary structure of proteins is the random sequence distribution of the 20 different amino acids concatenated in a polypeptide chain. Each of the 20 different amino acids consists of two parts: a backbone of protein and a unique side chain or "R group" that determines the physical and chemical properties of the amino acid. Each amino acid consists of an amine ($NH_2^+$) and carboxylic acid moiety ($COO^-$). The general formula of the 20 different amino acids can be classified into four categories based upon the net charge on the protein molecule. These categories and the amino acids contained in these categories are

1. *Positively charged basic amino acids*: lysine (Lys), argentine (Arg), and histidine (His)

$$NH_2$$
$$|$$
$$H_2N\text{-}C_4H_8\text{-}CH\text{-}COOH \quad (Lysine)$$
$$NH \qquad\qquad NH_2$$
$$\parallel \qquad\qquad\quad |$$
$$NH_2\text{-}C\text{-}NH\text{-}C_3H_6\text{-} CH\text{-}COOH \; (Arginine)$$

$$NH_2$$
$$|$$
$$H_2N \qquad NH\text{-}CH_2\text{-}CH\text{-}COOH \; (Histidine)$$
$$\backslash \quad /$$
$$\backslash \; /$$
$$C$$

2. *Negatively charged acidic amino acids*: aspartic acid (Asp), glutamic acid (Glu)

$$NH_2$$
$$|$$
$$HOOC\text{-}CH_2\text{-}CH\text{-}COOH \; (Aspartic Acid)$$

$$NH_2$$
$$|$$
$$HOOC\text{-}C_2H_4\text{-}CH\text{-}COOH \; (Glutamic Acid)$$

3. *Polar amino acids*: glycine (Gly), serine (Ser), threonine (Thr), cysteine (Cys), tyrosine (Tyr), glutamine (Gln), asparagine (Asn),

CH$_2$-COOH  (Glycine)
|
NH$_2$
OH- CH$_2$CH-COOH  (Serine)
|
NH$_2$
CH$_3$-CH-CH-COOH  (Threonine)
|   |
OH  NH$_2$

HS-CH$_2$-CH-COOH  (Cysteine)
|
NH$_2$
OH-φ-CH$_2$-CH-COOH  (Tyrosine)
|
NH$_2$
NH$_2$-C-C$_2$H$_4$-CH-COOH  (Glutamine)
||        |
O        NH$_2$

NH$_2$-C-CH$_2$-CH-COOH  (Asparagine)
||        |
O        NH$_2$

4. *Nonpolar amino acids*: alanine (Ala), valine (Val), leucine (Leu), isoleucine (Ile), proline (Pro), methionine (Met), phenylalanine (Phe), tryptophan (Trp)

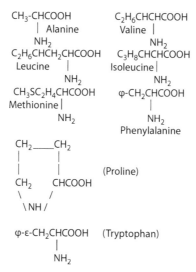

CH$_3$-CHCOOH                   C$_2$H$_6$CHCHCOOH
|  Alanine                       Valine  |
NH$_2$                          NH$_2$
C$_2$H$_6$CHCH$_2$CHCOOH        C$_3$H$_8$CHCHCOOH
Leucine  |                       Isoleucine|
NH$_2$                          NH$_2$
CH$_3$SC$_2$H$_4$CHCOOH         φ-CH$_2$CHCOOH
Methionine|                      |
NH$_2$                          NH$_2$
                                 Phenylalanine

CH$_2$____CH$_2$
|        |
|        |        (Proline)
CH$_2$    CHCOOH
\   /
\ NH /

φ-ε-CH$_2$CHCOOH   (Tryptophan)
|
NH$_2$

## 1.5.2 SECONDARY STRUCTURE

The polypeptide backbone exists in different sections of the protein as an α-helix, β-pleated sheet, or random γ-coil. The study of the protein secondary structure has attracted a lot of attention in the literature. As will be discussed in later chapters, protein secondary structures can be constructed from the primary structure chain sequence distribution. The secondary structure pertains to the *stereoisomerism* exhibited by the polypeptide chain. The problem of secondary structure prediction is one of hydrogen bonding. The polar groups present in the backbone of the polypeptide chain, C = O, N–H, are capable of hydrogen bond formation. The two structures that solve the problem are the α-helix and β-pleated sheet, in which extended polypeptide backbones are side by side. These structures are stable. They can occur at the exterior of the proteins with appropriate hydrophilic side chains or in the hydrophobic interior of proteins with appropriate hydrophobic side chains.

In the α-helix, the polypeptide backbone is twisted into a right-hand helix. The structure was first recognized in α-keratin by Pauling. For l-amino acids, the right-handed helix is more stable than the left-handed one. The structure has a pitch of 3.6 amino acids per turn. This results in the C = O of each peptide bond being aligned to form a hydrogen bond with the peptide bond NH of the fourth distant amino acid residue. The C = O groups point in the direction of the axis of the helix and are aimed at the NH groups with which they hydrogen bond, giving maximum bond strength and making the α-helix a stable structure. Thus, every C = O and NH group of the polypeptide backbone are hydrogen bonded in pairs forming a stable, cylindrical, rodlike structure. Amino acids vary in their tendency to form α-helices. Proteins are made of mixtures of α-helix and β-sheet structures. This is also a stable structure in which the polar groups of the polypeptide backbone are hydrogen bonded to one another. The polypeptide chain lies in an extended or β form with the C = O and NH groups hydrogen bonded to those of a neighboring chain. The structure was first recognized in β-keratin. Several chains can form a sheet of polypeptide. It is pleated because successive α carbon atoms of the amino acid residues lie slightly above and below the plane of the β sheet, alternately. The adjacent polypeptide chains bonded together can run in the same direction (parallel) or opposite directions (antiparallel). In the latter case, a polypeptide may make tight β turns to fold the chain back on itself.

The random coil refers to a section of polypeptide in a protein whose conformation is not recognizable as one of the defined structures of helices and pleated sheets. It is determined by side-chain interactions and, within a given protein, is fixed rather than varying in a random way.

One good way to measure the protein secondary structure is by x-ray crystallography. In addition, the techniques of neutron diffraction and NMR can be used to measure the protein secondary structure.

### 1.5.3 TERTIARY STRUCTURE

The folding of the secondary structure into a macrostructure such as globules is called the *tertiary structure* of proteins. A given protein in a physiological environment can have a complex three-dimensional structure. The amino acid "backbone" of a protein can rotate freely, allowing amino acids from distal protein domains to come into close contact with each other. As these regions of the protein interact with one another, they will create and stabilize a particular protein conformation. Disulfide bond creation between cysteine residues is one of the primary stabilizing mechanisms.

### 1.5.4 QUATERNARY STRUCTURE

Two polypeptide chains connected by hydrogen bonding form the quaternary structure of protein.

The protein chemist is confronted with the problem of isolating, purifying, and characterizing the protein. Identification of a suitable source such as fresh tissue, is the first step for purification of proteins. This is subjected to the action of a blender for grinding to obtain a homogenate that is rich in protein as well as contaminating material. Proteins are temperature sensitive and fragile. The homogenate is filtered and freed from unwanted material by treatment with suitable solvents. Denaturation of proteins is avoided by control of pH and temperature. As a general rule, purification of proteins is carried out at temperatures close to the freezing point.

*Salt precipitation* is used to effect separation by addition of ammonium sulfate so that the desired protein remains either in the supernatant or in the precipitate. A mixture of proteins is passed down an *ion-exchange column* and separated by binding to the column. The bed is regenerated by eluting agents of varying pH. By increasing the pH of the effluent, different fractions of proteins are obtained. The protein solution binds to ion-exchange materials such as cellulosic polymers.

The molecular weight of the proteins can be determined by the use of *gels*. The discovery of the *ultracentrifuge* in the early twentieth century was an advancement to obtain precise molecular weights. Svedberg [24] won the Nobel Prize in physics in 1926 for his efforts in the development of the use of the ultracentrifuge. *Sedimentation* is used to measure the molecular weights of proteins and in order to study protein–protein interactions. Sucrose-density gradients can be used to separate molecular fragments.

*Electrophoresis* is a method with superior resolution used to separate macromolecules from complex mixtures by the application of an electric field. The macromolecules, placed at one end of the matrix and called the "gel," are subjected to an electric field. Different macromolecules in the gel will migrate at different speeds, depending on the nature of the gel and the characteristics of the macromolecules. Electrophoretic techniques can be used to separate any biomacromolecule such as nucleic acids, polypetides, and carbohydrates. Tiselius [25] won the Nobel Prize in chemistry in 1948 for his work on the development of electrophoresis as a technique to separate and characterize proteins from complex mixtures.

The use of polyacrylamide gel electrophoresis (PAGE) had a major impact on the capability for isolation and characterization of proteins. Polyacrylamide gels are formed by cross-linking the acrylamide monomer with the chemical agent *N-N*-methylene bisacrylamide. The polymerization reaction proceeds as free radical catalysis with the use of ammonium persulfate and the base TEMED (*N-N-N-N*-tetramethylenediamine) as the initiator. PAGE can be used to resolve the ladders in DNA structure. They can be used to characterize proteins according to their size or charge. Three methods were developed to measure the *primary structure* of protein. These are the Sanger's method, dansyl chloride method, and Edman degradation techniques.

All three techniques are laborious. A lot of material and years of analysis are required to complete the analysis of even a short protein. In the Edman degradation method, the peptide fragment is treated with phenylisothiocynate at pH 8 to yield a phenylthiocarbamyl derivative at the N-terminal. The derivative is treated with acid in organic solvent so that the N-terminal amino acid undergoes cyclization to produce phenylthiohydantoin, which is cleaved from the peptide fragment. Thiohydantoin derivative can be identified using *paper chromatography*, and the peptide is further subjected to the same treatment every time, forming a thiohydantoin derivative from the amino end. Dansyl chloride reagent is used for determination of N-terminal residue in alkaline condition. The N-terminal residue forms a yellow fluorescent derivative that can be easily detected. Even small amounts of amino acids can be deducted. Indirect methods can be used to save time. The cDNA responsible for the creation of the protein can be cloned, and its sequence measured. Then, by deduction, the protein sequence can be obtained.

*Mass spectrometry* (MS) is a method where the mass of the molecules that have been ionized can be measured using a mass spectrometer. MS has become a key tool in proteomics research because it can analyze and identify compounds that are present at extremely low concentrations (as little as 1 pg) in very complex mixtures by analyzing its unique signature. A critical concern in MS is that the methods used for ionization can be so harsh that they may generate very little product to measure at the end. The development of "soft" desorption ionization methods by John Fenn and Koichi Tanaka [26], which allowed the application of MS to biomolecules on a wide scale, earned them a share of the Nobel Prize in chemistry in 2002.

*Isoelectric focusing* is a variation of electrophoresis and can be used for separating mixtures of protein. A column consists of gel having positive and negative charges. When the protein mixture is injected into the column, the molecules polarize in the electric field in such a way that the negatively charged ones move toward the anode and those with the positive charge move toward the cathode. At the isolectric point (i.e., the point in the tube where each protein attains a pH which is neutral), the driving force to migration stops. A sharp bend forms at this juncture.

*2D gel electrophoresis* is a way to couple different gel systems with different resolving powers to dramatically improve separations and resolution of complex mixtures of proteins. 2D gel electrophoresis is an incredibly useful analytic tool, and provides a foundation for what is now referred to as *proteomics*.

Chromatography is extensively used for separating different molecular species, including proteins. Different types of chromatography are recognized, such as *adsorption chromatography, ion-exchange chromatography,* and *partition chromatography.* Paper chromatography is suitable for separation of small amounts of low molecular weight compounds that are soluble in the liquid phase. The liquid phase is water and the mobile phase consists of a mixture of organic solvents. The paper is spotted with the substance to be separated and immersed in a trough containing the mobile phase. The spots on the paper are developed using a suitable developing reagent. The partition coefficient, Rf, is the ratio between the distance run by the compound and the distance traveled by the solvent. Ion exchange chromatography can be employed for separation and purification of proteins. The protein sample is prepared in the right type of buffer and then applied to the ion exchange column. Molecules possessing no charge will easily pass through, while charged molecules will interact with the exchanger and get adsorbed. Proteins can then be eluted from the exchangers.

*Gel filtration chromatography* is a technique for purification and separation of macromolecules based upon their molecular size. Gel permeation, gel exclusion, and molecular sieving are similar methods. Gel filtration media include polydextran gels, polyacrylamide gels, agarose gels, and controlled pore glass beads. The porosity of gel beads in a column is controlled depending on the problem of separation at hand. Larger molecules elute out and smaller molecules diffuse through the pores of beads. Later, this can be eluted by using a buffer. Desaltation of a sample can be affected. A biospecific interaction of a protein with a specific ligand is used in *affinity chromatography*. The chromatographic column utilizes an inert matrix or support medium that offers binding sites for the desired protein to be purified. The adsorbate has to be specific, and the adsorbent can be porous, hydrophilic, and capable of covalent binding. Agarose gel, polyacrylamide, and controlled pore-sized glass are examples of adsorbents used by this method. A *thin-layer chromatography* (TLC), the chromatogram, can be developed a number of times with different solvents with good separation.

Quantitative analysis of multiple components can be affected in one hour's time using this method. A binding medium such as calcium sulfate is used in TLC. The adsorbent is activated, and then spots are generated and dried. Spots in the form of colored zones may be observed by ultraviolet light. *High performance liquid chromatography* (HPLC) can be used to isolate, purify, and identify compounds in a mixture. Rapid analysis of nonvolatile, ionic, thermally labile compounds that were previously difficult to separate can be measured using HPLC. Molecular components of the cell can be determined with high sensitivity, speed, accuracy, and resolution. The mobile phase is a liquid. The solute needs to be soluble in the mobile phase. The mobile phase is forced under high pressure of more than 6000 psi into the column. Normal-phase chromatography, bounded-phase chromatography, and reverse-phase chromatography are three kinds of HPLC methods. Separation of ribosomal proteins can be effected in this manner. The effect of pore size, pore volume, silica density, and surface area on a given separation is complex.

Crick, Watson, and Wilkins [27] were awarded the Nobel Prize in medicine in 1962 for their work on the molecular configuration of nucleic acids, the genetic code, and the involvement of RNA in the synthesis of proteins and its significance for information transfer in living material. Nucleic acids are long-chain polymers of nucleotides. Each nucleotide consists of three parts: sugar (ribose or deoxyribose), phosphoric acid, and a nitrogenous base. Four different nitrogenous bases are adenine, guanine, cytosine, and thymine. Adenine (A) and guanine (G) are purines, and cytosine (C) and thymine (T) are pyrimidines. Uracil (U) is another nitrogenous base found in RNA in place of thymine (T) in DNA. The double helix, three-dimensional structure of DNA was discovered Watson, Crick, and Wilkins. They used x-ray crystallography to determine the structure. Rosaland Franklin was deceased by the time of the prize but also was one of the pioneers. DNA's discovery is hailed as the most important work in biology of the last 100 years, and the field it opened may be the next scientific frontier for the next 100 years. Outside the helix backbone of the ladder is the sugar phosphate chain. A complete turn of the ladder is called *pitch,* and is about 3–4 Å in length. The space between bases is 2–4 Å, and the diameter of the helix 20 Å.

The sequences of bases in DNA by a process called *translation* determine the sequence distribution of protein molecules. All genetic information in living organisms of any kind is carried by the nucleic acids, usually by the DNA. Certain small viruses use RNA as their genetic material. The four different bases in DNA can assume $4 \times 4 = 16$ combinatorial forms, 64 triplets, and 256 quartets. The set of bases of that code is called a *codon*. The two DNA strands are antiparallel. One strand runs in the $5'$–$3'$ direction and the other strand in the $3'$–$5'$ direction. The two polynucleotide chains of double helix interact with one another. The hypothesis that the linear sequence of nucleotides in DNA specifies the linear sequence of amino acids in proteins evolved over a period of time. Kornberg and coworkers [28] discovered and characterized the enzyme polymerase. This was followed a few years later by RNA polymerase characterization. They clarified the manner by which information in DNA is transcribed into an RNA, which is now referred to as messenger RNA. Kornberg's son, Roger [29], received the Nobel Prize in chemistry in 2006 for his studies in the molecular basis for eukaryotic transcription.

## 1.6  POLYMER ALLOYS

A polymer *alloy* is a commercial polymer blend with improvement in property balance with the use of *compatibilizer* interface stabilization. Dow Chemical (Midland, Michigan) has patented [30] compatibilized blends of polycarbonate and linear polyethylene. A graft copolymer such as ethylene propylene diene monomer grafted styrene acrylonitrile (EPDM-g-SAN) can be used as a compatibilizing agent. Molded articles of compatibilized EPDM/SAN blends display high impact resistance, good melt processability, and weld-line strength. Molded articles of this blend display high impact resistance, good melt processability, and weld-line strength, and show a reduced tendency towards delamination. Products can be processed into films, fibers, extruded sheets, multilayer laminates, molded or shaped articles of instrument housings, data storage apparatus, and articles for electronic applications. The composition of the PC in the blend reported was 50–90% and linear PE from 1–40%.

Dow has also reported [31] polymer alloys made of S-AMS (styrene-alpha/methyl styrene copolymer), and PP (polypropylene), and a compatibilizer made up of styrene-grafted-olefin copolymer. The S-AMS copolymers were prepared using anionic polymerization. Poly AMS has a low ceiling temperature of 61°C. The ceiling temperature is the temperature above which depolymerization reactions are favorable compared with molecular weight-building monomer to polymer reactions. Monomers used in anionic polymerization are purified by distillation or by use of molecular sieves. Highly purified monomers, along with an inert solvent, are charged to a reactor vessel and reactions performed at temperatures above the ceiling temperature of the monomer, AMS. The copolymer ceiling temperature is much higher. This way, the depolymerization reactions are circumvented in preparation of the polymer.

Polymer alloys made of nylon and ABS engineering thermoplastic were sold by Monsanto under the name of Triax 1000. The compatibilizer used was a terpolymer of styrene and acrylonitrile and glycidyl methacrylate. The effect of the compatibilizer was a step change improvement in the performance property of the polymers. Golf balls can be made from the product. PC/ABS blends are also extensively marketed. The enthalpy of mixing is nearly zero for the PC/SAN system at the AN composition close to that used to prepare ABS engineering thermoplastic. An amine-functional SAN polymer was proposed by Wildes et al. [32], as a reactive compatibilizer for PC–ABS blends. This polymer was expected to be miscible with the SAN copolymer matrix of ABS engineering thermoplastic, and the pendant secondary amine groups should react with PC at the carbonate linkage to form a SAN-g-PC copolymer at the PC–ABS interface. The copolymer molecules ought to provide improved morphological stability by suppressing phase coalescence. They examined the morphology of the blends using transmission electron microscopy (TEM).

Often, even immiscible and/or incompatible polymers are also made compatible by addition of a compatibilizer. Compatibilizers are believed to primarily reside at the polymer/polymer interfaces. The compatibilizer can be presynthesized or formed by in situ polymerization. Often the product engineer can make a judicious choice of compatibilizer resulting in improvement of mechanical properties, sometimes with synergistic effects. The compatibilizer is believed to "stitch" itself across the polymer/polymer interfaces. The addition of a compatibilizer lowers the interfacial tension between the two immiscible phases. The compatibilizer may sometimes be made of block copolymer composed of two different components. One of the block components may be miscible with polymer A, and the block component may be miscible with polymer B. Even if a polymer A and polymer B are immiscible, such a block copolymer would *compatibilize* the blend of A and B.

The addition of a compatibilizer may result in other favorable effects, such as retardation of dispersed phase coalescence via steric stabilization. This promotes the formation of a distribution of small particles of the dispersed phase in the matrix. The interfacial adhesion is improved between the blend components and reduces the possibility of interfacial failure.

Reactive compatibilization has been used in the development of polymer blends with polyamide as a constituent. Blends of ABS and polyamide are of significant interest commercially. Maleic anhydride has been grafted into ABS. Terpolymers of styrene acrylonitrile and maleic anhydride with a small amount of methyl methacrylate have been patented as compatibilizer by Monsanto (now Bayer) for polyamide/ABS blends. Kudva et al. [33] studied the use of glycidyl methacrylate/methyl methacrylate monomers to form copolymers in order to control the morphological, rheological, and mechanical characteristics of polyamide/ABS blends. They found that addition of glycydyl methacrylate/methyl methacrylate (GMA/MMA) copolymers to PA/ABS blends reduced the dispersed-phase domain size of the blend. The changes in the dispersed-phase morphology were found to stem from reactions between the nylon-6 end groups and the copolymer. This occurred to some measure by steric stabilization of the SAN domains against coalescence. At low GMA content in the copolymer, the notched Izod impact strength was found to become worse than that of binary nylon-6/ABS blends. The addition of the compatibilizer was found to break-up the co-continuity of the ABS phase and promote the formation of two populations of ABS domains.

## 1.7 SUMMARY

Polymer blends offer improved performance/cost ratio and flexibility to tailor-made products. Compatible, incompatible, miscible, and immiscible blends are distinguished from each other. Polymer alloys are a commercial blends with improved property balance by use of a compatibilizer. By 1992, the worldwide production of plastics was twice that of steel. The manufacture of polymers would grow by a factor of 10 prior to market saturation. Table 1.1 provided a chronology of events in technology development in the field of polymer blends since the nitrile rubber/gutta percha was blended in 1846 first through the development of PPS/PBT in 1990.

According to P. Flory, finding thermodynamically miscible polymers is difficult to do unless they have specific interactions with each other. Very few miscible polymer blends of nonpolar or weakly polar homopolymers were identified during a 30-year period following the publication of Flory's book. During the 1980s, a steadily increasing number of miscible systems has been reported. Krause found 282 chemically dissimilar polymer pairs that appeared to be miscible. Of these, 75% were miscible on account of specific interactions, and 15% of miscible pairs were due to clever choice of copolymer composition. Miscibility can be predicted using;

$$\Delta G = \Delta H - T\Delta S.$$

The spinodal and binodal curves can be estimated from Equations (1.4) and (1.5). Typical LCST, UCST, and a spinodal curve are shown in Figure 1.1.

A miscible polymer blend is expected to exhibit one-glass transition temperature. Partial miscibility is indicated by two-glass transition temperatures. PC and aliphatic esters were found to form a miscible polymer blend. A copolymer of styrene and bromostyrene formed a miscible pair with the copolymer of xylenyl and bromoxylenyl ether.

Several polyaryl ether-sulfones were found to be miscible with each other (Figure 1.2). AMS–AN copolymers have been found to be miscible with PMMA and with PEMA, and the blends exhibit LCST-type behavior. PS forms a miscible blend with polyxylenyl ether and with tetramethyl bisphenol polycarbonate, and PPO and P-εCL (poly-epsilon caprolcatone). SAN forms miscible blends with PMMA and PVC.

Specific interactions between chain segments can be by

Hydrogen bonding
Acid/base interactions
Dipole/dipole interactions
Ion/dipole interactions

The two polymer blend $T_g$s that are only partially miscible can be calculated from a quadratic equation as shown by Sharma ([12–22]);

$$AT_g^2 + BT_g + C = 0.$$

Insulin was among the first primary protein structures identified. The sequence of amino acids in a protein is defined by genes and encoded by the genetic code.

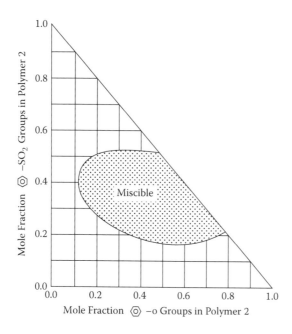

**FIGURE 1.2**   Miscible region of poly(aryl ether sulfone) polymer–polymer blend.

Primary structure, secondary structure, tertiary structure, and quaternary structure are the four different structure types of proteins. The primary structure of protein consists of a random polymer chain sequence distribution of amino acids. There exist 20 different amino acids. Each amino acid consists of an amine and carboxylic acid group. The 20 different amino acids are lysine, arginine, and histidine, which are basic; aspartic acid and glutamic acid, which are acidic; glycine, serine, threonine, cysteine, tyrosine, glutamine, and asparagines, which are polar; and alanine, valine, leucine, isoleucine, proline, methionine, phenylalanine, and tryptophan, which are nonpolar. Because of hydrogen bond formation, a secondary structure of protein is formed, depending on the primary sequence structure of protein. The forms take α-helix, β-sheet, and γ-coil shapes. F. Sanger won two Nobel Prizes: one for discovering the primary chain sequence distribution structure of insulin and another for discovering the chain sequence distribution of nucleic acid. He developed a molecular labeling method. Paper chromatography and gel acrylamide electrophoresis are used to separate the molecular fragments and are sequence deduced.

There are a number of bioseparation techniques. These are salt precipitation, ion-exchange, ultracentrifuge, sedimentation, poly(acrylamide) gel electrophoresis, Sanger's method, dansyl chloride method, Edman degradation technique, paper chromatography, mass spectrometry, isoelectric focusing, 2D gel electrophoresis, adsorption chromatography, partition chromatography, ion-exchange chromatography, gel-filtration chromatography, affinity chromatography, thin-layer chromatography, and high-performance liquid chromatography (HPLC).

Nucleic acids consist of a ribose sugar or deoxyribose, phosphoric acid, and nitrogenous bases. Adenine, guanine, cytosine, tymine, and uracil are nitrogenous bases. DNA has a double helix structure and two antiparallel strands. The linear sequence of protein is specified by the linear sequence of nucleic acids by a process of translation and transcription. Har Gobind Khorana received the Nobel Prize for nucleic acid synthesis and the genetic code. The flow of information in gene expression from DNA to mRNA is called *transcription*, and from mRNA to protein is referred to as *translation*. Exons are the coding regions of DNA and intergenic regions, the introns. Splicing occurs when exons are linked together and introns eliminated.

Dow has prepared a compatibilized blend of PC and linear PE. The compatibilizer used was EPDM grafted with SAN. The product has high impact strength and good melt processability. Polymer alloys with S-AMS copolymer and PP with styrene-grafted polyolefin copolymer have been reported. Triax 1000 of Monsanto is a blend of nylon and ABS compatibilized with styrene-acrylonitrile and glycidyl methacrylate terpolymer. The compatibilizer often improves the property balance of an immiscible blend. Reactive compatibilization is an emerging technique.

## EXERCISES

1. Do polymer blends possess properties superior to the individual polymer constituents of the blend or possess properties between the properties of the individual polymer constituents?
2. What is the difference between a compatible and a miscible blend?
3. What is the role of a compatibilizer in a blend?
4. What is the difference between a polymer alloy and a polymer solution?
5. What are the differences between an immiscible and an incompatible blend?
6. Can a blend be immiscible and compatible?
7. Can a blend be incompatible and miscible?
8. What is meant by a partially miscible blend?
9. How is IPN different from polymer nanocomposites?
10. Why is morphology of a blend an important consideration in product analysis?
11. What are the differences between agitation and mixing?
12. Why are screw extruders preferred to a Banbury mixer?
13. How was nylon/ABS blends compatibilized in Triax 1000?
14. Discuss the features of PA/PVC blends.
15. Discuss the features of PVC/ABS blends.
16. Can ABS engineering thermoplastic be considered a polymer blend?
17. What did Nobel laureate Flory allude to as "difficult" in attempting to form miscible polymer blends?
18. What did Krause's study find?
19. What is the difference between spinodal and binodal curve?
20. How can thermodynamic miscibility be studied using Gibbs free energy of mixing?
21. What do LCST and UCST mean?
22. Give examples of miscible polymer blends.
23. What are phase diagrams used for?
24. What are the four types of specific interactions?
25. What is the form of quadratic equation to predict the two glass transition temperatures observed in partially miscible polymer blends?

26. What is the difference between polar amino acids and negatively charged amino acids?
27. What is the significance of the β-sheet in the secondary structure of protein?
28. How is ultracentrifuge used in the separation of proteins?
29. Where is the PAGE technique used?
30. What is meant by affinity chromatography?
31. How did Dow compatibilize PC and polyethylene blends?
32. How was the S-AMS copolymer compatibilized with polypropylene?

## REFERENCES

1. L. A. Utracki, History of commercial polymer alloys and blends (From a perspective of the patent literature), *Polym. Eng. Sci.*, Vol. 35, 1–17, 1995.
2. J. Baker, Improvement in Grain-Binders, US Patent 93,035, 1869.
3. P. Pfleiderer, Machine for Mixing, Kneading, or Otherwise Treating Plastic Masses, US Patent 534,968, 1895.
4. P. J. Flory, *Principles of Polymer Chemistry*, Cornell University Press, Ithaca, New York, 1953.
5. S. Krause, Polymer-polymer miscibility, *Pure Appl. Chem.*, Vol. 58, 12, 1553–1560, 1986.
6. R. P. Kambour, J. T. Bendler, and R. C. Bopp, Phase behavior of polystryene, poly 2,6-dimethyl-1,4-phenylene oxide and their brominated derivatives, *Macromolecules*, Vol. 16, 753–757, 1983.
7. J. E. Harris and L. M. Robeson, Miscible Blends of Poly (Aryl Ether Sulfones), US Patent, 4,804, 723, Amoco, Chicago, IL, 1989.
8. T. Nishi and T. Kwei, Cloud point curves for poly(vinyl methyl ether) and monodisperse polystyrene mixtures, *Polymer*, Vol. 16, 285–290, 1975.
9. S. Cimmino, E. D. Pace, E. Martuscelli, C. Silvestre, D. M. Rice, and F. E. Karasz, Miscibility of syndiotactic polystyrene/poly(vinyl methyl ether) blends, *Polymer*, Vol. 34, 214–217, 1993.
10. S. H. Goh, D. R. Paul, and J. W. Barlow, Miscibility of poly(alphamethyl styrene)-co-acrylonitrile with polyacrylates and polymethacrylates, *Polym. Eng. Sci.*, Vol. 22, 1, 34–39, 1982.
11. L. M. Robeson, *Polymer Blends: A Comprehensive Review,* Hanser Gardner, Munich, Germany, 2007.
12. K. R. Sharma, Quantitation of Partially Miscible Systems in Polymer Blends—Entropy of Mixing Change at Glass Transition, 37th Annual Convention of Chemists, Haridwar, India, November 2000.
13. K. R. Sharma, Mathematical Modeling of Partially Miscible Copolymers in Blends, 215th ACS National Meeting, Dallas, TX, March 1998.
14. K. R. Sharma, *Applied Polymer Thermodynamics*, CIEC, Tampa, FL, January 1997.
15. K. R. Sharma, *Course Development in Applied Polymer Thermodynamics*, ASEE Annual Conference & Exposition, Milwaukee, WI, June 1997.
16. K. R. Sharma, *Ternary Polymer Blend for Hydrogen Production vs. Inorganic High Surface Area Materials*, Interpack 99, Mauii, HI, July 1999.
17. K. R. Sharma, Phase Computer to Predict Polymer Phase Behavior, ASEE North Central Regional Meeting, Erie, PA, April 1999.
18. K. R. Sharma, Miscibility and Moisture Sensitivity of Starch Based Binary Blends, 51st Southeast Regional Meeting of the ACS, SERMACS, Knoxville, TN, October 1999.
19. K. R. Sharma, Predictive Design of Starch Based Binary Blends, 91st AIChE Annual Meeting, Dallas, TX, October/November 1999.

20. K. R. Sharma, Ternary Polymer Blends for Hydrogen Production vs Inorganic High Surface Area Zeolite Materials, 218th ACS National Meeting, New Orleans, LA, August 1999.

21. K. R. Sharma, Adhesion Issues in Starch Based Binary Blends, 218th ACS National Meeting, New Orleans, LA, August 1999.

22. K. R. Sharma, Mesoscopic Mixing in Miscible SAN/PMMA Blend, CHEMCON 2001, Chemical Engineering Congress, Chennai, India, December, 2001.

23. F. Sanger, The Chemistry of Insulin, Nobel Lecture, 1958, in Nobel Lectures Elsevier, Amsterdam, Netherlands, 1964.

24. T. Svedberg, The ultracentrifure, *Les Prix Nobel*, Stockholm, Sweden, 1928.

25. A. Tiselius, Electrophoresis and adsorption analysis as aids in investigations of large molecular weight substances and their breakdown products, *Les Prix Nobel*, Stockholm, Sweden, 1948.

26. K. Tanaka, The origin of macromolecule ionization by laser irradiation, *Les Prix Nobel*, Stockholm, Sweden, 2002.

27. F. H. C. Crick, On the genetic code, *Les Prix Nobel*, Stockholm, Sweden, 1962.

28. A. Kornberg, Bertsch, L. L., Jackson, J. F. and H. Khorana, *Proc. Natl. Acad. Sci.,* 61, 315, 1964.

29. R. Kornberg, The Molecular Basis of Eukaryotic Transcription, Les Prix Nobel, Stockholm, Sweden, 2006.

30. H. T. Pham, S. P. Namhata, and C. P. Bosnyak, Compositions for Tough and Easy Melt Processible Polycarbonate/Polyolefin Blend Resin, US Patent 6,025,420, Dow Chemical Co., Midland, MI, 2000.

31. V. I. W. Stuart and D. B. Priddy, Blends of Polypropylene and Vinylaromatic/α-Methylstyrene Copolymers, US Patent 4,704,431, Dow Chemical Co., Midland, MI, 1987.

32. S. S. Wildes, T. Harada, H. Keskkula, D. R. Paul, V. Janarthanan, and A. R. Padwa, Compatibilization of PC/ABS Blends Using an Amine Functional SAN, *ANTEC Conference Proceedings*, 2438–2442, 1998.

33. R. A. Kudva, H. Keskkula, and D. R. Paul, Compatibilization of nylon 6/ABS blends using glycidyl methacrylate/methyl methacrylate copolymers, *Polymer*, Vol. 39, 12, 2447–2460, 1998.

# 2 Equation of State Theories for Polymers

## LEARNING OBJECTIVES

- Small molecules versus macromolecules
- Kinetic representation of pressure
- Derivation of ideal gas law
- PT diagram of small molecule pure substance
- PT diagram of polymer
- van der Waals cubic equation of state
- Virial equation of state
- Redlich–Kwong equation of state and Soave modification
- Peng–Robinson equation of state
- Tait equation for polymer liquids
- Flory, Orwoll, and Vrij models
- Prigogine square-well cell model
- Sanchez–Lacombe lattice fluid theory
- Materials with negative coefficient of thermal expansion and second law of thermodynamics

## 2.1 SMALL MOLECULES AND LARGE MOLECULES

The knowledge of thermodynamic properties such as internal energy, enthalpy, free energy, entropy, and Helmholtz free energy is important in the design and use of any substance in practice. For fluids, these properties are usually evaluated from measurements of molar volume as a function of temperature and pressure, yielding, pressure–volume–temperature (PVT), relations which may be expressed mathematically as *equations of state*. For small molecules such as a monatomic gas, it has been found that the ideal gas law is a good representation of the PVT behavior. The ideal gas law can be written as

$$PV = nRT. \tag{2.1}$$

where $P$ is the pressure with units of $Nm^{-2}$; $V$ $(m^3)$ is the volume; $R$ is the universal molar gas constant $(Jmole^{-1}K^{-1})$; $n$ is the number of moles of the substance; and $T$ $(K)$ is the temperature. The ideal gas law can be derived as follows.

### 2.1.1 KINETIC REPRESENTATION OF PRESSURE

Consider a box of gas molecules. Let each of the molecules have a velocity $v$, with three components, $v_x$, $v_y$, and $v_z$. Let the box be a cube of side 1 m. When one of the gas molecules collides with one of the walls of the container, assuming an elastic collision, the momentum change during collision may be given [1] as

$$\text{Rate of momentum change due to one collision} = mv_x - (-mv_x) = 2mv_x, \quad (2.2)$$

where $m$ is the mass of a molecule.

Assuming a round trip of 2l, the time taken between two collisions of the same molecule with the same wall appears as follows:

$$\text{Between collisions} = \frac{2l}{v_x}; \quad (2.3)$$

$$\text{The frequency of collisions on account of 1 molecule} = \frac{v_x}{2l}, \quad (2.4)$$

$$\text{and the rate of change of momentum at the wall} = \frac{v_x}{2l}\frac{2mv_x}{}. \quad (2.5)$$

The rate of change of momentum at the wall on account of $N$ molecules is shown as

$$\text{Molecules} = \frac{mv_{x1}^2}{l} + \frac{mv_{x2}^2}{l} + \frac{mv_{x3}^2}{l} + .... + \frac{mv_{xN}^2}{l}. \quad (2.6)$$

The force exerted by $N$ molecules at the wall is equal to the rate of change of momentum from Newton's second law. The pressure exerted by the fluid is $F/A$, and hence,

$$P = \frac{mv_{x1}^2}{l^3} + \frac{mv_{x2}^2}{l^3} + \frac{mv_{x3}^2}{l^3} + ... + \frac{mv_{xN}^2}{l^3}. \quad (2.7)$$

The root mean square velocity of the molecule is defined as

$$N<v^2> = v_1^2 + v_2^2 + v_3^2 + ... + v_N^2. \quad (2.8)$$

And, accounting for the motion of molecules in three dimensions, we combine Equations (2.7) and (2.8) as

$$P = \frac{mN<v^2>}{3l^3} = \rho<v^2>, \quad (2.9)$$

where the density of the fluid, $\rho$, can be seen to be $mN/l^3$. Equation (2.9) gives the kinematic representation of pressure.

## 2.1.2   DERIVATION OF IDEAL GAS LAW

From the Boltzmann equipartition energy theorem, the temperature of the fluid can be written as

$$\frac{mv^2}{2} = \frac{3k_B T}{2}.$$  (2.10)

Combining Equations (2.9) and (2.10) and multiplying and dividing the numerator and denominator by the Avogadro number, $A_N$,

$$P = \frac{mN3k_B A_N <T>}{A_N 3l^3} = \frac{RT}{V}.$$  (2.11)

where $V$ = molar volume in m³/mole.

Thus, the relationship $PV = RT$ for one mole of the gas can be derived. This is the ideal gas law. The assumptions in the box of molecules were in elastic collision, and the gas molecule occupies negligible volume compared to the volume of the container. In Equation (2.11) it can be seen that $A_N$ is the Avogadro number. $A_N k_B$ yields the universal gas constant $R$ (J/mole/K). Further, $mN/A_N$ gives the number of moles of gas $n$ present in the box. Also, $A_N l^3/mN$ gives the molar volume of the gas.

## Example 2.1   Show for an ideal gas that $C_p - C_v = R$

From Appendix A,

$$H = U + PV.$$  (2.12)

For an ideal gas, $PV = RT$.
   Hence, Equation (2.12) becomes

$$H = U + RT.$$  (2.13)

Differentiating Equation (2.13) with respect to $T$,

$$\frac{\partial H}{\partial T} = \frac{\partial U}{\partial T} + R.$$  (2.14)

It can be seen that

$$\left(\frac{\partial H}{\partial T}\right)_P = C_p \; and \; \left(\frac{\partial U}{\partial T}\right)_v = C_v.$$  (2.15)

Combining Equations (2.14)/(2.15):

$$C_p - C_v = R \text{ for an ideal gas.}$$

Polymers are large macromolecules. They behave differently compared to the small molecules. The typical pressure temperature (PT) diagram for a small molecule is shown in Figure 2.1.

Per the phase rule for a pure substance, the number of degrees of freedom can be written as

$$\Im = C - P + 2. \tag{2.16}$$

where $C$ is the number of components, and $P$ the number of phases. At the triple point, as shown in Figure 2.1, the number of degrees of freedom would be $1 - 3 + 2 = 0$. This means that the temperature, pressure, and molar volume at the triple point are fixed quantities for a pure substance. At low pressures and high temperatures, the substance exists as a gas. At lower temperatures, the substance exists as vapor. The curve between A and the triple point in Figure 2.1 is the *sublimation curve* and represents the equilibrium curve between the solid phase and vapor phase of 1 mole of the substance. The curve between the triple point and point B is the *fusion curve*. It separates the solid phase from the liquid phase of the substance.

At very low pressures and temperatures close to 0 K, a fourth state of matter has been discovered. This is the Bose–Einstein condensate. Eric A. Cornell obtained the Nobel prize in 2001 for his work on the Bose–Einstein condensate and fourth state of matter. The critical point of the pure substance can be seen in Figure 2.1. Beyond this point, the liquid and gas are indistinguishable from each other and exist as a fluid. Are there similar critical points at the end of the fusion curve and sublimation curves? The PVT and other phenomena at very low pressures and temperatures are subjects of exploratory research. The critical point may also be viewed as the highest pressure and temperature at which a pure chemical species is observed to exist in vapor/liquid equilibrium [2]. The vertical line in Figure 2.1 is an *isotherm*, and a horizontal line in Figure 2.1 is an *isobar*. The solid lines in Figure 2.1 indicate a

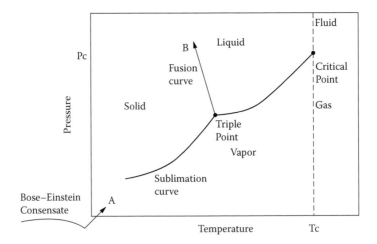

**FIGURE 2.1** PT diagram of a pure substance (small molecule).

change of state. When the curve is traversed, an abrupt change in properties of the fluid occurs, that is, solid becomes liquid, etc.

**Example 2.2   Evaluate the Difference** $\left(\dfrac{\partial U}{\partial T}\right)_P - \left(\dfrac{\partial U}{\partial T}\right)_V$

Assumption: Fluid is an ideal gas.

$$C_V = \left(\frac{\partial U}{\partial T}\right)_V \tag{2.17}$$

$$\left(\frac{\partial U}{\partial T}\right)_P = \left(\frac{\partial (H - PV)}{\partial T}\right)_P = C_P - P\left(\frac{\partial V}{\partial T}\right)_P \tag{2.18}$$

$$\left(\frac{\partial U}{\partial T}\right)_P - \left(\frac{\partial U}{\partial T}\right)_V = Cp - Cv - PV\beta = R(1 - \beta T) \tag{2.19}$$

$$\beta V = \left(\frac{\partial V}{\partial T}\right)_P, \tag{2.20}$$

where $\beta$ is the volume expansivity.

An isothermal compressibility can be defined as

$$\kappa = -\frac{1}{V}\left(\frac{\partial V}{\partial P}\right)_T. \tag{2.21}$$

For the regions of the diagram in Figure 2.1 where only a single phase exists, a relation is implied between the three quantities P, V, and T. Such a relation is referred to as the *PVT* equation of state. It relates pressure, molar volume, and temperature for a pure, one-component substance in the equilibrium state. An equation of state may be used to solve for any one of the three quantities P, V, and T as a function of the other two. For instance, V can be viewed as a function of temperature and pressure, V = f(T,P). Thus:

$$dV = \left(\frac{\partial V}{\partial T}\right)_P dT + \left(\frac{\partial V}{\partial P}\right)_T dP = \beta dT - \kappa dP. \tag{2.22}$$

When pure substances are found to deviate from ideal gas behavior, equations of state other than the ideal gas law can be used to describe the substances. Some of the equations of state used for substances whose molecules are small are described in the following sections.

### 2.1.3   VAN DER WAALS CUBIC EQUATION OF STATE

$$\left(P + \frac{a}{V^2}\right)(V - b) = RT, \tag{2.23}$$

where $b$ is the molecular volume per mole and $a$ represents the attraction forces between the molecules. The finite volume occupied by molecules in the box are accounted for by a parameter $b$. The intermolecular interaction forces are also accounted for by another parameter $a$. Johannes van der Waals [3] was awarded the Nobel Prize for his work.

### 2.1.4 VIRIAL EQUATION OF STATE

$$Z = \frac{PV}{RT} = 1 + \frac{B}{V} + \frac{C}{V^2} + \frac{D}{V^3} + \dots\dots$$

$$Z = 1 + B'P + C'P^2 + D'P^3 + \dots$$

(2.24)

The compressibility factor $Z$ is expressed as a sum of infinite series. The constants in the virial equation carry physical significance, and correlations such as Pitzer's are available for them. Kamerlingh Onnes [4], a Nobel laureate for his work at low temperatures and liquid helium, proposed Equation (2.24).

### 2.1.5 REDLICH AND KWONG EQUATION OF STATE AND SOAVE MODIFICATION

$$P = \frac{RT}{V - b} - \frac{a}{V(V + b)}.$$

(2.25)

The cubic equation of state was modernized by the introduction of the RK equation [5] as given by Equation (2.25) is a function of temperature. As follows, $a$ and $b$ are given as functions of temperature and other parameters:

$$a = \frac{\psi \alpha R^2 T_c^2}{P_c}$$

$$b = \frac{\Omega R T_c}{P_c},$$

(2.26)

where $T_c$ and $P_c$ are critical temperature and critical pressure, $\psi$ and $\Omega$ are pure numbers independent of substance and determined for a particular equation of state. The parameter $\alpha$ is a function of the reduced temperature $Tr = (T/T_c)$. The RK equation is a two-parameter corresponding states correlation. The Soave, Redlich, and Kwong (SRK) equation [6] is a modification of the RK equation and is a three-parameter corresponding state correlation. The Peng–Robinson (PR) equation of state [7] has an additional parameter in the form of an acentric factor. The RK equation was an improvement over other equations of their time, but it performed poorly with respect to the liquid phase.

Large molecules such as polymers behave differently from small molecules. A PT diagram of a polymer [8] poly(4-methyl-1-pentene) is shown in Figure 2.2.

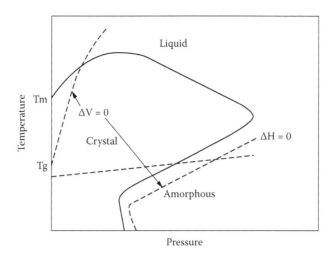

**FIGURE 2.2**   Temperature–pressure diagram for poly(4-methyl-1-pentene).

As can be seen in this figure, a polymer behaves differently compared with substances having small molecules. There is a crystalline phase in addition to the
amorphous solid and liquid phases. By the time the temperatures that are needed
to vaporize the polymer are achieved, in practice they may degrade or depolymerize. The rest of the chapter is devoted to the study of the equations of state
that can be used to describe the behavior of polymer systems. Negative pressures
must be watched for in these systems. One feature that comes out clearly from
the analysis presented in Sections 2.1.1 and 2.1.2 is that pressure in a box filled
with gas molecules is the force exerted by the gas molecules on the walls of the
container. The minimum pressure achievable is zero $Nm^{-2}$. This can happen at
very low temperatures when the velocity of the molecules reaches a state of rest
or zero velocity. At this juncture, or at some time prior to this juncture, the force
exerted by the molecules on the walls of the container will be zero and hence
the pressure is zero. Any value of pressure lower than 0, such as "negative pressure," cannot accurately describe real substances. It could be that it derives from
a mathematical model that is no longer valid to describe the system in that range
of conditions.

For example, a cubic equation of state can make a positive "x intercept" in the
pressure–temperature diagram. Such an *isochor* or phase boundary envelope can
imply an existence of negative pressure. But time and again it turns out that, at very
low temperatures, no equation of state is available that depicts real substances well.
So the positive x intercept has to be interpreted as an incorrect description of the
cubic equation of state. Experimental data is needed to define the envelope of appli-
cability of the contours predicted by equations of state or phase boundaries developed from sound theories. An example of such an envelope is shown in Figure 2.2 as
the dotted-line contours of $\Delta V = 0$ for the polymeric system. The isotherms at zero
pressure can be noted in equations of states, EOS, that are polynomials of higher
order.

## 2.2  PVT RELATIONS FOR POLYMERIC LIQUIDS

Numerous equations of state for polymer liquids have been developed. These equations provide valuable thermodynamic information that can be used to predict properties of polymer blends and polymer solutions. The predictions of phase behavior of polymer blends vary considerably from one equation of state to another. EOS theories for polymer liquids can be roughly grouped into three kinds:

1. Lattice-fluid theory
2. Hole models
3. Cell models

The three categories differ in how they represent the compressibility and expansion of the polymer systems under scrutiny. Volumetric changes are restricted to a change in cell volume in cell models. Lattice vacancies are allowed in lattice-fluid theory, and the cell volume is assumed constant. Cell expansion and lattice vacancies are allowed by hole models. The models also differ on the lattice type, such as a face-centered cubic, orthorhombic, hexagonal, and also in their selection of interpolymer/interoligomer potential such as Lennard–Jones potential, hard-sphere, or square-well.

Six of the commonly used EOS for polymers were reviewed by Rodgers [9]. The six EOS are discussed in the following sections.

## 2.3  TAIT EQUATION

An empirical representation of PVT data of polymer systems is given by the Tait equation. It can be viewed as an isothermal compressibility model. The general form of the Tait equation can be written as follows:

$$V = V_0\left(1 - C\ln\left(1 + \frac{P}{B}\right)\right). \tag{2.27}$$

The Tait equation was first developed over 120 years ago [10] in order to fit compressibility data of freshwater and seawater. Modifications of the Tait equation are used for fitting liquid density data worldwide. The constant $C$ in Equation (2.27) is taken as 0.0894 and is a universal constant. The *zero pressure isotherms* are given by

$$V_0 = V'e^{(\alpha T)}. \tag{2.28}$$

where $\alpha$ is the thermal expansion coefficient. The Tait parameter, $B$, can be expressed as

$$B = B_0 e^{-B_1 T}. \tag{2.29}$$

The Tait equation is a four-parameter representation of the $P$-$V$-$T$ behavior of polymers. The four parameters are the zero pressure isotherm, $V_0$, Tait parameter, $B$, thermal expansion coefficient, $\alpha$, and activation energy, $B_1$. In some industrial systems polynomial expressions are used for the zero-pressure isotherm and Tait parameter as follows:

$$V_0 = a_0 + a_1T + a_2T^2$$
$$B = b_0 + b_1T + b_2T^2. \tag{2.30}$$

The zero-isotherm and Tait parameter for some polymer systems were provided for 56 polymers by Rodgers [9]. The values for some systems are given in Table 2.1. For a few polymers, the temperature, $T$, is expressed as °C.

In Figure 2.3 the isotherms are shown at 25°C and 100°C for polystyrene in the pressure volume diagram for polystyrene as predicted by the Tait equation given in Table 2.1. It can be seen that the model does not predict the change at the glass transition temperature of the polymer. Further, the effects of molecular weight of the polymer are also not taken into account. The specific volume as a function of pressure for polystyrene in shown in Figure 2.3.

**TABLE 2.1**
**Tait EOS Parameters for Some Polymer Liquids**

| Polymer | Zero-Pressure Isotherm $V_0$ | Tait Parameter $B$ |
|---|---|---|
| Low density polyethylene | $1.19 + 2.841 \times 10^{-4}T + 1.87 \times 10^{-6}T^2$ | $2022e^{-5.243\times10^{-3}T}$ |
| PS | $0.9287e^{5.131\times10^{-4}T}$ (°C) | $2169e^{-3.319\times10^{-3}T}$ |
| PMMA | $0.8254 + 2.84 \times 10^{-4}T + 7.8 \times 10^{-7}T^2$ | $2875e^{-4.146\times10^{-3}T}$ |
| Polydimethylsiloxane | $1.0079e^{9.121\times10^{-4}T}$ | $894e^{-5.701\times10^{-3}T}$ |
| Polytetrafluoroethylene | $0.32 - 9.59 \times 10^{-4}T$ | $4252e^{-9.380\times10^{-3}T}$ |
| Polybutadiene | $1.0970e^{6.600\times10^{-4}T}$ | $1777e^{-3.593\times10^{-3}T}$ |
| PEO | $0.8766e^{7.0871\times10^{-4}T}$ | $2077e^{-3.947\times10^{-3}T}$ |
| Isoptatic polypropylene | $1.1606e^{6.7\times10^{-4}T}$ | $1491e^{-4.177\times10^{-3}T}$ |
| PET | $0.6883 + 5.9 \times 10^{-4}T$ | $3697e^{-4.1503\times10^{-3}T}$ |
| Polyphenylene oxide | $0.781e^{2.151\times10^{-5}T^{3/2}}$ (Kelvin) | $2278e^{-4.290\times10^{-3}T}$ |
| PC | $0.7357e^{1.859\times10^{-5}T^{3/2}}$ (Kelvin) | $3100e^{-4.078\times10^{-3}T}$ |
| Polyether ether ketone | $0.7158e^{6.690\times10^{-4}T}$ | $3880e^{-4.214\times10^{-3}T}$ |
| Polyvinylmethyl ether | $0.9585e^{6.653\times10^{-4}T}$ | $2158e^{-4.588\times10^{-3}T}$ |
| Polyamide-6 | $0.7597e^{4.7011\times10^{-4}T}$ | $3767e^{-4.660\times10^{-3}T}$ |
| Tetramethyl polycarbonate | $0.8497 + 5.073 \times 10^{-4}T + 3.83 \times 10^{-7}T^2$ | $2314e^{-4.242\times10^{-3}T}$ |
| Ethyl vinylacetate | $1.024e^{2.173\times10^{-5}T^{3/2}}$ (Kelvin) | $1822e^{-4.537\times10^{-3}T}$ |

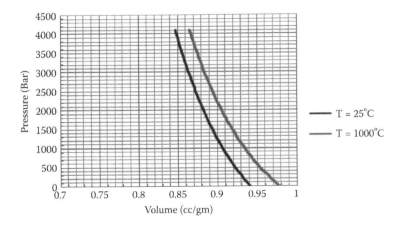

**FIGURE 2.3**   P–V diagrams for polystyrene.

The equation of state parameters for the Tait equation may be calculated as follows:

$$\left(\frac{\partial P}{\partial V}\right)_{T,cr} = 0$$

$$\left(\frac{\partial^2 P}{\partial V^2}\right)_{T,cr} = 0$$

(2.31)

The form of the Tait equation allows for Equation (2.31) to be met at infinite pressure.

## 2.4   FLORY, ORWOLL, AND VRIJ (FOV) MODEL

Au contraire to the empirical equation of Tait for EOS predictions, theoretical models can be used but generally require an understanding of forces between the molecules. These laws, strictly speaking, need be derived from quantum mechanics. However, Lenard–Jones potential and hard-sphere law can be used. The use of *statistical mechanics* is an intermediate solution between quantum and continuum mechanics. A *canonical partition function* can be formulated as a sum of Boltzmann's distribution of energies over all possible states of the system. Necessary assumptions are made during the development of the partition function. The thermodynamic quantities can be obtained by use of differential calculus. For instance, the thermodynamic pressure can be obtained from the partition function $Q$ as follows:

$$P = k_B T \frac{\partial \ln Q}{\partial V}$$

(2.32)

where $k_B$ is the Boltzmann constant, $Jmole^{-1}K^{-1}$. The resulting expression for pressure, $P$, as a function of temperature and specific volume is the EOS.

The principle of corresponding states can be used to express the pressure, temperature, and specific volume in terms of *reduced variables*. Experimental observations reveal that the compressibility factor, $Z$ Equation (2.24), for different fluids exhibits similar behavior when correlated as a function of reduced temperature, $T_r$ and reduced pressure, $P_r$. The reduced variables may be defined with respect to some characteristic quantity. For example, they can be defined as follows with respect to critical temperature and critical pressure:

$$T_r = \frac{T}{T_c}$$
$$P_r = \frac{P}{P_c}. \qquad (2.33)$$

The theorem of corresponding states is simply as follows:

> All fluids, when compared at the same reduced temperature and reduced pressure, have approximately the same compressibility factor, and all deviate from ideal-gas behavior to about the same degree.

Corresponding-states correlations of Z based on this theorem are referred to as two-parameter correlations. They require the use of two reducing parameters, $T_c$ and $P_c$. These correlations are nearly exact when used to describe noble gases such as argon, krypton, and xenon. When complex fluids were encountered, a three-parameter corresponding states parameter was found to be needed. An acentric factor, $\omega$, can be defined as introduced by Pitzer [11]. The acentric factor for a pure substance is defined with respect to its vapor pressure;

$$\frac{d\log_{10}\left(P_r^{sat}\right)}{d\left(\dfrac{1}{T_r}\right)} = Slope. \qquad (2.34)$$

The logarithm of the vapor pressure of a pure fluid is linear with respect to the reciprocal of absolute temperature. The slope is expected to be the same for all pure fluids if the two-parameter theorem of corresponding states is valid. When experimental measurements are obtained this is found not to be case. The slope changes with the change in fluid type. This can serve as the third corresponding state parameter. Pitzer [11] noted that all vapor pressure data for pure fluids such as Ar, Kr, and Xe lie on the same line when plotted as $\log_{10}(P_r^{sat})$ versus $1/T_r$, and that line passes through $\log_{10}(P_r^{sat})$ and $T_r = 0.7$. Data from other fluids define locations that can be fixed in relation to the line for the simple fluids, SF, by the difference:

$$\log_{10}P_r^{sat}(SF) - \log_{10}P_r^{sat}{}_{(T_r=0.7)}. \qquad (2.35)$$

The acentric factor is defined as this difference evaluated at $T = 0.7$:

$$\omega = -1.0 - \log_{10} P_r^{sat}{}_{(T_r=0.7)}. \tag{2.36}$$

Therefore, $\omega$ can be determined for any fluid from $T_c$, $P_c$, and a single vapor pressure measurement made at $T_r = 0.7$. Values of $\omega$ for different substances are available in Tables 2.2 and 2.3.

The thermodynamic variables are rendered dimensionless by using characteristic parameters for polymers. The characteristic parameters need be chosen allowing the equation to obey the corresponding states principle. The FOV EOS is one that is extensively used. The FOV equation of state can be written as

$$\frac{P_r V_r}{T_r} = \frac{V_r^{1/3}}{(V_r^{1/3} - 1)} - \frac{1}{T_r V_r}, \tag{2.37}$$

where $P_r = P/P^*$; $T_r = T/T^*$; $V_r = V/V^*$. The characteristic parameters are given by

$$T^* = \frac{s\eta}{ck_B}$$

$$P^* = \frac{ck_B T^*}{V^*} \tag{2.38}$$

$$V^* = \sigma^3,$$

where $s$ is the contacts per segment, $\sigma$ is the "hard-sphere" radius, and $\eta$ is the intersegment interaction energy. $V^*$ can be seen to be the hard-core cell volume. The characteristic parameters, $P^*$, $T^*$, and $V^*$, for some industrial polymer systems are given in Table 2.2 [9].

Negative pressure has no meaning physically. When Equation (2.37) is applied to real systems, the predictions may indicate negative values for pressure. Absolute pressure at the minimum can be 0 Nm$^2$ but no lower. It is the force per unit area exerted by the molecules on the walls of the container. In Section 2.1.1, the kinetic representation of pressure was derived for ideal gas. The zero pressure isotherms can be written for FOV as follows:

$$\frac{V_r^{4/3}}{(V_r^{1/3} - 1)} = \frac{1}{T_r} \, or \, T_r = \frac{1}{V_r}\left(1 - \frac{1}{\sqrt[3]{V_r}}\right). \tag{2.39}$$

In Figure 2.4 the pressure-volume isotherm is shown at 25°C for polystyrene using the FOV equation of state. The characteristic parameters were obtained from Table 2.2. The EOS does not explicitly account for molecular weight and glass transition temperature effects.

**TABLE 2.2**

**Characteristic Temperature, Pressure, and Specific Volume for Use in FOV**

| Polymer | P*(bar) | T*(K) | V*(cc/gm) |
|---|---|---|---|
| LDPE | 5292 | 6356 | 0.9794 |
| PS | 4052 | 8118 | 0.8277 |
| PMMA | 5688 | 7717 | 0.7204 |
| PDMS | 3269 | 5184 | 0.8264 |
| PTFE | 4049 | 7088 | 0.4215 |
| PBD | 4544 | 5522 | 0.9173 |
| PEO | 6016 | 7147 | 0.7719 |
| i-PP | 3974 | 7011 | 1.0072 |
| PET | 8510 | 8215 | 0.6452 |
| PPO | 6509 | 7360 | 0.7472 |
| PC | 6710 | 8039 | 0.7070 |
| PEEK | 8329 | 8667 | 0.6642 |
| PVME | 5128 | 6607 | 0.8266 |
| PA6 | 4110 | 9182 | 0.6896 |
| TMPC | 5142 | 8156 | 0.7720 |
| EVA18 | 4536 | 6870 | 0.9724 |

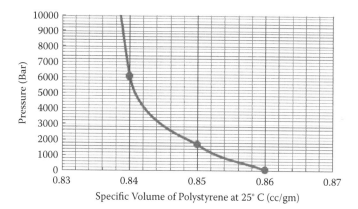

**FIGURE 2.4**   Pressure–volume isotherm for polystyrene at 25°C using FOV.

## 2.5  PRIGOGINE SQUARE-WELL CELL MODEL

Polymer molecules are modeled as having two distinct sets of modes contributing to the partition function in the cell models. The two modes are internal and external modes. The internal modes are used to represent the internal motions of the molecules, and the intermolecular interactions are accounted for by the external modes. Prigogine and Kondepudi [12] proposed the conceptual separation of the two modes. The PVT properties of the polymer systems will be affected by the external modes. A polymer molecule is divided into *r* repeat units. Each repeat unit has 3 degrees of freedom. The changes in cell volume can be used to explain the compressibility and thermal expansivity of the polymer system. Different models developed based on this approach differ in their form of intermolecular potential used and the contact geometry of segments of the polymer. The Lennard–Jones potential and cell model for liquids are applied to polymer liquids [12].

Two forms of the cell model (CM) are then developed: harmonic oscillator approximation and square-well approximation. Both forms assume hexagonal closed packing (HCP) lattice structure for the cell geometry. The model developed by Paul and Di Benedetto [13] assumes that the chain segments interact with a cylindrical symmetric square-well potential. The FOV model discussed in the earlier section uses a hard-sphere type repulsive potential along with a simple cubic (SC) lattice structure. The square-well cell model by Prigogine was modified by Dee and Walsh [14]. They introduced a numerical factor to decouple the potential from the choice of lattice structure. A universal constant for several polymers was added and the modified cell model (MCM) was a three-parameter model. The Prigogine cell EOS model can be written as follows.

$$\frac{P_r V_r}{T_r} = \frac{V_r^{1/3}}{\left(V_r^{1/3} - 0.8909\right)} - \frac{2}{T_r}\left(\frac{1.2045}{V_r^2} - \frac{1.011}{V_r^4}\right). \tag{2.40}$$

The variables used in Equation (2.40) are similar to those used in the FOV model. The HCP lattice structure leads to the 0.8909. The Lenard–Jones/612 potential is assumed.

**TABLE 2.3**

**Characteristic Pressure, Temperature for Prigogine CM**

| Polymer | P*(bar) | T*(K) | V*(cc/gm) |
|---------|---------|-------|-----------|
| LDPE    | 5917    | 5007  | 1.1606    |
| PS      | 5990    | 5167  | 0.9148    |
| PMMA    | 7650    | 4783  | 0.7941    |
| PDMS    | 3880    | 3254  | 0.9180    |
| PTFE    | 3239    | 4488  | 0.4829    |
| PBD     | 6815    | 3811  | 1.0269    |
| PEO     | 6968    | 4226  | 0.8447    |
| i-PP    | 4258    | 4712  | 1.1470    |
| PET     | 8688    | 5070  | 0.7193    |
| PPO     | 6685    | 4585  | 0.8355    |
| PC      | 7747    | 4969  | 0.7835    |
| PEEK    | 7970    | 5418  | 0.7465    |
| PVME    | 6770    | 4298  | 0.9210    |
| PA6     | 4654    | 7048  | 0.7956    |
| TMPC    | 6094    | 4875  | 0.8473    |
| EVA18   | 5445    | 4434  | 1.0874    |

The MCM model proposed by Dee and Walsh [14] can be written as follows:

$$\frac{P_r V_r}{T_r} = \frac{V_r^{1/3}}{(V_r^{1/3} - 0.8909q)} - \frac{2}{T_r}\left(\frac{1.2045}{V_r^2} - \frac{1.011}{V_r^4}\right). \tag{2.41}$$

They noted that the potential and hard-core cell volume were coupled by the choice of cell lattice structure. In order to decouple from a specific lattice they introduced a quantitative factor that can be used to scale the hard-core cell volume in the free volume term. The factor 1 was found to be nearly constant for several polymers and falls out as 1.07. The reduced variables and characteristic parameters used in Equation (2.41) are the same as those used for the CM model.

## 2.6 LATTICE FLUID MODEL OF SANCHEZ AND LACOMBE [15]

Sanchez and Lacombe [15] developed a molecular theory of classical fluids based on a well-defined statistical mechanical model. The model fluid reduces to the classical lattice gas in one special case. It can be characterized as an Ising or lattice fluid. The model fluid undergoes a liquid–vapor transition. Only three parameters are required to describe a fluid. These parameters have been determined and tabulated for several fluids.

The Gibbs free energy, $G$, can be expressed in terms of the configurational partition function $Z$ in the pressure ensemble by

$$G = -k_B T \ln(Z)$$

$$Z = \sum_V \sum_E \Omega e^{-\beta(E + PV)}, \tag{2.42}$$

where $\Omega$ is the number of configurations available to a system of $N$ molecules whose configurational potential energy and volume are $E$ and $V$, respectively. The summations extend over all values of $E$ and $V$. The fundamental problem is to determine $\Omega$. In the lattice fluid theory, the problem is to determine the number of configurations available to a system of $N$ molecules, each of which occupied $r$ sites (an $r$-sized oligomer) and $N_0$ vacant lattice sites (holes). This problem is yet to be solved completely for the simple case of dimers and holes on a rectangular lattice. Guggenheim [27] obtained an approximate value of $\Omega$ for a multicomponent mixture of $r$-mer on a lattice. Guggenheim's solution for a binary mixture of $r$-length oligomers and monomers (holes) shall be used to evaluate the partition function using a *mean field approximation*. The number of lattice sites for a binary mixture of $N_r$ length oligomers and $N_0$ empty sites is given by

$$N_r = N_0 + rN. \tag{2.43}$$

The coordination number of the lattice Z is given by Equation (2.44). Each interior mer of a linear chain is surrounded by the Z−2 nearest nonbonded neighbors and two bonded neighbors.

$$qZ = (r-2)(Z-2) + 2(Z-1) = r(Z-2) + 2. \tag{2.44}$$

The total number of nearest neighbor pairs in the system is $(Z/2)N_r$. Only $(Z/2)N_q$ are nonbonded pairs where

$$N_q = N_0 + qN. \tag{2.45}$$

The numbers $N_r$, $N_q$, and $N_0$ are related by

$$\frac{Z}{2}N_q = N_r\left(\frac{Z}{2}-1\right) + N + N_0. \tag{2.46}$$

An $r$ repeat oligomer is characterized by a symmetry number $\sigma$. It is also characterized by a flexibility parameter $\delta$. It is equal to the number of ways in which the $r$ repeat oligomer can be arranged on a lattice after one of its repeat units has been fixed on a lattice site. It is a measure of the $r$ length oligomers' internal degrees of freedom: $\delta_{max} = Z(Z-1)^{r-2}$. The number of configurations available to a system of $Nr$ length oligomers and $N_0$ empty sites is given by

$$\Omega = \left(\frac{\delta}{\sigma}\right)^N \frac{N_r!}{N_0!N!}\left(\frac{N_q!}{N_r!}\right)^{Z/2}. \tag{2.47}$$

Using Stirling's approximation,

$$n! \cong \left(\frac{n}{e}\right)^n \tag{2.48}$$

$$Lt_{Z\to\infty}\Omega = \left(\frac{1}{f_0}\right)^{N_0}\left(\frac{\omega}{f}\right)^N,$$

where $\omega = \delta r / \sigma e^{r-1}$. The empty site fraction $f_0$ and the fraction of occupied sites $f$ are given by

$$f_0 = \frac{N_0}{N_r}$$

$$f = \frac{rN}{N_r}. \tag{2.49}$$

The large Z limit equation is the Flory approximation. The parameter $\delta$ is independent of temperature and pressure. The closed packed volume $rV*$ of a molecule is independent of temperature and pressure. The closed packed volume of a repeat unit is $V*$. It is also the volume of a lattice site. The closed packed volume of $N_r$ repeat oligomers (no holes) is

$$V* = N(r,V*).$$ (2.50)

If $\rho*$ is the closed packed mass density, then the closed packed molecular volume is given by

$$rV* = \frac{M}{\rho*},$$ (2.51)

where $M$ is the molecular weight and $\rho*$ is the crystal density as a first approximation. The volume is the product of number of sites times the site volume:

$$V = (N_0 + rN)V* = N_rV* = \frac{V*}{f}$$

$$E = -\frac{Z}{2}N_r \sum_i \sum_j p(i,j)\varepsilon_{ij}.$$ (2.52)

The energy of lattice depends on nearest neighbor interactions: $\varepsilon_{ij}$ is the pair interaction energy between components $i$ and $j$, and $p(i,j)$ is the pair joint probability of an $(i,j)$ pair in the system. In the present case, there are only two component "holes" and "mers." The only nonzero pair interaction energy is the one associated with nonbonded $r$ repeat oligomer–$r$ repeat oligomer interactions. Hole–hole, hole–repeat unit(s), and bonded repeat unit(s)–repeat unit(s) are assigned zero energy. If a random mixing of holes and molecules is assumed, then,

$$p(i,j) = \frac{qN^2}{N_q N_r}.$$ (2.53)

Equation (2.53) becomes in the large Z limit

$$p(i,j) = \left(\frac{rN}{N_r}\right)^2 = f^2$$

$$E = -N_r f^2 \left(\frac{Z\varepsilon}{2}\right)f^2 \quad or \quad \frac{E}{rN} = -\frac{\varepsilon * V*}{V} = -\varepsilon * f,$$ (2.54)

where $\varepsilon$ is the nonbonded repeat unit–repeat unit interaction energy and $\varepsilon* = Z\varepsilon/2$ is the total interaction energy per repeat unit. The quantity $r\varepsilon*$ is the characteristic interaction energy per molecule in the absence of holes: $\varepsilon*$ is also the energy required to create a lattice vacancy (hole). Since $E$ and $\Omega$ are functions of a single parameter, the number of holes in the lattice, the double sum over $E$ and $V$ required in the evaluation of the partition function, can be replaced by a single sum over $N_0$:

$$Z = \sum_{0}^{\infty} \Omega e^{-\beta(E+PV)}$$

(2.55)

$$G = E + PV - k_B T \ln \Omega.$$

The general procedure in statistical mechanics is to approximate the above sum by its maximum term. The maximum term is overwhelmingly larger than any other for a macroscopic system. Mathematically, this is equivalent to equating the free energy to the logarithm of the generic term in the partition function and then finding the minimum value of the free energy. In terms of dimensionless variables,

$$\frac{G}{Nr\varepsilon*} \cong \bar{G} = -\rho_r + P_r V_r + T_r \left( (V_r - 1)\ln(1-\rho_r) + \frac{1}{r}\ln\left(\frac{\rho_r}{w}\right) \right).$$

(2.56)

The $T_r$, $P_r$, $V_r$, and $\rho_r$ are reduced variables defined with respect to the characteristic parameters. Thus:

$$P_r = \frac{P}{P*}; \; T_r = \frac{T}{T*}$$

$$V_r = \frac{V}{V*}; \; T* = \frac{\varepsilon*}{k_B}; \; P* = \frac{\varepsilon*}{V*}; \; V* = n(rv*).$$

(2.57)

The fraction of occupied sites f and empty site fraction $f_0$ are related to the mass density, $\rho$, and the close packed mass density, $\rho*$, by

$$f = \frac{\rho}{\rho*}; f_0 = 1 - \rho_r.$$

(2.58)

The free energy is minimized as in other models:

$$\rho_r^2 + P_r + T_r \left( \ln(1-\rho_r) + \rho_r \left( 1 - \frac{1}{r} \right) \right) = 0.$$

(2.59)

The equation of state is given by Equation (2.59). It has the capability of accounting for molecular weight effects. The complete thermodynamic description of the model is given by

$$G = E + PV - k_B T \ln \Omega$$

$$\frac{G}{Nr\varepsilon^*} \cong G_r = -\rho_r + P_r V_r + T_r\left((V_r - 1)\ln(1 - \rho_r) + \frac{1}{r}\ln\left(\frac{\rho_r}{w}\right)\right). \qquad (2.60)$$

All other thermodynamic properties follow from the standard thermodynamic formulas. The volume expansivity, $\beta$, and isothermal compressibility, $\kappa$, can be written as follows:

$$T\beta = \frac{1 + P_r V_r^2}{T_r V_r\left(\dfrac{1}{V_r - 1} + \dfrac{1}{r}\right) - 2}$$

$$P\kappa = \frac{P_r V_r^2}{T_r V_r\left(\dfrac{1}{V_r - 1} + \dfrac{1}{r}\right) - 2}. \qquad (2.61)$$

If the pressure is increased isothermally, the vaporlike minimum will disappear. If the pressure is decreased isothermally, the liquidlike minimum will eventually disappear. The model fluid undergoes a liquid–vapor transition. The spinodal and binodal curves can be calculated from the Gibbs free energy expression. If at a given pressure and temperature, two minima appear, the higher free energy minimum corresponds to a *metastable state*. The locus of the pressures and temperatures that define the metastability limit of a phase is called the *spinodal*. The discussion about spinodal and binodal coinciding is controversial. The spinodal condition can be obtained from Equation (2.60) and the tangency condition is

$$\frac{\partial \ln \rho_r}{\partial \rho_r} = 1. \qquad (2.62)$$

The reduced spinodal temperature can be expressed in terms of spinodal density:

$$T_r^s = \frac{2\rho_r^s(1 - \rho_r^s)}{\left(\rho_r^s + (1 - \rho_r^s)\dfrac{1}{r}\right)}. \qquad (2.63)$$

The spinodal pressure can be expressed as

$$P_r^s = \frac{-2\rho_r^s(1-\rho_r^s)\ln(1-\rho_r^s)+\left(1-\frac{1}{r}\right)\rho_r^{s2}(1-\rho_r^s)+\rho_r^{s2}}{\rho_r^s+(1-\rho_r^s)\frac{1}{r}}.$$  (2.64)

Equation (2.64) is valid for all values of spinodal reduced density, $\rho_r^s$, for which the spinodal pressure is greater than 0, $P_r^s > 0$. The critical point is obtained when the density of the two phases become equal so that the critical point not only satisfies Equation (2.62) but also the condition

$$\frac{\partial^2 \ln\rho_r}{\partial\rho_r^2} = 0.$$  (2.65)

A pure fluid is completely characterized by the three molecular parameters, $v^*$, $r$, and $\varepsilon^*$, and the scale factors $T^*$, $P^*$, and $\rho^*.P^*v^*/RT^* = 1$ and $M/r = v^*\rho^* = RT^*\rho^*/P^*$. In principle, any thermodynamic property can be utilized to determine these parameters. Saturated vapor pressure data is useful as they are readily available for a variety of fluids. A compendium of such data is available for organic liquids. The lattice fluid theory of Sanchez and Lacombe as described is similar to the van der Waals EOS as discussed in the earlier section for small molecules. The virial form of the EOS of Sanchez and Lacombe can be written as follows:

$$\frac{PV}{Nk_BT} = \frac{rP_rV_r}{T_r} = 1+r\rho_r\left(\frac{1}{2}-\frac{1}{T_r}\right)+\frac{r}{3}\rho_r^2+\frac{r}{4}\rho_r^3+......$$  (2.66)

The characteristic parameters for Sanchez and Lacombe free volume EOS for selected polymers are given in Table 2.4.

The EOS predictions of Sanchez and Lacombe are shown in Figure 2.5 for PMMA at 25°C at infinite molecular weight. The Sanchez and Lacombe (SL) EOS allows for the effect of polymer molecular weight in the EOS. The characteristic temperature values in Table 2.4 for the same set of polymers as indicated before for SL, EOS, are a lot lower compared to other EOS values. The proximity of the characteristic temperatures of polymers to operating temperatures in the molding and other polymer processing operations may indicate that this model may be a better representation of polymer PVT behavior compared with other models.

**TABLE 2.4**

**Characteristic Pressure, Temperature for Sanchez–Lacombe Lattice Fluid Theory**

| Polymer | P*(bar) | T*(K) | ρ*(gm/cc) |
|---|---|---|---|
| LDPE | 4679 | 615 | 0.9137 |
| PS | 3715 | 688 | 1.1199 |
| PMMA | 5169 | 668 | 1.2812 |
| PDMS | 2885 | 466 | 1.1084 |
| PTFE | 3572 | 630 | 2.2150 |
| PBD | 4402 | 462 | 1.0151 |
| PEO | 4922 | 656 | 1.1776 |
| i-PP | 3664 | 633 | 0.9126 |
| PET | 7261 | 761 | 1.4081 |
| PPO | 5541 | 681 | 1.2166 |
| PC | 5744 | 728 | 1.2925 |
| PEEK | 7137 | 804 | 1.3679 |
| PVME | 4630 | 567 | 1.1198 |
| PA6 | 4225 | 785 | 1.3502 |
| TMPC | 4324 | 752 | 1.1774 |
| EVA18 | 4077 | 612 | 0.9452 |

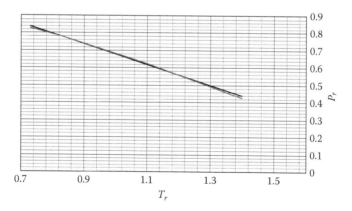

**FIGURE 2.5** Reduced density versus temperature for poly(methyl methacrylate) (PMMA) at 25°C by Sanchez–Lacombe lattice fluid theory.

## 2.7  NEGATIVE COEFFICIENT OF THERMAL EXPANSION

The volume expansivity of pure substances as given by Equation (2.20) is

$$\beta = \frac{1}{V}\left(\frac{\partial V}{\partial T}\right)_P. \tag{2.67}$$

It is a parameter that is used to measure the volume expansivity of pure substances and is defined at constant pressure, P. In the field of materials science, the property of *linear coefficient of thermal expansion* is an important consideration in materials selection and design of products. This property is used to account for the change in volume when the temperature of the material is changed. The linear coefficient of thermal expansion is defined as

$$\alpha = \left(\frac{l_f - l_0}{l_0\left(T_f - T_0\right)}\right) = \frac{\Delta l}{l_0 \Delta T} = \frac{\varepsilon}{\Delta T}. \tag{2.68}$$

For isotropic materials, $\beta = 3\alpha$. Instruments such as dilatometers and x-ray diffraction (XRD) can be used to measure the thermal expansion coefficient. Typical values of volume expansivity for selected isotropic materials at room temperature are provided in Table 2.5.

**TABLE 2.5**
**Volume Expansivity of Selected Materials at Room Temperature**

| # | Material | $\beta$, Volume expansivity ($*10^{-6}$ K$^{-1}$) |
|---|----------|---------------------------------------------------|
| 1 | Aluminum | 75.0 |
| 2 | Copper | 49.8 |
| 3 | Iron | 36.0 |
| 4 | Silicon | 9.0 |
| 5 | 1020 Steel | 36.0 |
| 6 | Stainless steel | 51.9 |
| 7 | Epoxy | 165.0 |
| 8 | Nylon 6,6 | 240 |
| 9 | Polyethylene | 300.0 |
| 10 | Polystyrene | 210 |
| 11 | Partially stabilized $ZrO_2$ | 31.8 |
| 12 | Soda–lime glass | 27.0 |
| 13 | Zirconium tungstate $ZrW_2O_8$ | −27.0 |
| 14 | Faujasite | −12.0 |
| 15 | Water (0–4 K) | Negative |
| 16 | Honey | Negative |

As can be seen from Table 2.5, the volume expansivity for pure substances is usually positive. In some cases it can be negative. Examples of materials with negative values for volume expansivity are water in the temperature range of O–4 K, honey, monoclinic selenium (Se), tellurium (Te), quartz, glass, faujasite, and cubic zirconium tungstate (alpha-$ZrW_2O_8$) [16] in the temperature range of 0.3–1050 K. Zerodur™ [17] is a glass-ceramic material that can be controlled to have a zero or slightly negative thermal coefficient of expansion and was developed by Schott Glass Technologies. It consists of a 70–80 wt% crystalline phase that has a high-quartz structure. The rest of the material is a glassy phase. The negative thermal expansion coefficient of the glassy phase and the positive thermal expansion coefficient of the crystalline phase are expected to cancel each other out, leading to a zero thermal coefficient material. Zerodur has been used as the mirror substrate on the Hubble telescope and the Chandra x-ray telescope. A dense, optically transparent and zero-thermal expansion material is necessary in these applications, because any changes in dimensions as a result of the changes in the temperature in space will make it difficult to focus the telescopes appropriately. Material scientists have developed ceramic materials based on sodium zirconium phosphate (NZP) that have a near-zero thermal-expansion coefficient.

The occurrence of negative values of volume expansivity quid pro quo is a violation of the second law of thermodynamics, according to some investigators such as Stepanov [18]. They propose that the first law of thermodynamics be changed from $dU = dQ - PdV$ to $dU = dQ + PdV$ in order to work with materials with a zero or negative coefficient of thermal expansion. This study examines where the second law of thermodynamics is violated, and proposes an isentropic volume expansivity definition in place of an isobaric volume expansivity as shown in Equation (2.1). In a famous problem, such as the development of the theory of the velocity of sound, such a change from defining an isothermal compressibility to isentropic compressibility brought the experimental observations closer to theory [19–21]. The proposal from this study is in part motivated by the work of Laplace (see the following section).

## 2.7.1 HISTORICAL NOTE

By the seventeenth century, it was realized that sound propagates through air at some finite velocity. Artillery tests have indicated that the speed of sound is approximately 1140 ft/s. These tests were performed by standing a known large distance away from a cannon, and noting the time delay between the light flash from the muzzle and the sound of the discharge. In Proposition 50, Book II of his *Principia*, Newton [20] theorized that the speed of sound was related to the *elasticity* of the air and could be given by the reciprocal of the compressibility. He assumed that the sound wave propagation was an isothermal process. He proposed the following expression for the speed of sound:

$$c = \sqrt{\frac{1}{\rho \kappa_T}}, \tag{2.69}$$

where the isothermal compressibility is given by

$$\kappa_T = -\frac{1}{V}\left(\frac{\partial V}{\partial P}\right)_T. \tag{2.70}$$

Newton calculated a value of 979 ft/s from this expression to interpret the artillery test results. The value was 15% lower than the gunshot data. He attributed the difference between experiment and theory to the existence of dust particles and moisture in the atmosphere. A century later Laplace [21] corrected the theory by assuming that the sound wave was isentropic and not isothermal. He derived the expression used to this day to instruct senior-level students in gas dynamics [19] for the speed of sound as

$$c = \sqrt{\frac{1}{\rho\kappa_s}} \tag{2.71}$$

$$\kappa_s = -\frac{1}{V}\left(\frac{\partial V}{\partial P}\right)_s. \tag{2.72}$$

By the time of the demise of Napoleon Bonaparte, the relation between propagation of sound in gas was better understood.

### 2.7.2  Violation of Second Law of Thermodynamics

From Equation (2.67),

$$\beta V = \left(\frac{\partial V}{\partial T}\right)_P. \tag{2.73}$$

From Maxwell Relations (Appendix A),

$$\left(\frac{\partial V}{\partial T}\right)_P = -\left(\frac{\partial S}{\partial P}\right)_T. \tag{2.74}$$

Combining Equation (2.73) and Equation (2.74), we get

$$\beta V = -\left(\frac{\partial S}{\partial P}\right)_T. \tag{2.75}$$

For a reversible process, the combined statement of the first and second laws can be written as

$$dH = C_p dT = TdS + VdP. \tag{2.76}$$

At constant temperature, Equation (2.76) becomes

$$TdS = -VdP.$$
(2.77)

From Equation (2.77)

$$\left(\frac{\partial S}{\partial P}\right)_T = -\frac{V}{T}.$$
(2.78)

Combining Equation (2.78) and Equation (2.75),

$$\beta = \frac{1}{T}.$$
(2.79)

From Equation (2.79), it can be seen that $\beta$ can never be negative, as the temperature is always positive, as stated by the third law of thermodynamics. So, materials with negative values for $\beta$, quid pro quo, violate the combined statement of the first and second laws of thermodynamics. Negative temperatures are not possible for vibration and rotational degrees of freedom. A freely moving particle or a harmonic oscillator cannot have negative temperatures, for there is no upper bound on their energies. Nuclear spin orientation in a magnetic field is needed for negative temperatures [22]. This is not applicable for engineering.

### 2.7.3 PROPOSED ISENTROPIC EXPANSIVITY

Along similar lines to the improvement given by Laplace to the theory of the speed of sound as developed by Newton (as discussed in Section 2.7.1) an *isentropic volume expansivity* is proposed:

$$\beta_s = \frac{1}{V}\left(\frac{\partial V}{\partial T}\right)_s.$$
(2.80)

Using the rules of partial differential for three variables, any function $f$ in variables $(x,y,z)$ it can be seen that

$$\left(\frac{\partial f}{\partial x}\right)_z = \left(\frac{\partial f}{\partial x}\right)_y + \left(\frac{\partial f}{\partial y}\right)_x\left(\frac{\partial y}{\partial x}\right)_z.$$
(2.81)

Thus,

$$\left(\frac{\partial V}{\partial T}\right)_s = \left(\frac{\partial V}{\partial T}\right)_P + \left(\frac{\partial V}{\partial P}\right)_T\left(\frac{\partial P}{\partial T}\right)_s.$$
(2.82)

For isentropic process, Equation (2.76) can be written as

$$C_p dT = V dP. \tag{2.83}$$

From Equation (2.83), it can be seen that

$$\left(\frac{\partial P}{\partial T}\right)_s = \frac{C_p}{V}. \tag{2.84}$$

Plugging Equation (2.84) and Equation (2.70) into Equation (2.82), we get

$$\beta_s V = \left(\frac{\partial V}{\partial T}\right)_s = \left(\frac{\partial V}{\partial T}\right)_P - C_p \kappa_T. \tag{2.85}$$

At constant pressure,

$$dH - dU = PdV. \tag{2.86}$$

Equation (2.86) comes from $H = U + PV$ the definition of specific enthalpy in terms of specific internal energy, $U$, pressure and volume, $P$ and $V$, respectively.

Equation (2.86) can be written as

$$(C_P - C_v)dT = PdV. \tag{2.87}$$

It can be realized from Equation (2.87) that

$$\left(\frac{\partial V}{\partial T}\right)_P = V\beta_P = \frac{(C_P - C_v)}{P} \tag{2.88}$$

Plugging Equation (2.88) into Equation (2.85), we have

$$\beta_s V = \left(\frac{\partial V}{\partial T}\right)_s = \frac{(C_P - C_v)}{P} - C_p \kappa_T \tag{2.89}$$

or

$$\beta_s = \frac{1}{V}\left(\frac{\partial V}{\partial T}\right)_s = \frac{(C_P - C_v)}{PV} - \frac{C_p \kappa_T}{P}. \tag{2.90}$$

For substances with a negative coefficient of thermal expansion under the proposed definition of isentropic volume, expansivity, $\beta_s$, does not violate the laws of thermodynamics quid pro quo.

Considering the thermal expansion process for pure substances in general and materials with negative coefficient of thermal expansion in particular, the process is not isobaric. Pressure was shown to be related to the square of the velocity of the molecules (Section 2.1.1).

### 2.7.4  MEASUREMENTS OF VOLUME EXPANSIVITY NOT ISOBARIC

The process of measurement of volume expansivity cannot be isobaric in practice. When materials expand, the root mean square velocity of the molecules increases. For the materials with negative coefficient of thermal expansion when materials expand, the root mean square velocity of molecules is expected to decrease. In either case, forcing such a process as isobaric is not a good representation of theory with experiments. Such processes can even be reversible or isentropic. Experiments can be conducted in a careful manner and the energy needed supplied or energy released removed, as the case may be, in a reversible manner. Hence, it is proposed to define volume expansivity at constant entropy. This can keep the quantity per se from violating the laws of thermodynamics.

## 2.8  SUMMARY

Thermodynamic models derived for small molecules may not be applicable for large molecules such as for polymers. The kinetic representation of pressure was derived from the force exerted by gas molecules colliding with the walls of the container. The ideal gas law was derived. The pressure/temperature (P/T) diagram for small molecules was discussed. Compressibility and volume expansivity of pure substances were defined. The van der Waals cubic equation of state (EOS), virial equation of state, Redlich Kwong EOS, and Soave modification of Redlich Kwong EOS were presented. The *P/T* diagram of poly(4-methyl-1-pentene) was shown to distinguish between the behavior of small molecules and large molecules. Negative pressure values are spurious.

EOS theories for polymer liquids were grouped into three categories: (1) lattice-fluid theory, (2) the hole model, and (3) the cell model. The Tait equation was first developed 121 years ago in order to fit compressibility data of freshwater and seawater. The Tait equation is a four-parameter representation of PVT behavior of polymers. Expressions for zero pressure isotherms and Tait parameters were provided, and values of Tait parameters for 16 commonly used polymers were tabulated.

FOV EOS theory was developed by formulation of canonical function of the Boltzmann distribution of energies and derivation of thermodynamic pressure. The theorem of corresponding states says that the same compressibility factor can be expected for all fluids when compared at reduced temperature and pressure. A two-parameter correlation for compressibility factor, Z, can be derived using the theorem of corresponding states. FOV EOS obeys the corresponding state principle. Characteristic temperature, pressure, and specific volume used in FOV EOS are tabulated for 16 commonly used polymers.

The Prigogine square-well cell model is based upon conceptual separation of internal and external modes for polymers separately accounted for in the partition function. Two forms of cell model, i.e., harmonic oscillator approximation and square-well approximation, were developed. The CM model is written in reduced variables and assuming Lennard–Jones 6/12 potential. The characteristic pressure/temperature for the CM model for 16 commonly used polymers were tabulated.

Sanchez and Lacombe developed the lattice fluid EOS theory using statistical mechanics. Gibbs free energy can be expressed in terms of configurational partition function Z in the pressure ensemble. In the lattice fluid theory development the problem is to determine the number of configurations available for a system of $N$ molecules each of which occupies $r$ sites and $N_0$ vacant sites or holes. Mean field approximation was used to evaluate the partition function. The SL EOS has the capability to account for molecular weight effects, unlike other EOS theories. Characteristic lattice fluid EOS parameters were tabulated for 16 commonly used polymers.

The volume expansion coefficient, $\beta$, cannot take on negative values and stay within the bounds of the laws of thermodynamics. An isentropic volume expansivity is proposed for materials reported in the literature with "negative thermal coefficient" values to stay within the bounds of thermodynamics.

## PROBLEMS

1. Obtain the virial form of the FOV EOS theory.
2. Obtain the virial form of the Prigogine square-well model?
3. Obtain the virial form of Tait EOS.
4. Obtain the virial form of the Sanchez and Lacombe free volume theory.
5. Derive expression for volume expansivity, $\beta$, and compressibility, $\kappa$ for the FOV EOS.
6. Derive an expression for volume expansivity, $\beta$, and compressibility $\kappa$ for Prigogine square-well EOS.
7. Derive an expression for volume expansivity, $\beta$, and compressibility, $\kappa$ for Sanchez and Lacombe free volume EOS.
8. The EOS developed by Sanchez and Lacombe using lattice fluid theory can be written as

$$\rho_r^2 + P_r + T_r \left[ \ln\left(1 - \rho_r\right) + \rho_r \left(1 - \frac{1}{r}\right) \right] = 0.$$

Calculate the density of PMMA at a temperature of 29°F and atmospheric pressure for
   a.  Molecular weight of 100,000
   b.  Molecular weight of 250,000. Do you get one unique solution? If you obtain multiple solutions, what is the significance of the other values?

9. Donahue and Prausnitz [23] developed the perturbed-hard chain theory (PHCT) based on perturbed hard-sphere theory for small molecules and Prigogine's theory for chain molecules. In order to account for attractive and repulsive forces among molecules, empirical parameters such as the $c$ parameter were introduced by subsequent investigators. More accurate expressions for the repulsive and attractive forces were

attempted in the PHCT model. The PHCT equation of state can be written as follows:

$$\frac{PV}{RT} = Z = 1 + cZ^{CS} + c\sum \frac{A_{nm}}{V_r^m T_r^{n-1}}. \tag{2.91}$$

Where $Z^{cs}$ can be obtained from the Carnahan–Starling equation. Calculate the EOS parameters for the PHCT EOS. Obtain *the* expression for volume expansivity.

10. The Born–Green–Yvon (BGY) integral equation of state [24] was derived and shown to fit PVT experimental data for both small molecules and polymers. The fit parameters were then used to make predictions about thermodynamic properties for the system of interest. In the athermal limit the BGY treatment is equivalent to Guggenheim's approximation of random mixing. The EOS can be written as follows:

$$P_r + T_r\left[\ln(1-\phi) - \frac{z\ln(1-\phi)}{2(1-\xi)} + \left(1 - \frac{1}{r}\right)\phi\delta_{kr} + \frac{zJ\xi^2}{2}\right] = 0, \tag{2.92}$$

where, $\delta_{kr}$ is the Kronecker delta and $J$ is given by

$$J = \frac{e^{-\frac{2}{zT_r}} - 1}{1 - \xi\left(1 - e^{-\frac{2}{zT_r}}\right)} \tag{2.93}$$

Obtain the EOS parameter of BGY equation of state and the expression for volume expansivity. Does the volume expansivity take on negative values?

11. The lattice cluster (LC) equation of state model was developed for polymers by Bawendi and Freed [25] and is given as follows:

$$P_r + T_r\left[\ln(1-\phi) + \phi + \phi^2\left(\frac{1}{T_r} - \frac{4}{zT_r} + \frac{1}{zT_r^2} + \frac{1}{z} + \frac{3}{z^2}\right)\right.$$
$$\left. + \phi^3\left(\frac{4}{zT_r} - \frac{4}{zT_r^2} - \frac{20}{3z^2}\right) + + \phi^4\left(\frac{3}{zT_r^2} + \frac{6}{z^2}\right)\right] = 0. \tag{2.94}$$

Obtain the EOS parameters of the LC equation of state and the expression for volume expansivity. Does the volume expansivity take on negative values?

12. The EOS developed by Panayiotou and Vera (PV) [26] using some concepts introduced by Guggenheim [27] is given below:

$$P_r + T_r\left[\ln(1-\phi) - \frac{z\ln(1-\phi)}{2(1-\xi)}\right] + \xi^2 = 0. \tag{2.95}$$

This EOS can be obtained from the BGY equation by expanding $J$ to $0(\beta\varepsilon)$. Obtain the EOS parameter of PV equation of state and the expression for volume expansivity. Does the volume expansivity take on negative values?

13. The Simha–Somcynsky (SS) [28] EOS is given below:

$$\frac{P_r V_r}{T_r} = \left[1 - 2^{-\frac{1}{6}} y (y V_r)^{-\frac{1}{3}}\right]^{-1} + \frac{2y}{T_r}(y V_r)^{-2}\left[1.011(y V_r)^{-2} - 1.2045\right].$$

(2.96)

The reduced variables are different for the SS equation of state; $y$ is the fraction of occupied sites, depending here on $r$ and on an additional parameter $c$ defined such that there are $3c$ degrees of freedom per chain. The reduced variables are given by

$$V_r = \frac{V}{Nrv^*}; \ T_r = \frac{c}{qz\beta\varepsilon}; \ P_r = \frac{Pr v^*}{qz\varepsilon},$$

(2.97)

where $v^*$ is the characteristic volume per unit segment. Calculate the EOS parameters for the SS equation. Obtain the expression for volume expansivity and check under what conditions do they take on negative values.

14. Obtain the virial form of the SS equation of state given in Problem 13.
15. Obtain the virial form of the LC equation of state given in Problem 11.
16. Obtain the virial form of the BGY equation of state given in Problem 10.
17. Obtain the virial form of the PV equation of state given in Problem 12.
18. Obtain the virial form the PHCT equation of state given in Problem 9.
19. Prove the following

$$\left(\frac{\partial \beta}{\partial P}\right)_T = -\left(\frac{\partial \kappa}{\partial T}\right)_P$$

(2.98)

20. Find the *isentropic volume expansivity* for systems described using the FOV equation of state. The isentropic volume expansivity as defined by Equation (2.80) is as follows:

$$\beta_s = \frac{1}{V}\left(\frac{\partial V}{\partial T}\right)_s.$$

(2.99)

21. Find the *isentropic volume expansivity* for systems described using the Prigogine square-well cell equation of state. The isentropic volume expansivity as defined by Equation (2.80).

22. Find the *isentropic volume expansivity* for systems described using the Sanchez–Lacombe lattice fluid theory equation of state. The isentropic volume expansivity as defined by Equation (2.80).

23. Find the *isentropic volume expansivity* for systems described using the PHCT equation of state. The isentropic volume expansivity as defined by Equation (2.80).

24. Find the *isentropic volume expansivity* for systems described using the BGY equation of state. The isentropic volume expansivity as defined by Equation (2.80).

25. Find the *isentropic volume expansivity* for systems described using the LC equation of state. The isentropic volume expansivity as defined by Equation (2.80).

26. Find the *isentropic volume expansivity* for systems described using the PV equation of state. The isentropic volume expansivity as defined by Equation (2.80).

27. Find the *isentropic volume expansivity* for systems described using the SS equation of state. The isentropic volume expansivity as defined by Equation (2.80).

28. Find the *isentropic volume expansivity* for systems described using the Tait equation of state. The isentropic volume expansivity as defined by Equation (2.80).

29. Obtain the thermal pressure coefficient $(\partial P/\partial T)$ for the following EOS theories:
    a. Tait equation
    b. FOV EOS
    c. Prigogine Square–Well Model
    d. Sanchez–Lacombe lattice fluid theory
    e. PHCT equation of state
    f. LC equation of state
    g. PV equation of state
    h. BGY equation of state
    i. SS equation of state

30. The Boyle temperature is defined as

$$\underset{P \to 0}{Lt}\left(\frac{\partial Z}{\partial P}\right)_T = 0 \qquad (2.100)$$

Obtain the Boyle temperature for the nine EOS models mentioned in Problem 29.

31. The infinite pressure limit of EOS theories can be obtained from Equation (2.100) as

$$\underset{P \to \infty}{Lt}\left(\frac{\partial Z}{\partial P}\right)_T = 0. \qquad (2.101)$$

Obtain the infinite pressure limit for the nine EOS models listed in Problem 29.

32. In the nine EOS models listed in Problem 29 for values of $Z < 1$, are intermolecular attractive forces dominant?

33. In the nine EOS models listed in Problem 29 for values of $Z > 1$, are intermolecular repulsion forces dominant?

34. In the nine EOS models listed in Problem 29 for values of $Z = 1$, are the intermolecular attraction and repulsion forces "balanced?"

35. The general form of EOS in terms of compressibility factor can be written as follows:

$$Z = 1 + Z_{repulsion}(\rho) - Z_{attraction}(T, \rho). \qquad (2.102)$$

What are the repulsive and attractive contributions in the nine EOS for models listed in Problem 29?

## REVIEW QUESTIONS

1. Some EOS theories predict negative pressures for certain values of temperature and specific volume. Are "negative pressure" values spurious?
2. What is the significance of the mathematical model predictions of "negative pressure"?
3. Does a critical point exist between the solid and vapor phase in the P/T diagram of pure substances?
4. What is the "fourth state of matter"?
5. Can the volume expansivity take on negative values for polymeric liquids?
6. For certain models, the EOS theory predicts multiple volumes for a given pressure and temperature? How would you decide which is the right one?
7. How many parameters does the Tait equation have?
8. What is the significance of the parameters in the FOV equation of state?
9. What is the significance of the parameters of the Prigogine "square-well" cell model?
10. What is the significance of each parameter in the Dee and Walsh modified cell model, MCM?
11. What is the difference in approach between the lattice-fluid theories and hole theories that predict EOS of polymeric liquids?
12. Which of the EOS theories predict the glass transition behavior in polymers?
13. Which EOS theories allow for variation with molecular weight of the polymers?
14. Does a solid–liquid critical point exist for polymer liquids in the P/T diagram?
15. Can polymers exist in the vapor form?
16. Do the end-groups that are present at the terminal ends of the polymers have an effect on the EOS predictions?
17. When the polymers cross-link, what happens to the predictions from the EOS theories?
18. Is the chain entanglement phenomena accounted for in the EOS theories?
19. How would you extend the EOS for homopolymer to a (1) random copolymer, (2) alternating copolymer, (3) block copolymer?
20. Is the molecular architecture such as star-shaped polymers or comb-shaped polymers accounted for by the EOS theories?
21. What would be a good EOS for oligomers?: (1) one for small molecules, (2) one for macromolecules?
22. How would you extend the EOS for homopolymer to a polymer blend that is miscible?

23. Show how you would calculate the solubility parameter from the EOS theories.
24. Is the Flory–Huggins parameter a function of temperature?
25. Are the EOS theories discussed in this chapter applicable for polymer nanocomposites?

## REFERENCES

1. D. Halliday, R. Resnick, and J. Walker, *Fundamentals of Physics,* 9th edition, John Wiley & Sons, New York, 2004.
2. J. M. Smith, H. C. van Ness, and M. M. Abbott, *Introduction to Chemical Engineering Thermodynamics*, 7th edition, McGraw Hill, New York, 2005.
3. J. D. van der Waals, The Equation of State for Gases and Liquids, Nobel Lecture, Stockholm, Sweden, 1910.
4. H. Kammerling Onnes, Expression of the equation of state of gases and liquids by means of series, *Commun. Phys. Lab. Univ. Leiden,* 71, 1901.
5. O. Redlich and J. N. S. Kwong, On the Thermodynamics of Solutions. V. An equation of state. Fugacities of gaseous solutions, *Chem. Rev.*, Vol. 44, 233–244, 1949.
6. G. Soave, Equilibrium constants from a modified Redlich-Kwong equation of state, *Chem. Eng. Sci.*, Vol. 27, 1197–1203, 1972.
7. D. Y. Peng and D. B. Robinson, A new two-constant equation of state, *Ind. Eng. Chem.: Fund.*, Vol. 15, 59–64, 1976.
8. S. Rastogi, M. Newman, and A. Keller, Thermodynamic anomalies in the pressure-temperature phase diagram of poly-4-methyl pentene-1: Disordering of crystals on cooling, *Prog. Colloid Polym. Sci.*, Vol. 87, 39–41. 1992.
9. P. A. Rodgers, Pressure-volume-temperature relationships for polymeric liquids: A review of equations of state and their characteristic parameters for 56 polymers, *J. Appl. Polym. Sci.*, Vol. 48, 1061–1080, 1993.
10. P. G. Tait, *Phys. Chem.*, Vol. 2, 1, 1888.
11. K. S. Pitzer, *Thermodynamics*, 3rd Edition, McGraw Hill, New York, 1995.
12. I. Prigogine and D. Kondepudi, *Modern Thermodynamics: Heat Engines to Dissipative Structures*, John Wiley & Sons, New York, 1999.
13. D. R. Paul and A. T. Di Benedetto, Thermodynamic and molecular properties of amorphous high polymers, *J. Polym. Sci.*, Vol. C16, (1967), 1269.
14. G. T. Dee and D. J. Walsh, *Macromolecules*, Vol. 21, 815, 1988.
15. I. C. Sanchez and R. H. Lacombe, An elementary molecular theory of classical fluids, *J. Phys. Chem.*, Vol. 80, 21, 1976.
16. M. B. Jakubinek, C. A. Whitman, and M. A. White, Negative thermal expansion materials: Thermal properties and implications for composite materials, *J. Therm. Anal. Calorim.*, DOI 1.1007/s10973-009-0458-9, Japan Symposium, 2008.
17. D. R. Askleland and P. P. Phule, *The Science and Engineering of Materials*, Thomson, Toronto, Ontario, Canada, 2006.
18. L. A. Stepanov, Thermodynamics of substances with negative thermal expansion coefficient, *Computer Modelling New Technol.*, Vol. 4, 2, 72–74, 2000.
19. J. D. Anderson, *Modern Compressible Flow with Historical Perspective*, 3rd Edition, McGraw Hill Professional, New York, 2003.
20. Sir Isaac Newton, *Philosophiae Naturalis Principia Mathematica*, 1687.
21. P. S. M. de Laplace, Sur la vitesse du son dans l'aire et dan l'eau, *Ann. Chim. Phys.*, 1816.
22. C. Kittel and H. Kroemer, *Thermal Physics,* 2nd edition, Freeman & Co, New York, 1980.

23. M. D. Donohue and J. M. Prausnitz, Perturbed hard chain theory for fluid mixtures: Thermodynamic properties for mixtures in natural gas and petroleum technology, *AIChE J*, 24, 849, 1978.

24. J. E. G. Lipson and S. S. Andrews, A Born-Green-Yvon integral equation treatment of a compressible fluid, *J. Chem. Phys.*, 96, 2, 1426–1434, 1992.

25. M. G. Bawendi and K. F. Freed, Lattice theories of polymeric fluids, *J. Chem. Phys.*, 93, 2194, 1989.

26. C. Panayiotou and J. H. Vera, An improved lattice-fluid equation of state for pure component polymeric fluids, *Polym. Eng. Sci.*, 22, 6, 345, 1982.

27. E. A. Guggenheim, Statistical thermodynamics of mixtures with non-zero energies of mixing, *Proc. Royal Society, London*, Ser. A., 283, 203, 213, 1944.

28. R. Simha and T. Somcynsky, On the statistical thermodynamics of spherical and chain molecule fluids, *Macromolecules*, 2, 4, 342, 1969.

# 3 Binary Interaction Model

## LEARNING OBJECTIVES

- Binary interaction model
- Intramolecular repulsions
- Copolymer–homopolymer miscibility
- Copolymer–copolymer miscibility
- Terpolymer–terpolymer miscibility
- Terpolymer–homopolymer miscibility
- Mean–field approach to obtain interaction parameters
- Spinodal and phase separation
- Compositional mismatch in copolymer systems with common monomers

## 3.1 INTRODUCTION

The thermodynamic basis to explain miscibility in polymer blends is an *exothermic* heat of mixing, since *entropic* contributions are so small in such systems. One school of thought is that the specific intermolecular interactions are responsible for exothermic heats of mixing. Hydrogen bond formation, $\pi$–$\pi$ complex formation, and a variety of specific interactions play an important role in determining polymer blend miscibility. On the other hand, the observed exothermic heats of mixing for many low and high molecular weight systems are quite small, indicating that other mechanisms may be involved.

*Intramolecular repulsions* in many cases may be an important factor in realizing exothermic heats of mixing rather than specific intermolecular interactions as usually assumed. This point of view was proposed independently by Kambour, Bendler, and Bopp [1], Ten Brinke, Karasz, and MacKnight [2], and Paul and Barlow [3]. A simple binary interaction model is applied to blends involving copolymers, and it was suggested that a similar rationale can be extended to blends involving homologous series of homopolymers by appropriate subdivision of the repeating units. This model can explain numerous experimental observations and provides a framework for quantification of the thermodynamics for some blend systems.

Over 20 pairs of miscible polymer blends have been reported [4]. Substantial progress has been made in polymer science and engineering since the first review by Krause [5] in understanding the phase behavior of blends [6–10] and interpreting the technology practiced in the industry better. The application of thermodynamics is visible in the improved generation and stabilization of the morphology of blends.

A necessary requirement for miscibility of polymer blends is that the free energy of mixing be negative:

$$\Delta G_m = \Delta H_m - T\Delta S_m. \tag{3.1}$$

An additional condition for stability for binary mixtures,

$$\frac{\partial^2 \Delta G_m}{\partial \phi_i^2} > 0, \tag{3.2}$$

where $\phi_i$ is the volume fraction of either component; any suitable measure of the concentration can be used. This model further assumes that the heat of mixing is described by a van Laar expression [3]:

$$\Delta H_m = (V_A + V_B) B \phi_A \phi_B \tag{3.3}$$

where B is the binary interaction energy density. The B parameter is related to the Flory–Huggins interaction parameter, $\chi$ by

$$\frac{B}{RT} = \frac{\chi_A}{\tilde{V}_A} = \frac{\chi_B}{\tilde{V}_B} = \tilde{\chi}_{AB}. \tag{3.4}$$

B is preferred since its basis is always clearly a unit mixture of volume. The binary interaction model for the heat of mixing can be extended to multicomponent mixtures as follows:

$$\frac{\Delta H_m}{V} = \sum_{i>j} B_{ij} \phi_i \phi_j. \tag{3.5}$$

The sum in Equation (3.5) excludes terms with $i = j$, and obviates double counting of terms with $i \neq j$. Further, $B_{ij} = B_{ji}$.

The sign of combinatorial entropy always favors mixing; its value is diminished for molecular weights of the order of magnitude of millions for most important polymers. Thus, in the limit of high molecular weights, the conditions of miscibility can only be satisfied by a negative interaction parameter, leading to the conclusion that exothermic mixing is a requirement for miscibility in high molecular weight polymer blends.

This is a simple model and cannot account for all the issues of mixture thermodynamics. Interaction parameters deduced from various phase behavior information are often believed to include other effects than purely enthalpic ones. This way, the LCST (lower critical solution temperature) behavior observed in polymer blends can be explained and accounted for quantitatively. These theories refine the binary interaction parameter by removing extraneous effects. EOS effects do not favor phase stability, and the $B$ parameter must be negative to have miscibility in high molecular weight blends. Interaction parameters used in the ensuing sections are not limited to the Flory–Huggins framework and can be viewed as free of equation of state effects.

The role of intramolecular repulsions as a causative factor in driving blend miscibility can be seen readily by considering mixtures of copolymers with homopolymers. Reports in the literature indicate cases of miscibility involving copolymers when their corresponding homopolymers are not miscible. For instance, pure polystyrene and pure polyacrylonitrile are not miscible with poly(methyl methacrylate). But the copolymer styrene–acrylonitrile (SAN) is miscible with poly(methyl methacrylate) PMMA over a range of acrylonitrile (AN) compositions. This system is quantitatively discussed in Section 2.2. The same is the case with PEMA in place of PMMA. Ethylenevinyl acetate (EVA) copolymers are miscible with polyvinyl chloride (PVC) for a range of VA composition in the copolymer. Neither polyethylene nor polyvinyl acetate are miscible with PVC. In a similar fashion, butadiene–acrylonitrile copolymers are found to be miscible with PVC for a range of AN compositions. Similarly, poly(alpha-methyl styrene) (poly-AMS), AN copolymers, are miscible with PMMA and PEMA; poly (o-chlorostyrene-p-chlorostyrene) copolymers are miscible with polyphenylene oxide (PPO) over a range of p-chlorostyrene composition. The higher the phase separation temperature (LCST), the more negative is the binary interaction parameter.

## 3.2 COMPOSITIONAL WINDOW OF MISCIBILITY: COPOLYMER–HOMOPOLYMER

Consider a polymer blend of a copolymer and a homopolymer, for example, SAN copolymer and PMMA homopolymer.

Let the composition of AN (1) in SAN copolymer be x mole fraction. Then, the styrene (2) mole fraction in SAN would be 1-x. Let the volume fraction PMMA (3) in the blend be given by $\phi$. The volume fraction SAN in the blend would then be given by (1-$\phi$). The effective binary interaction parameter of the blend can be written as follows:

$$B = B_{13}x(1-\phi)\phi + B_{23}(1-x)(1-\phi)\phi - B_{12}x(1-x)(1-\phi)^2. \qquad (3.6)$$

From the binary interaction values given in [11], the $B_{ij}$ values are [11] S–AN: + 6.67 cal/cm$^{-3}$; S–MMA: + 0.18 cal/cm$^{-3}$; MMA–AN: + 4.11 (cal/cm$^{-3}$)

Equation (3.6) becomes

$$B = 4.11x(1-\phi)\phi + 0.18(1-x)(1-\phi)\phi - 6.74x(1-x)(1-\phi)^2. \qquad (3.7)$$

It can be seen that the miscible regions of the SAN copolymer–PMMA homopolymer blend depends on (1) the composition of AN in SAN (x) and (2) composition of PMMA in the blend ($\phi$). Equation (3.7) is plotted in Figure 3.1A for various values of the PMMA volume fraction in the blend, $\phi$. Thus, for volume fraction of 0.1,

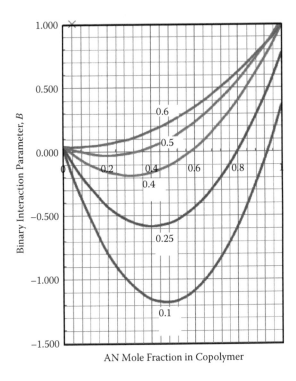

**FIGURE 3.1A**  Compositional window of miscibility in SAN–PMMA blends predicted by binary interaction model.

the PMMA is miscible in SAN for a wider compositional window of miscibility of AN composition. As the volume fraction of PMMA is increased, the compositional window of miscibility of AN composition decreases. For a PMMA with volume fraction $\phi = 0.5$, SAN with AN composition in mole fraction between 10–30% is found to have negative $B$ values, indicating possible miscibility with PMMA copolymer. This has decreased from a AN composition range in SAN of 0.05–0.90 for a blend with PMMA volume fraction, $\phi = 0.1$. The binary interaction values for five different volume fractions of PMMA in the blend and 20 different mole fractions of AN in SAN were calculated in an Excel spreadsheet. The actual values are shown in Table 3.1. The miscibility of SAN/PMMA blend systems as a function of volume fraction for various copolymers of SAN at different AN compositions is shown in Figure 3.1B. It can be seen that at low volume fraction of SAN in the blend miscibility is expected. At high AN composition in the SAN copolymer, immiscible blends are expected.

## 3.3  COMPOSITIONAL WINDOW OF MISCIBILITY: COPOLYMERS WITH COMMON MONOMERS

During the high volume manufacture of copolymers such as SAN, and subsequent processing of SANs with different compositions of AN, it is important to calculate

TABLE 3.1
Binary Interaction Parameter, B Values, for SAN–PMMA Blend

| AN X | B ($\phi = 0.1$) | B ($\phi = 0.25$) | B ($\phi = 0.4$) | B ($\phi = 0.5$) | B ($\phi = 0.6$) |
|---|---|---|---|---|---|
| 0 | 0.016 | 0.034 | 0.043 | 0.045 | 0.043 |
| 0.05 | −0.225 | −0.109 | −0.025 | 0.014 | 0.039 |
| 0.1 | −0.440 | −0.234 | −0.081 | −0.008 | 0.040 |
| 0.15 | −0.627 | −0.339 | −0.125 | −0.022 | 0.047 |
| 0.2 | −0.787 | −0.425 | −0.156 | −0.028 | 0.059 |
| 0.25 | −0.919 | −0.493 | −0.176 | −0.025 | 0.077 |
| 0.3 | −1.024 | −0.541 | −0.183 | −0.014 | 0.100 |
| 0.35 | −1.102 | −0.571 | −0.179 | 0.006 | 0.128 |
| 0.4 | −1.153 | −0.581 | −0.162 | 0.034 | 0.162 |
| 0.45 | −1.176 | −0.573 | −0.133 | 0.070 | 0.201 |
| 0.5 | −1.172 | −0.546 | −0.092 | 0.115 | 0.245 |
| 0.55 | −1.140 | −0.499 | −0.039 | 0.168 | 0.295 |
| 0.6 | −1.082 | −0.434 | 0.027 | 0.230 | 0.350 |
| 0.65 | −0.996 | −0.350 | 0.104 | 0.300 | 0.411 |
| 0.7 | −0.883 | −0.247 | 0.194 | 0.379 | 0.477 |
| 0.75 | −0.742 | −0.124 | 0.296 | 0.466 | 0.548 |
| 0.8 | −0.574 | 0.017 | 0.410 | 0.561 | 0.625 |
| 0.85 | −0.379 | 0.177 | 0.536 | 0.665 | 0.707 |
| 0.9 | −0.157 | 0.356 | 0.674 | 0.778 | 0.795 |
| 0.95 | 0.093 | 0.554 | 0.824 | 0.898 | 0.888 |
| 1 | 0.370 | 0.771 | 0.986 | 1.028 | 0.986 |

the compositional window of miscibility between two copolymers A and B, both made of the same comonomers but with different compositions. For example, will SAN with 25% weight fraction AN and SAN with 60 wt% AN when blended together be miscible? In order to calculate the compositional window of miscibility among copolymers with common monomers, the effective binary interaction parameter B is written for such systems as follows:

$$B = B_{12}x_A(1-\phi)(1-x_B)\phi - B_{12}x_A(1-\phi)^2(1-x_A).$$ (3.8)

Let the compositional mismatch between the copolymer A and copolymer B be given by

$$x_A - x_B = \xi.$$ (3.9)

For the SAN–SAN system, say $\phi = 0.5$, $x_A = 0.4$, Equation (3.9) becomes the following, realizing that $B_{12} = 6.74$ cal/cc for the system to stay miscible:

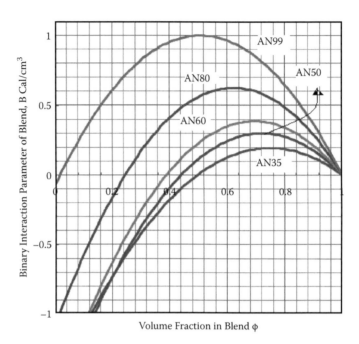

**FIGURE 3.1B** Effect of volume fraction on SAN–PMMA blend miscibility predicted by binary interaction model.

$$\frac{1-\phi}{\phi} > \frac{\xi}{x_A(1-x_A)} \tag{3.10}$$

$$\xi < 0.24. \tag{3.11}$$

Thus, a compositional mismatch in AN of 24 mol% can be tolerated at 40% AN containing SAN before the system of SAN–SAN becomes immiscible. Thus, the miscible window is within a 12% mole fraction of AN mismatch between SAN–SAN blends, where the system will be miscible. This turns out to be a 14% by weight mismatch in AN composition between the two copolymers SAN and SAN.

## 3.4  COMPOSITIONAL WINDOW OF MISCIBILITY: TERPOLYMER SYSTEM WITH COMMON MONOMERS

The methylmethacrylate–alphamethylstyrene–acrylonitrile terpolymer system is an attractive component in compatible blends with acrylonitrile butadiene styrene (ABS). The compositional window of terpolymer and terpolymer miscibility is of interest during manufacture and during blending. The common monomers in the terpolymer A, terpolymer B are α-methylstyrene, acrylonitrile, and methyl methacrylate. The *B* values [11] are α-MS–AN +7.96 cal/cc, AMS-MMA +0.12, MMA–AN +4.32 cal/cc. The binary interaction parameter for the terpolymer terpolymer blend with common monomers can be written as

$$B = B_{12}(X)(Y) + B_{13}(X)(Z) + B_{23}(Y)(Z). \tag{3.12}$$

where,

$$X = (\phi_1' - \phi_1'')$$
$$Y = (\phi_2' - \phi_2'') \tag{3.13, 3.14}$$
$$Z = (\phi_3' - \phi_3'')$$

X, Y, and Z are the compositional mismatch [12,13] between a components in polymer A versus the same component in polymer B.

$$\text{For miscibility } B \leq 0 = f(X,Y,Z). \tag{3.15}$$

In a Quattro Pro spreadsheet values for X were generated using random number generator between (−1,+1). The @RAND key is used and the range is confined to (−1,1) with the values generated being fractional numbers. The same for Y and Z, making sure that

$$\sum_{i=1}^{3} \phi_i' = \sum_{i=1}^{3} \phi_i'' = 1. \tag{3.16}$$

The results are contoured for X = const and plotted as B versus Y in Figure 3.2. The cost of computation without the definition for compositional mismatch would be n × m × k × 1 or $n^4$ where n and m are the composition of component 1 and 2 in the terpolymer A and k, and 1 is the composition of component 1 and 2 in the terpolymer B. By using the compositional mismatch definition, the cost of computation is 2 × n or n. Thus, this procedure is more efficient in searching for the compositional window of miscibility.

In a similar manner, the blend of terpolymer A, styrene–methyl methacrylate–acrylonitrile, and polymer B, a different terpolymer with monomers styrene, methyl methacrylate, and acrylonitrile, can be studied using the binary interaction model and miscibility window obtained:

$$B = B_{12}(X)(Y) + B_{13}(X)(Z) + B_{23}(Y)(Z). \tag{3.17}$$

The B values are [11] S–AN: +6.74 cal/cm$^{-3}$; S–MMA: +0.18, cal/cm$^{-3}$; MMA–AN: + 4.11 cal/cm$^{-3}$.

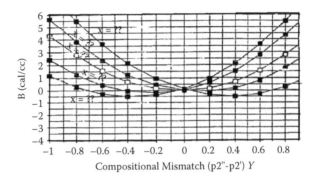

FIGURE 3.2 **FIGURE 3.2**   Miscibility of terpolymer–terpolymer with common monomers predicted by binary interaction model.

## 3.5   COMPOSITIONAL WINDOW OF MISCIBILITY: TERPOLYMER AND HOMOPOLYMER SYSTEM WITHOUT COMMON MONOMERS

A number of technologists have attempted to obtain miscible systems with polycarbonate (PC) as one of the blend constituents. Given the high performance and cost associated with PC polymers, blends of PC and multicomponent copolymers may get the job done for less cost. PC/ABS blends have been reported in the product literature. Although they were not compatibilized, they exhibit good balance of properties. In this section a terpolymer made of three comonomers, (1) α-methlystyrene (AMS), (2) acrylonitrile (AN), and (3) methyl methacrylate (MMA) can be evaluated for miscible regions in the copolymer compositional space and blend composition with (4) tetramethyl polycarbonate (TMPC) homopolymer. The binary interaction parameter for the terpolymer/homopolymer system can be written as follows:

$$B = B_{AD}x_A\phi(1-\phi) + B_{BD}(x_B)\phi(1-\phi) + B_{CD}(1-x_A-x_B)\phi(1-\phi) - B_{AB}x_Ax_B\phi^2$$

$$-B_{AC}x_A(1-x_A-x_B)\phi^2 - B_{BC}x_B(1-x_A-(x_B)\phi^2,$$

(3.17a)

where $x_A$ and $x_B$ are the terpolymer composition of comonomer $A$ and $B$, and $\phi$ is the volume fraction of terpolymer 1 in the two-component blend of terpolymer/homopolymer.

The binary interaction parameter B values [11] are AMS–AN: +7.96 cal/cm$^{-3}$; AMS-MMA; +0.12: MMA-AN +4.32 cal/cm$^{-3}$. The interaction parameters for TMPC/AMS, TMPC/AN, and TMPC/MMA may be taken as [17] 0.18 cal/cm$^{-3}$, 5.57 cal/cm$^{-3}$, and 1 cal/cm$^{-3}$, respectively. The miscibility regions for a 70% volume fraction terpolymer 1 with homopolymer 2 is shown in Figure 3.3. The $B$ values for the blend vs. composition are plotted and contoured for constant copolymer composition of AN. $B$ is the plotted versus composition of AMS. The $B$ values from the calculation are tabulated in Table 3.2.

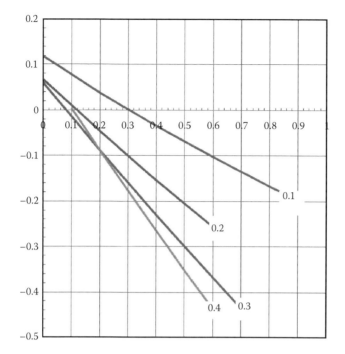

**FIGURE 3.3** Miscibility compositional window for terpolymer–homopolymer miscibility.

## 3.6 SPINODAL CURVE FROM *B* VALUES AND EOS

Once the binary interaction parameters for the blend system are known, EOS theory can be used to predict phase separation behavior. Lower critical solution temperature (LCST) is the temperature above which a miscible system becomes immiscible. Upper critical solution temperature (UCST) is the temperature above which an immiscible polymer blend system becomes miscible. Some polymer–polymer systems exhibit either LCST or UCST or both or neither. Another set of phase separation can be obtained as shown in the copolymer–homopolymer example in Section 3.2 by varying the blend volume fraction. The Gibbs free energy of mixing per unit volume for a binary system of two polymers can be written as

$$\Delta G_{mix} = \Delta H - T\Delta S = B\phi_A\phi_B + TR\left(\frac{\rho_A\phi_A \ln\phi_A}{MW_A} + \frac{\rho_B\phi_B \ln\phi_B}{MW_B}\right), \quad (3.18)$$

where $\rho$, $\phi$, and $MW$ are the density, volume fraction, and molecular weight with the considered species shown as a subscript. $R$ is the universal molar gas constant; $T$ is the absolute temperature in Kelvin, and $B$ in the binary interaction energy (cal/cc). The interaction energy parameter $B$ lumps the heat of mixing and other noncombinatorial

**TABLE 3.2**

**Terpolymer–Homopolymer Blend at $\phi = 0.7$**

| AMS ($x_A$) | AN ($x_B$) | MMA ($x_C$) | $B_{blend}$ (Cal/cm³) |
|---|---|---|---|
| 0 | 0.1 | 0.9 | 0.117558 |
| 0.2 | 0.1 | 0.7 | 0.039214 |
| 0.4 | 0.1 | 0.5 | −0.03443 |
| 0.6 | 0.1 | 0.3 | −0.10336 |
| 0.8 | 0.1 | 0.1 | −0.16759 |
| 0.9 | 0.1 | 0 | −0.19795 |
| 0 | 0.2 | 0.8 | 0.067452 |
| 0.2 | 0.2 | 0.6 | −0.04539 |
| 0.4 | 0.2 | 0.4 | −0.15352 |
| 0.6 | 0.2 | 0.2 | −0.25696 |
| 0 | 0.3 | 0.7 | 0.059682 |
| 0.2 | 0.3 | 0.5 | −0.08765 |
| 0.4 | 0.3 | 0.3 | −0.23029 |
| 0.6 | 0.3 | 0.1 | −0.36821 |
| 0.7 | 0.3 | 0 | −0.43541 |
| 0.1 | 0.4 | 0.5 | 0.002744 |
| 0.15 | 0.4 | 0.45 | −0.04257 |
| 0.2 | 0.4 | 0.4 | −0.08758 |
| 0.3 | 0.4 | 0.3 | −0.17674 |
| 0.4 | 0.4 | 0.2 | −0.26471 |
| 0.5 | 0.4 | 0.1 | −0.35151 |
| 0.6 | 0.4 | 0 | −0.43714 |

effects. Assuming that $B$ is independent of composition, a spinodal can be derived from Equation (3.18), subject to the criteria

$$\frac{\partial^2 \Delta G_{mix}}{\partial \phi_i^2} = 0 = TR \left( \frac{\rho_A}{\phi_A MW_A} + \frac{\rho_B}{\phi_B MW_B} \right) - 2B_{spinodal} = 0. \tag{3.19}$$

A spinodal condition is defined by Equation (3.19). A $B_{crit}$ was obtained by Merfeld and Paul [14] by obtaining the third derivative of $\Delta G_{mix}$ with respect to the composition, which is zero. This is where the boundary between the miscible and immiscible lies and is supposed to be a balance between the combinatorial entropy and component interactions:

$$B_{crit} = \frac{TR}{2} \left( \sqrt{\frac{\rho_A}{MW_A}} + \sqrt{\frac{\rho_B}{MW_B}} \right). \tag{3.20}$$

Miscibility is expected when $B$ is more favorable than $B_{crit}$. This model can only predict the UCST-type phase boundaries. Although when $B$ is allowed to vary with temperature in an empirical manner, the LCST can be captured. Most polymer blends show LCST-type phase behavior. EOS theories account for the compressible nature of the polymer mixtures and can be used to predict LCST-type phase separation. The lattice fluid theory of Sanchez and Lacombe was discussed in Chapter 2. The EOS from the discussion can be written as

$$\rho_r^2 + P_r + T_r\left(\ln(1-\rho_r) + \rho_r\left(1-\frac{1}{r}\right)\right) = 0. \tag{3.21}$$

The spinodal condition can be derived as

$$\frac{\partial^2 \Delta G_{mix}}{\partial \phi_i^2} = 0 = \frac{1}{2}\left(\frac{1}{r_1\phi_1} + \frac{1}{r_2\phi_2}\right) - \rho_r\left(\frac{\Delta P^* v^*}{k_B T} + \frac{1\psi^2 T_r P^*\beta}{2}\right). \tag{3.22}$$

where $\beta$ is the isothermal compressibility and $\psi$ is a dimensionless function. The $\Delta P^*$ is the bare interaction density and is used instead of the $B$ values in the equations discussed in the earlier sections. If the volume fractions used in the Flory–Huggins theory are assumed to be equivalent to those used in the Sanchez–Lacombe theory, the Flory–Huggins interaction energy and the bare interaction energy can be related:

$$B_{spinodal} = \rho_r \Delta P^* + \left\{[P_2^* - P_1^* + \Delta P^*(\phi_2 - \phi_1)] + \frac{TR}{\rho_r}\left(\frac{1}{r_1^0 v_1^*} - \frac{1}{r_2^0 v_2^*}\right)\right.$$

$$\left. -TR\left(\frac{1}{\rho_r} + \frac{\ln(1-\rho_r)}{\rho_r^2}\right)\left(\frac{1}{v_1^*} - \frac{1}{v_2^*}\right)\right\}^2 \left\{\left(\frac{2TR}{v^*}\right)\left(\frac{2\ln(1-\rho_r)}{\rho_r^3}\right) + \frac{1}{\rho_r^2(1-\rho_r)} + \frac{1-1/r}{\rho_r^2}\right\}^{-1}. \tag{3.23}$$

where $r_i^0$ is the number of sites occupied by molecule $i$ in the pure close-packed state such that $r_i^0 v_i^* = r_i v^*$.

## 3.7   COPOLYMER/HOMOPOLYMER BLENDS OF AMS–AN/PVC

PVC/SAN homopolymer–copolymer blend systems have been reported to be miscible over some composition range of AN in the copolymer and blend composition. Zhang et al. [15] found that the value of LCST increases as the strength of interaction between the component polymer increases. The miscibility behavior of PVC/AMS–AN blends was found to depend on the AMS content in AMS–AN copolymer using dynamical mechanical analysis (DMA) and scanning electron microscopy (SEM). The PVC was found to be immiscible with AMS–AN copolymer as the AMS content in AMS–AN is less than 15 wt% by melt mixing.

However, the miscibility of PVC/AMS–AN blends is substantially improved with the increase of AMS content in AMS–AN copolymer containing identical

AN content, and it develops homogeneous morphology when the AMS content in AMS–AN is more than 22.5 wt%. So the miscibility of PVC/AMS–AN blends depends not only on the AN content in AMS–AN copolymer but also on the different AMS content in AMS–AN copolymer. The miscibility of PVC/AMS–AN-37.5 blends has been investigated by changing the composition of the blends. It was found that the PVC/AMS–AN-37.5 blend system is a miscible system in the whole composition. The presence of plasticizer has a substantial effect on the blends. A high miscibility is founded in PVC/AMS–AN/DOP blends when 10~20 wt% DOP are added, and SEM pictures show a true homogeneous phase.

## 3.8 COPOLYMER/HOMOPOLYMER BLENDS OF AMS–AN WITH OTHER COPOLYMERS

The phase behavior of blends of poly(methyl methacrylate) (PMMA), polyvinyl chloride (PVC), tetramethylbisphenol A polycarbonate (TMPC), poly(2,6-dimethyl-1-4, phenylene oxide) (PPO), polycaprolactone (PCL), and low molecular weight polycarbonate PC with AMS–AN copolymers was studied by Gan, Paul, and Padwa [18]. The copolymerization of AMS with AN was carried out in bulk at 40°C using mixtures of *t*-butylperoxy-2-ethylhexanoate (TBPEH) and *t*-butyl peroxyneodecanoate (TBPOND) from Lucidol, Atomchem, Buffalo, New York). For copolymers with less than 25 wt% AN, 0.05 wt% (TBPEH) and 0.008 wt% TBPOND were used. For copolymers containing greater than 25 wt% AN, 0.1% TBPOND initiator was used. The monomers used were commercial polymerization grade. The inhibitors were removed using activated alumina (Scientific Polymer Products, Ontario, NY). The polymerization was carried out in 1-liter capped bottles in which the monomers were nitrogen purged by bubbling at room temperature for about 5 min. After the required time to achieve approximately 1–2% conversion, the solution was concentrated on a rotary evaporator at 65°C using increasing vacuum levels to control the rate of evaporation.

When most of the monomers were removed, the resulting viscous solution was diluted to an estimated 5 wt% concentration with 2-butanone, and reprecipitated by dropwise addition into five volumes of methanol in a high-speed blender. The vacuum-filtered moist polymer was redissolved to about 5 wt% concentrations and precipitated again. The resulting polymer was dried overnight at 80°C in a vacuum oven with a nitrogen bleed.

AN content was estimated by automated Carlo–Erba CHN (carbon, hydrogen, nitrogen) elemental analysis. Molecular weights for copolymers containing less than 45 wt% AN were estimated by gel permeation chromatography (GPC) in tetrahydrofuran (THF), using four columns containing a mixed pore-size bed of 5 μm beads (Polymer Laboratories, Amherst, Massachusetts). The flow rate was 1 ml/min at 35°C with a refractive index detector. Molecular weights are reported in polystyrene equivalents. The glass transition temperatures and degree of polymerization of the samples used in the study by Gan, Paul, and Padwa [18] are shown in Table 3.3.

All blends of AMS–AN copolymers were prepared by solution casting from THF onto a glass plate at 60°C. The cast films were dried at 60°C for about 10 min before

## TABLE 3.3
## Polymer Used in the Study by Gan, Paul, and Padwa [18]

| Copolymer | AN (wt%) | $T_g$ (°C) | $M_n$ |
|---|---|---|---|
| AMS–AN4 | 4 | 152 | 10070 |
| AMS–AN6.5 | 6.5 | 151 | 13520 |
| AMS–AN9.4 | 9.4 | 149 | 17260 |
| AMS–AN11.9 | 11.9 | 148 | 25420 |
| AMS–AN16.5 | 16.5 | 148 | 37840 |
| AMS–AN19.4 | 19.4 | 147 | 49200 |
| AMS–AN23.4 | 23.4 | 140 | 100,300 |
| AMS–AN30 | 30.0 | 127 | 57,000 |
| AMS–AN38 | 38.1 | 121 | 81,230 |
| AMS–AN41.4 | 41.4 | 115 | 75,300 |
| AMS–AN48.5 | 48.5 | 116 | 62,500 |
| AMS–AN57.6 | 57.6 | 114 | 75,000 |
| SAN18 | 18.0 | 106 | 76,100 |
| SAN20 | 20.0 | 106 | 88,120 |
| SAN24 | 23.5 | 108 | 72,070 |
| SAN26 | 26.0 | 106 | 57,400 |
| SAN29 | 29.0 | 104 | 44,000 |
| SAN30 | 30.0 | 105 | 168,000 |
| PMMA | | 108 | 52,900 |
| PVC | | 85 | 99,000 |
| TMPC | | 190 | 33,000 |
| PPO | | 218 | 29,400 |
| PCL | | −74 | 15,500 |
| PC | | 127 | 4,504 |
| Poly AMS | | 169 | 53,398 |

further drying in the vacuum oven at a temperature of 20–30°C higher than the glass transition of the blend, $T_g$. Glass transition temperatures were measured using a Perkin–Elmer differential scanning calorimeter, DSC-7, at a scanning rate of 20°C/min. The onset of the change in heat capacity was defined as the $T_g$. The temperatures at which phase separation occurs on heating (i.e., LCST behavior) were measured using both DSC and optical methods.

In the DSC thermal methods, the samples were isothermally annealed in the sample holder at various temperatures for 4–5 min and then quickly quenched to room temperature before increasing the temperature at a scanning rate of 20°C/min in the DSC-7 to look at glass transition behavior. In the optical methods, films were converted with a glass slide and annealed on a hot stage at various constant temperatures around the phase separation temperature for a certain period of time. The phase separation temperature was taken as the lowest annealing temperature at which the film was found to be cloudy.

**TABLE 3.4**

**Equation of State Parameters for Sanchez–Lacombe Lattice Fluid Theory (Temperature Range 220–270°C)**

| Polymer | $\rho^*$(gm/cc) | P*(bar) | T*(K) |
|---|---|---|---|
| Poly-AMS | 1.1268 | 4258 | 827 |
| Poly–AN | 1.2080 | 5192 | 909 |
| AMS–AN6.5 | 1.1321 | 4319 | 833 |
| AMS–AN9.4 | 1.1345 | 4346 | 835 |
| AMS–AN11.9 | 1.1365 | 4369 | 837 |
| AMS–AN16.5 | 1.1402 | 4412 | 841 |
| AMS–AN19.4 | 1.1426 | 4439 | 843 |
| AMS–AN23.4 | 1.1458 | 4477 | 846 |
| AMS–AN30 | 1.1512 | 4538 | 852 |
| AMS–AN41.4 | 1.1604 | 4645 | 861 |
| PMMA | 1.2564 | 5090 | 742 |
| TMPC | 1.1854 | 4395 | 729 |
| PPO | 1.1732 | 4083 | 758 |
| PCL | | 4212 | 736 |
| PVC | | | |

The PVT properties were obtained using the Gnomix PVT apparatus by measurement of the change in specific volume as a function of temperature. The density at 30°C and 1 atm was determined in a density gradient column. The EOS characteristic parameters calculated for AMS–AN copolymers and a range of homopolymers are in Table 3.4.

Blends of various homopolymers with AMS–AN copolymers were systematically examined for miscibility and phase separation temperatures in cases where LCST behavior was detected. The experimental data was used to calculate the interaction energy using the Sanchez–Lacombe lattice fluid equation of state theory. The analysis assumes that the experimental phase separation temperatures are represented using the spinodal curve and that the "bare" interaction energy density $\Delta P^*$ was found to be independent of temperature. Any dependence on the $B$ interaction parameter with temperature stems from compressibility effects. $\Delta P^*$ was determined as a function of copolymer composition. $\Delta P_{ij}^*$ values obtained for blends of the various homopolymers with AMS–AN copolymers were then compared with corresponding ones obtained from SAN copolymers. Flory–Huggins $B_{ij}$ values were calculated from the experimental miscibility limits using the binary interaction model for comparison with $\Delta P_{ij}^*$ values.

PMMA was blended with AMS–AN copolymers with different AN contents as shown in Table 3.3. Blend films were prepared by hot casting. They were found to be transparent and exhibited a single-composition dependent $T_g$ as expected for miscible blends for a wide range of AMS–AN copolymers. Cloudy films were obtained when PMMA was blended with AMS–AN copolymers with 38 wt% AN content higher.

The LCST phase separation temperatures for PMMA–AMS–AN copolymer blends for two different AN compositions at 9.4 wt% and 30 wt% is show in Figure 3.4. It

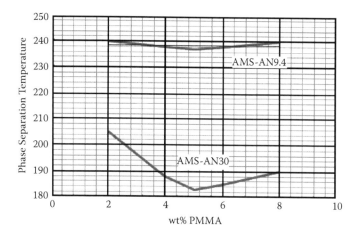

**FIGURE 3.4** Phase separation temperatures in miscible copolymer–homopolymer blends of AMS–AN/PMMA.

was found that below 150°C, PMMA forms miscible blends with AMS–AN copolymers with AN composition from 4–30 wt% AN. The phase separation temperatures experimentally observed characteristic parameters of AMS–AN copolymers from Table 3.4 were used to calculate the $\Delta P^*$ values. It was found that

$$\Delta P_{AMS-AN}{}^* = 7.87, \ \Delta P_{AMS-MMA}{}^* = 0.02 \ \text{and} \ \Delta P_{MMA-AN}{}^* = 5.06 \ \text{cal/cc.}$$

The calculated spinodal curve and the experimentally observed phase separation temperatures agree with each other well. The AMS–AN repeat units repel each other in the copolymer. This drives miscibility. The compressibility and EOS effects play a role in driving the miscibility. The $T_1{}^*$ and $\rho_{1r}$ values become larger with increasing AN content in the AMS–AN copolymer as can be seen from Table 3.4. The smaller compressibility of AMS–AN copolymers caused by $\rho_{1r}$ increasing is favorable for phase stability. However, as AN content increases $(T_1{}^* - T_2{}^*)^2$ increases, which strongly decreases blend stability. Therefore, as AN content increases, competing terms from the equations lead to immiscibility at higher contents of AN in AMS–AN copolymers.

The interaction energies $\Delta P^*$ deduced from the analysis can be converted into the corresponding quantities $B$ or $B_{ij}$ in the Flory–Huggins theory as shown by Kim and Paul [19]:

$$B_{sc} = \rho_r \Delta P^* +$$

$$\frac{[(P_2^* - P_1^* + (\phi_2 - \phi_1)\Delta P^*) + \dfrac{RT}{\rho_r}\left(\dfrac{1}{r_1^0 v_1^*} - \dfrac{1}{r_2^0 v_2^*}\right) - RT\left(\dfrac{\ln(1-\rho_r)}{\rho_r^2} + \dfrac{1}{\rho_r}\right)\left(\dfrac{1}{v_1^*} - \dfrac{2}{v_2^*}\right)]^2}{\dfrac{2RT}{v^*}\left(\dfrac{2\ln(1-\rho_r)}{\rho_r^3} + \dfrac{1}{\rho_r^2(1-\rho_r)} + \dfrac{(1-1/r)}{\rho_r^2}\right)}.$$

(3.24)

The interaction energy densities at the drying temperature of 150°C from the above analysis are $B_{AMS-MMA} = 0.12$ cal/cc; $B_{MMA-AN} = 4.32$ cal/cc; $B_{AMS-AN} = 7.96$ cal/cc. The larger values of $B_{ij}$ compared with $\Delta P_{ij}^*$ values indicate possible EOS contributions. The compositional window of miscibility of copolymer–homopolymer blends of AMS–AN copolymer and PMMA homopolymer for the binary interaction parameter values obtained is shown in Figure 3.5. The $B$ values are plotted as a function of AN composition in the copolymer for different volume fractions of the blend.

PMMA homopolymer was found miscible with SAN copolymers containing AN content from 9.5–28 wt%. The phase separation temperatures for these systems are usually found above their thermal decomposition temperatures. The Flory–Huggins interaction energy density $B_{ij}$ calculated at the drying temperature of 120°C from the above analysis is $B_{S,MMA} = 0.23$ cal/cc, $B_{MMA-AN} = 4.49$ cal/cc, and $B_{S,AN} = 7.02$ cal/cc. Replacing AMS for styrene in copolymers with AN in the minority was found to broaden the miscibility window with PMMA homopolymer. It was found to lower the phase separation temperature at the mid AN range. The change in phase behavior observed is due to changes in $\Delta P_{ij}^*$ and $\Delta T_{ij}^*$ values.

Blends of tetramethyl bisphenol-A polycarbonate, TMPC with AMS–AN copolymers were prepared by precipitation from THF into methanol. Miscible regions were found for copolymers of AMS–AN with a range of AN compositions from 4.0–16.5 wt%. Phase separation temperatures were determined by examination of the glass

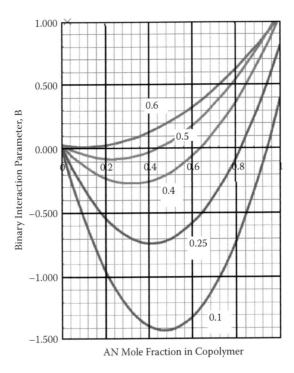

**FIGURE 3.5**  Binary interaction parameters for AMS–AN/PMMA blend for various AN fractions in the copolymer and various volume fractions in the blend.

transition temperature, $T_g$, behavior of blends annealed 4–5 min at various temperatures. Phase separated blends based on AMS–AN copolymers containing 9.4–16.5 wt% AN became homogenous again when annealed at 170°C for 2 days. Thus, the LCST behavior was captured. The measured phase separation temperatures and characteristic parameters from EOS as given in Table 3.4 are used to obtain the $\Delta P^*$ values for the blend system. The $\Delta P_{TMPC-AN}^* = 5.92$ cal/cc; $\Delta P_{TMPC-AMS}^* = 0.005$; $\Delta P_{AMS-AN}^* = 8.59$ cal/cc. TMPC was found not miscible with AMS–AN copolymer systems with AN composition of 17–18 wt%. This may be due to the large $(T_1^* - T_2^*)$ that grows with AN content. The Flory–Huggins interaction parameter calculated from the above analysis at 180°C are as follows: $B_{AMS-TMPC} = 0.18$ cal/cc; $B_{AN,-TMPC} = 5.67$ cal/cc; and $B_{AN-AMS} = 7.84$ cal/cc. The compositional window of miscibility for the copolymer–homopolymer system consisting of TMPC/AMS–AN copolymer for the calculated values of the binary interaction parameters are shown in Figure 3.6. The B values of the blend are plotted as a function of AN copolymer composition for various values of the volume fraction of the blend.

It can be seen that the replacement of styrene with AMS in SAN copolymers decreases the compositional window of miscibility. The phase separation temperatures are lowered. Homopolymer blends of polystyrene and TMPC have been found miscible with each other. Blends of poly(alpha-methylstyrene) and TMPC have been found to be immiscible.

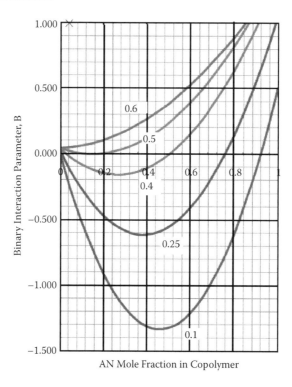

FIGURE 3.6   Binary interaction parameters for AMS–AN copolymer with TMPC homopolymer as a function of AN composition for various values of volume fraction of the blend.

Blends of AMS–AN copolymers with PVC are used to increase the heat resistance of the product. Blends of PVC and AMS–AN copolymers were formed by precipitation with THF solution into methanol. Miscible blends were found for copolymers of AMS–AN with AN composition containing 11.9–30.0 wt%. Phase separation temperatures were noted by annealing and DSC methods. Phase separation temperatures below 130°C were found. Using the characteristic EOS parameters listed in Table 3.4 and the phase separation temperatures that were measured, the $\Delta P^*$ values were calculated for the blend system. These values were found to be $\Delta P_{AN,VC}^* = 4.30$ cal/cc; $\Delta P_{AMSVC}^* = 0.26$ cal/cc; and $\Delta P_{AN,AMS}^* = 8.6$ cal/cc. The Flory–Huggins interaction parameters at 130°C were calculated and $B_{AN,VC} = 4.22$ cal/cc; $B_{AMS,VC} = 0.37$ cal/cc; $B_{AMS,AN} = 8.04$ cal/cc. The binary interaction parameter of the copolymer–homopolymer blend for AMS–AN copolymer and PVC homopolymer for various AN compositions in the copolymer at various volume fractions in the blend is shown in Figure 3.7.

It can be seen that the use of AMS as replacement for styrene in SAN copolymers widens the compositional window of miscibility and increases the phase separation temperature.

In a similar fashion the compositional window of miscibility for AMS–AN copolymer with poly(2,6-dimethyl-1,4-phenylene oxide) (PPO) and AMS–AN copolymer, and poly(ε-caprolactone) (PCL) is shown in Figure 3.8 and Figure 3.9, respectively.

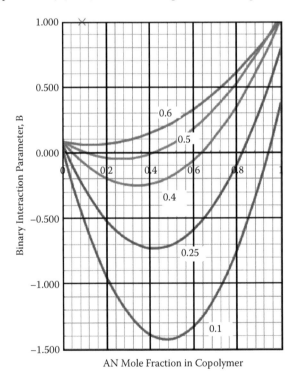

**FIGURE 3.7**  Binary interaction parameters for AMS–AN copolymer with PVC homopolymer as a function of AN composition for various values of volume fraction of the blend.

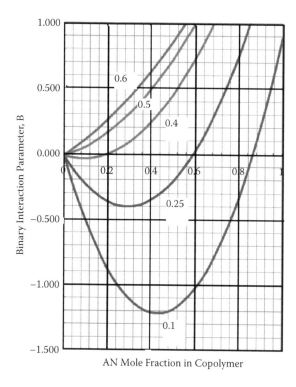

**FIGURE 3.8** Binary interaction parameters for AMS–AN copolymer with PPO homopolymer as a function of AN composition for various values of volume fraction of the blend.

The calculated $B_{ij}$ values for the systems studied in Reference 18 are listed below in Table 3.5.

## 3.9 INTRAMOLECULAR REPULSION AS DRIVING FORCE FOR MISCIBILITY–MEAN FIELD APPROACH

For several years, polymers were expected to be miscible only when there was favorable specific interaction between them. Gibbs free energy of mixing was developed and shown to contain three different contributions: combinatorial entropy of mixing, exchange interaction, and a free volume term. Specific interactions are a prerequisite for the miscibility of polymers. This was the drift that the Nobel laureate P. Flory indicated as discussed in Chapter 1. A number of exceptions to this rule have been found. A number of miscible systems have been found to have a *random copolymer* as one of the blend constituents. Both the copolymers that were found miscible with random microstructure are butadiene–styrene copolymer and vinyl chloride and vinyl acetate copolymers. The homopolymers PBd, PS, PVC, and PVA can form six binary blends. This is from $^4C_2$ = 6. They are not miscible in any of the six binary blends that can be formed from four homopolymers. This would mean that there are no specific interactions

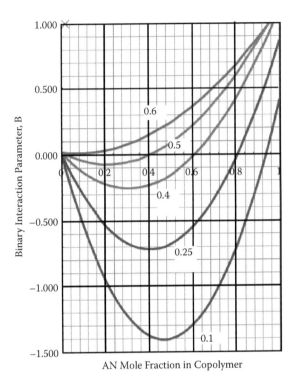

**FIGURE 3.9** Binary interaction parameters for AMS–AN copolymer with PCL homopolymer as a function of AN composition for various values of volume fraction of the blend.

between the blend constituents even when the copolymers are found to be miscible. Miscibility [2] was found in $o$-chlorostyrene and $p$-chlorostyrene copolymer; the poly(2,6-dimethyl-1,4-phenylene oxide) and PPO system; the $o$-fluorostyrene and PPO system. The miscibility was found for a certain range of copolymer compositions. Miscibility was found only up to a certain temperature at which phase separation occurs. LCST-type behavior was detected.

A miscibility window was identified in the temperature-composition plane. The miscibility windows of polyphenylene oxide/orthochlorostyrene/parachlrostyrene (PPO/$o$ClS-pClS) and polyphenylene oxide/orthoflurostyrene-parafluorostyrene (PPO/$o$FS-pFS) were compared with each other. The maxima in the miscibility window of PPO/$o$ClS-pClS were found at the center of the composition axis, and the maxima in the PPO/$o$FS-pFS system were found skewed to the $o$-FS-rich side. The miscibility window was not observed for the PPO/$o$-bromostyrene and $p$-bromostyrene copolymer blend system. Kambour et al. [1] formulated a Flory–Huggins type theory for mixtures of homopolymers and random copolymers. They argued that such a system can be miscible for a suitable choice of the copolymer composition, without the presence of any specific interaction, because of a so-called repulsion between the two different monomers comprising the copolymer.

### TABLE 3.5
### Binary Interaction Parameters and Blend Characteristic Pressures for Certain Polymer Systems [18]

| Comonomer pairs | $B_{ij}$ (cal/cc) | $\Delta P_{ij}^*$ (cal/cc) |
|---|---|---|
| S–AN | 7.02 | 7.37 |
| AMS–AN | 7.96 | 8.6 |
| S-MMA | 0.23 | 0.23 |
| AMS-MMA | 0.12 | 0.02 |
| S-TMPC | −0.14 | −0.17 |
| AMS–TMPC | 0.18 | 0.01 |
| S–VC | 0.16–0.38 | 0.17–0.40 |
| AMS–VC | 0.37 | 0.26 |
| S–PPO | −0.41 | −0.42 |
| AMS-PPO | −0.06 | -0.10 |
| S–PCL | 0.09 | 0.07 |
| AMS–PCL | 0.09 | 0.07 |
| MMA–AN | 4.32 | 4.36 |
| TMPC–AN | 5.67 | 5.92 |
| VC–AN | 4.24 | 4.30 |
| PPO–AN | 9.91 | 10.5 |
| PCL–AN | 4.62 | 4.44 |

A quasi-lattice model can be used to represent polymer systems. A polymer molecule is assumed to consist of a number of segments. Each segment is assumed to occupy one lattice site. The excluded volume effect is taken into account by allowing each site to be occupied only once. The number of different ways a polymer chain may be arranged in a lattice arrangement can be calculated. Approximate solutions to this problem were developed by Flory [16] and Huggins [17]. They obtained mathematical expressions for the free energy of mixing by combining the derived combinatorial entropy of mixing with a Hildebrand–van Laar–Scatchard enthalpy of mixing. The enthalpy of mixing is characterized by an interaction parameter $\chi$. The $\chi$ parameter was reinterpreted to obtain good agreement between theory and experimental results. EOS theories such as FOV and the lattice fluid of Sanchez and Lacombe were developed with one of the goals to explain the LCS-type behavior observed in some miscible polymer systems. These EOS theories were discussed in Chapter 2 and can be used to obtain explicit mathematical expressions for $\chi$ as a function of temperature and composition.

Ten Brinke et al. [2] proposed a six $\chi$ parameters expression for free energy change upon blending two copolymers with four different comonomers. Let the two copolymers be indicated by 1 with comonomers $A$ and $B$, and 2 with comonomers $C$ and $D$, respectively. The monomer composition of $A$ in copolymer 1 is given by $x_A$ and monomer $B$ is $(1-x_A)$, respectively; the monomer $C$ composition in copolymer 2 be given by $x_C$ and monomer $D$ composition $(1-x_C)$, respectively. The expression can be written as

$$\frac{\Delta G_{mix}}{RT} = \frac{\phi_1}{N_1}\ln\phi_1 + \frac{\phi_2}{N_2}\ln\phi_2 + \phi_1\phi_2\{(x_A x_C)\chi_{AC} + (1-x_A)(x_C)\chi_{BC} + x_A(1-x_C)\chi_{AD}$$

$$+(1-x_A)(1-x_C)\chi_{BD} - x_A(1-x_A)\chi_{AB} - x_C(1-x_C)\chi_{CD}\}.$$

(3.25)

This expression is similar to the Flory–Huggins expression for binary of homopoly-mers provided a $\chi_{blend}$ can be defined as follows:

$$\chi_{blend} = (x_A x_C)\chi_{AC} + (1-x_A)(x_C)\chi_{BC} + x_A(1-x_C)\chi_{AD} +$$
$$(1-x_A)(1-x_C)\chi_{BD} - x_A(1-x_A)\chi_{AB} - x_C(1-x_C)\chi_{CD}.$$

(3.26)

For a polyblend of copolymer 1 with comonomers A and B, and homopolymer 2 with monomer C, the expression for $\chi_{blend}$ can be written as

$$\chi_{blend} = x_A\chi_{AC} + (1-x_A)\chi_{BC} - x_A(1-x_A)\chi_{AB}.$$

(3.27)

Consider a polyblend of a terpolymer and a homopolymer. The terpolymer consists of 3 monomers: (i) methacrylonitrile, (ii) alphamethyl styrene, and (iii) acrylonitrile. The homopolymer is polycarbonate. The expression for $\chi_{blend}$ can be written as

$$\chi_{blend} = x_A\chi_{AD} + x_B\chi_{BD} + (1-x_A-x_B)\chi_{CD} - x_A(1-x_A-x_B)\chi_{AC} - x_A x_B\chi_{AB} - x_B x_C\chi_{BC}.$$

(3.28)

Consider two copolymers 1 and 2 made up of the same comonomers A and B. They differ only in their composition. Let the mismatch in the composition of comonomer A between copolymer 1 and copolymer 2 be given by; $\xi = x_A - x_C$. The expression for $\chi_{blend}$ can be written as

$$\chi_{blend} = +(1-x_A)(x_A - \xi)\chi_{AB} + x_A(1-x_A + \xi)\chi_{AB} +$$
$$-x_A(1-x_A)\chi_{AB} - (x_A - \xi)(1-x_A + \xi)\chi_{AB} = +\xi^2\chi_{AB}.$$

(3.29)

Many of the polymer blend systems that were found miscible were also found to exhibit a LCST-type phase separation at elevated temperatures. The $\chi$ interaction parameters can be expected to increase with increase in temperature. In order to predict this LCST phenomena using the theoretical approach discussed in the above sections, consider a copolymer and homopolymer blend system. The $\chi$ interaction parameter of a two-component blend comprises two contributions—an exchange interaction and free volume term. Equation (3.27) can be rewritten including the temperature effects as

$$\chi_{blend} = \frac{c_1}{v_r T}(x_A \chi_{AC} + (1 - x_A)\chi_{BC} - x_A(1 - x_A)\chi_{AB}). \tag{3.30}$$

where $v_r$ is the reduced volume and $c_1$ an integration constant. The effective interaction parameter can be written as

$$\chi_{eff} = x_A \chi_{AC} + (1 - x_A)\chi_{BC} - x_A(1 - x_A)\chi_{AB}. \tag{3.31}$$

As $\chi_{eff} > 0$, $\chi_{blend}$ decreases as a function of temperature. When $\chi_{eff} < 0$, $\chi_{blend}$ increases as a function of temperature. For different positive values of the three interaction parameters, $\chi_{AC}$, $\chi_{BC}$, and $\chi_{AB}$, either of these cases may be found. The different possibilities are shown in Figure 3.10.

The $\chi$ interaction parameter of a copolymer with two comonomers and a homopolymer with one monomer given by Equation (3.26) can be rearranged and written as

$$\chi_{blend} = x_A^2 \chi_{AB} + x_A(\chi_{AC} - \chi_{BC} - \chi_{AB}) + \chi_{BC}. \tag{3.32}$$

It can be seen that the interaction parameter of the blend is a quadratic function of the copolymer 1 composition. As can be seen from Figure 3.4, a number of different cases can result from different systems;

> *Positive*: $\chi_{AC}$, $\chi_{BC}$, $\chi_{AB}$—In this case $\chi_{blend}$ is a concave function of the copolymer composition, $x_A$. It may contain two, one, or zero zeros at a prescribed temperature. For values of copolymer composition falling between the two

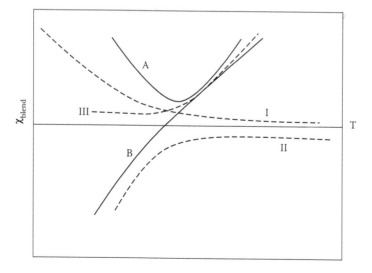

**FIGURE 3.10** Blend interaction parameter as a function of temperature [2].

zeros, the interaction parameter of the blend is an increasing function of temperature. Outside of this region, the interaction parameter of the blend decreases with increase in temperature.

*Negative:* $\chi_{AC}$, $\chi_{BC}$, $\chi_{AB}$—In this case, $\chi_{blend}$ is a convex function of the copolymer composition, $x_A$. It may contain, two, one or zero zeros at a prescribed temperature. Values of copolymer composition between the two zeros represent immiscible regions. A window of miscibility is obtained upon taking into account the effect of temperature.

*Negative:* $\chi_{AC}$; positive: $\chi_{AB}$—The $\chi_{blend}$ is negative when $\chi_{AC}$ is negative, and the system is miscible even at copolymer composition, $x_A$ close to 1.

## 3.10  SUMMARY

The thermodynamic basis to explain miscibility in polymer blends is the exothermic heat involved in mixing, as entropic contributions are small for such systems. Intramolecular repulsions may be an important factor in realizing exothermic heats of mixing. This approach was independently presented by Kambour, Bendler, and Bopp [1]; Brinke, Karasz, and MacKnight [2]; and Paul and Barlow [3].

The heat of mixing is described by the van Laar expression given in Equation (3.3) where $B$ is the binary interaction energy density. The relation between the $B$ parameter and Flory–Huggin's interaction parameter $\chi$ is given by Equation (3.4). The binary interaction model for heat of mixing can be extended to multicomponent mixtures as given by Equation (3.4). The binary interaction model for heat of mixing can be extended to multicomponent mixtures as given by Equation (3.5).

The compositional window of miscibility for systems with copolymer and homopolymer blends was developed using a binary interaction model. When Equation (3.6) becomes negative, the free energy of mixing becomes negative, driving miscibility. The system SAN copolymer–PMMA homopolymer was illustrated in Figure 3.1. The composition of AN and volume fraction of a blend where miscibility can be expected was calculated using an MS Excel spreadsheet.

The binary interaction model was applied to a system of copolymer–copolymer binary blend with common monomers in Section 3.3. The SAN/SAN blend, the compositional mismatch that can be tolerated to stay miscible, was calculated. A method to calculate the compositional window of miscibility for a terpolymer–terpolymer blend with common monomers was illustrated in Section 3.4. The @ RAND key can be used in an MS Excel spreadsheet to arrive at the compositional window of miscibility in computational time. The miscible regions for terpolymer–terpolymer binary blends with common monomers are shown in Figure 3.2. The compositional window of miscibility for terpolymer–homopolymer miscibility without any common monomers was calculated using a binary interaction model in Section 3.5. The system chosen for the illustration was TMPC/AMS–AN–MMA.

The expression to define the spinodal curves in the phase diagram of binary blends was obtained from the criteria for equilibrium and stability and from the Gibbs free energy expression. EOS theory such as the lattice fluid theory of Sanchez–Lacombe can be used to capture the temperature variation of the binary interaction parameter $B$ values. The copolymer–homopolymer blends of AMS–AN/PVC were discussed. Copolymer–homopolymer blends of AMS–AN copolymer with homopolymers such as PMMA, TMPC, PPO, and PCL were derived. The LCST–phase separation temperature for PMMA/AMS–AN for two different AN compositions are shown in Figure 3.4.

Ten Brinke et al. [2] proposed a 6 $\chi$ parameter expression for free energy in mixing for two copolymers with four monomers. Different possibilities are shown in Figure 3.10. The interaction parameter variation with copolymer composition may take a convex or concave curvature, depending on the values.

## EXERCISES

1. As can be seen from the illustrations, negative values of the $B_{ij}$ parameter drive miscibility. So can a copolymer/homopolymer be miscibilized by addition of a termonomer to the copolymer at similar composition.
2. How do $B_{ij}$ values change with molecular weight?
3. Can $B_{ij}$ values be estimated from fundamental theories and models?
4. What is the difference between intramolecular repulsion and intermolecular attraction?
5. Can the entropy of mixing be negative?
6. Does the criteria for equilibrium and stability applied to first and second derivatives of the Gibbs free energy apply to the Helmholtz free energy, A, in a similar fashion?
7. What is the difference between a spinodal and binodal condition?
8. Are the volume fractions a function of temperature?
9. What is the relation between the binary interaction parameter values and heat capacity, $C_p$, values?
10. Why is $B_{ij} = B_{ji}$?
11. Is the van Laar mixing model applicable for protein polymers during folding?
12. How would binary interaction model expect to fare for star polymers?
13. Does chain branching affect the binary interaction parameters?
14. SAN graft chains at 25% AN and SAN matrix chains at 25% AN have the same binary interaction parameter from the model. Would the intramolecular repulsion phenomena be the same between graft–graft chains, graft–matrix chains, and matrix–matrix chains?
15. What is the role of the chain sequence distribution on the binary interaction model?
16. Can the binary interaction model be applied to block copolymer systems as it is applied to random copolymer systems?
17. What is the difference between LCST and UCST?
18. How many monomers can go into a multicomponent copolymer before they have no impact on the blend thermodynamics?
19. Why are the bare interaction densities $P_{ij}$ preferred to $B_{ij}$ values in deriving spinodals in binary blends?
20. What are the differences between a spinodal represented by a straight line and a spinodal that is curvilinear?
21. What are metastable blends?

22. In a similar manner to finding a lower consolute solution temperature, can a lower consolute solution volume fraction be found?

23. In a similar manner to identifying an upper consolute solution temperature, can a upper consolute solution volume fraction be found?

24. The EOS of Sanchez and Lacombe in terms of reduced pressure, temperature, and density is a transcendental equation. In combining this equation with the free energy of mixing equation from the binary interaction model, is a numerical solution needed or can an analytical solution be obtained?

25. Polymer $A$ and polymer $B$ and polymer $A$ and polymer $C$ are found to be miscible when binary blends were made from them. Can polymer $B$ and polymer $C$ be expected to be miscible when binary blends are made using them.

## PROBLEMS

1. Show that, for systems blended in a *reversible* manner with negligible change of volume upon mixing,

$$\Delta G_{mix} = V\Delta P_{mix}. \tag{3.33}$$

2. Estimate the binary interaction density, $B_{ij}$ for a styrene/maleic anhydride (S/MA) binary pair from the data provided in the following equation, using the binary interaction model. Use the *critical molecular weight technique* as discussed in [14]. This technique is one where the change in interaction energy density with molecular weight can be tapped into. The entropic contributions to the free energy of mixing changes with the molecular weight. Show that by combining Equation (3.20) for $B_{crit}$ with the binary interaction energy for blends with common monomers, the following expression can be written:

$$\phi_2 = \sqrt{\frac{RT}{2B_{21}}} \left[ \sqrt{\frac{\rho_{21}}{MW_{21}}} + \sqrt{\frac{\rho_3}{MW_3}} \right], \tag{3.34}$$

where 2 refers to MA in SMA copolymer; 21 refers to SMA copolymer; 3 refers to PS homopolymer. The blend compositions were 50/50 and obtained by solvent casting and annealing at 170°C.

   As can be seen from Equation (3.34), a plot of $\phi_2$ versus $\sqrt{\rho_{21}/MW_{21}}$ can be used to obtain $B_{21}$ from the slope of the curve. The phase boundary of such a plot is tabulated in Table 3.6.

3. Equation (3.12) is written for the binary interaction energy density $B$ for a terpolymer $A$–terpolymer blend $B$ with all three common monomers. The $X$, $Y$, and $Z$ are in terms of $(\phi_1'-\phi_1'')$, $(\phi_2'-\phi_2'')$, and $(\phi_3'-\phi_3'')$. $\phi_1'$ is the composition of monomer 1 in terpolymer $A$, $\phi_{1A}$ multiplied with the volume fraction of terpolymer $A$, $\phi_A$ in the blend (i.e., $\phi_1' = \phi_{1A}\phi_A$). Rewrite Equation (3.12) in terms of the terpolymer compositional mismatch and volume fraction of the two terpolymers in the blend. For a certain fixed compositional mismatch, evaluate the variation of the binary interaction energy as a function of the *volume fraction* of the polymer in the blend.

**TABLE 3.6**

**Experimental Cloud Point Curve for SMA–PS Binary Blend**

| Wt% MA composition in SMA copolymer | $MW_3$ (polystyrene) |
|:---:|:---:|
| 6.0 | 60,000 |
| 7.0 | 30,000 |
| 8.0 | 20,000 |
| 9.0 | 15,000 |
| 10.0 | 10,000 |
| 12.0 | 5,000 |
| 14.0 | 2,500 |

4. Consider a binary blend of two terpolymers with *two common monomers*. Develop the binary interaction energy for such a blend. Discuss the miscible regions in terms of the compositional mismatch and *volume fraction* in the blend.

5. Consider a binary blend of two terpolymers with *one common monomer*. Develop the binary interaction energy for such a blend. Discuss the miscible regions in terms of the one composition mismatch and four other terpolymer compositions and *volume fraction* of the terpolymers in the blend.

6. Effect of volume fraction of blend on terpolymer/homopolymer miscibility In Section 3.5 a terpolymer made of three comonomers, (A) alphamethlystyrene (AMS), (B) acrylonitrile–AN, and (C) methyl methacrylate (MMA), was evaluated for miscible regions when blended with (D) tetramethyl polycarbonate (TMPC) homopolymer as a function of the terpolymer composition and blend volume fraction. The miscibility regions for a 70% volume fraction terpolymer 1 with homopolymer 2 is shown in Figure 3.3. The $B$ values for the blend are contoured for constant copolymer composition of AN. The $B$ is plotted versus composition of AMS. The $B$ values from the calculation are tabulated in Table 3.2. Obtain a similar plot as Figure 3.3 for miscible regions in the blend as a function of volume fraction of the polymers in the blend. Are these contours also linear?

7. *Solvent effect:* Phase separation can be seen when a miscible pair is cast from a common solvent. Show that this effect is caused by an asymmetry in the polymer–solvent interactions leading to a closed region of immiscibility in the ternary solvent–polymer–polymer phase diagram [20].

8. The isothermal miscibility map for 50/50 blends of SMA copolymer with TMPC–PC copolymer was given in [14]. It is shown below in Table 3.7. Obtain the $B$ values for
   a. TMPC–PC
   b. TMPC–PS
   c. TMPC–MA
   d. PC–PS
   e. PC–MA
   f. PS–MA

**TABLE 3.7**

**Experimental Cloud Point Curve for SMA/TMPC_PC Binary Blend**

| #  | Weight % MA in SMA | Weight % PC in TMPC–PC |
|----|--------------------|------------------------|
| 1  | 17.0               | 0.0                    |
| 2  | 15.0               | 10.0                   |
| 3  | 12.0               | 19.0                   |
| 4  | 9.0                | 22.0                   |
| 5  | 6.0                | 21.0                   |
| 6  | 3.0                | 17.0                   |
| 7  | 0.0                | 10.5                   |

9. The isothermal miscibility map as shown in Table 3.7 delineates the miscibility from the immiscibility regions. Develop the binary interaction energy for the binary blend between two copolymers without any common monomers. Use linear regression and obtain the best-fit parameters for the six binary interaction parameter $B_{ij}$ pairs using the information provided. Interpolate the isothermal miscibility map if necessary.

10. *Ternary blend using binary interaction model*: Gan et al. [18] found for certain copolymer compositions and volume fractions the ternary blend system of styrene acrylonitrile copolymer (SAN), polycarbonate (PC) homopolymer and polycaprolactone (PCL) was completely miscible. Develop the expression for binary interaction energy $B$ for the ternary blend using binary interaction model. Is the intramolecular repulsion in the copolymer sufficient to drive miscibility with two other homopolymers without any common monomers?

11. The interaction energy densities at the drying temperature of 150°C for the monomers alpha-methylstyrene (AMS), methyl methacrylate (MMA), and acrylonitrile (AN) pairwise were found to be $B_{AMS,MMA} = 0.12$ cal/cc; $B_{MMA,AN} = 4.32$ cal/cc; $B_{AMS,AN} = 7.96$ cal/cc. The larger values of $B_{ij}$ compared with $\Delta P_{ij}^{*}$ values indicate possible EOS contributions. The compositional window of miscibility of copolymer/homopolymer blends of AMS–AN copolymer and PMMA homopolymer for the binary interaction parameter values obtained is shown in Figure 3.6. The $B$ values are plotted as a function of AN composition in the copolymer for different volume fractions of the blend. In a similar manner, construct a plot of binary interaction energy density as a function of volume fraction of the blend contoured for various values of AN copolymer composition.

12. The binary interaction parameter values pairwise for AMS, TMPC, and AN comonomers were measured [18] as $B_{AMS,TMPC} = 0.18$ cal/cc; $B_{AN,\ TMPC} = 5.67$ cal/cc; $B_{AN,AMS} = 7.84$ cal/cc. The compositional window of miscibility for the copolymer/homopolymer system consisting of TMPC/AMS–AN copolymer for the calculated values of the binary interaction parameters are shown in Figure 3.7. The $B$ values of the blend are plotted as a function of AN copolymer composition for various values of the volume fraction of the blend. In a similar manner,

construct a plot of binary interaction energy density as a function of volume fraction of the blend contoured for various values of AN copolymer composition.

13. The binary interaction parameter values pairwise for AMS, vinyl chloride (VC), and AN, were measured at 130°C [18] as $B_{AN,VC}$ = 4.22 cal/cc; $B_{AMS,VC}$ = 0.37 cal/cc; and $B_{AMS,AN}$ = 8.04 cal/cc. The binary interaction parameter of the copolymer/homopolymer blend for AMS–AN copolymer and PVC homopolymer for various AN compositions in the copolymer at various volume fractions in the blend is shown in Figure 3.8. The $B$ values of the blend are plotted as a function of AN copolymer composition for various values of the volume fraction of the blend. In a similar manner, construct a plot of binary interaction energy density as a function of volume fraction of the blend contoured for various values of AN copolymer composition.

14.. In Figure 3.9, the binary interaction parameters for AMS–AN copolymer with PPO homopolymer are plotted as a function of AN composition for various values of volume fraction of the blend. In a similar manner, construct a plot of binary interaction energy density as a function of volume fraction of the blend contoured for various values of AN copolymer composition. Use the $B_{ij}$ values provided in Table 3.5.

15.. In Figure 3.10 the binary interaction parameters for AMS–AN copolymer with PCL homopolymer are plotted as a function of an AN composition for various values of volume fraction of the blend. In a similar manner, construct a plot of binary interaction energy density as a function of volume fraction of the blend contoured for various values of AN copolymer composition. Use the $B_{ij}$ values provided in Table 3.5.

16. Use the $B_{ij}$ values provided in Table 3.5 and obtain the compositional window of miscibility of PPO–AN and SAN copolymers. Did you obtain a wider window of miscibility?

# REFERENCES

1. R. P. Kambour, J. T. Bendler, and R. C. Bopp, Phase behavior of polystyrene, poly(2,6-dimethyl-1,4-phenylene oxide), and their brominated derivatives, *Macromolecules*, 16, 753, 1983.

2. G. Ten Brinke, F. E. Karasz, and W. J. Macknight, Phase behavior in copolymer blends: Poly(2,6-dimethyl-1,4-phenylene oxide) and halogen-substituted styrene copolymers, *Macromolecules*, 16, 1827, 1983.

3. D. R. Paul and J. W. Barlow, A binary interaction model for miscibility of copolymers in blends, *Polymer*, 25, 487, 1984.

4. K. R. Sharma, *Course Notes, Applied Polymer Thermodynamics—Chemical Engineering Thermodynamics*, Ch.E. 344, Chemical Engineering, West Virginia University, Morgantown, WV, 1995.

5. S. Krause, Polymer-polymer miscibility, *Pure & Appl. Chem.*, 58, 12, 1553–1560, 1986.

6. D. R. Paul, J. W. Barlow, and H. Keskkula, Polymer blends, in *Encylcopedia Polymer Science and Technology*, 2nd edition, volume 3, 758, 1988.

7. M. M. Coleman, J. J. Graf, and P.C. Painter, *Specific Interactions and the Miscibility of Polymer Blends,* Technomic Press, Lancaster, PA, 1991.

8. I. C. Sanchez, Relationships between polymer interaction parameters, *Polymer*, 30, 471, 1989.

9. O. Olabisi, L. M. Robeson, and M. T. Shaw, *Polymer-Polymer Miscibility*, Academic Press, New York, 1979.

10. D. J. Walsh, W. W. Grassley, S. Datta, D. J. Lohse, and L. J. Fetters, *Macromolecules*, 25, 5236, 1992.

11. P. P. Gan, D. R. Paul, and A. R. Padwa, *Polymer*, 35, 7, 1487, 1994.

12. K. R. Sharma, Miscibility Compositional Window in Terpolymers with Common Monomers using @RAND in Spreadsheet, 30th IEEE Southeastern Symposium on System Theory, Morgantown, WV, March 1998.

13. K. R. Sharma, Course Development in Applied Polymer Thermodynamics, ASEE Annual Conference and Exposition, Milwaukee, WI, June 1997.

14. G. D. Merfeld and D. R. Paul, Binary interaction parameters from blends of SMA copolymers with TMPC-PC copolycarbonates, *Polymer*, 39, 10, 1999–2009, 1998.

15. L. Zhang, S. Sun, X. Li, M. Zhang, and H. Zhang, Effect of $\alpha$-MSt content in $\alpha$-MSAN copolymer on the miscibility of PVC/$\alpha$-MSAN blends, *Polym. Bull.*, 61, 99–106, 2008.

16. P. J. Flory, *Principles of Polymer Chemistry*, Cornell University Press, Ithaca, New York, 1953.

17. M. L. Huggins, Thermodynamic properties of solutions of long-chain compounds, *Ann. New York Acad. Sci.*, 41, 1, 1942.

18. P. P. Gan, D. R. Paul, and A. R. Padwa, Phase behavior of blends of homopolymers with alphamethyl styrene/acrylonitrile/copolymers, *Polymer*, 35, 1487–1502, 1994.

19. C. K. Kim and D. R. Paul, Interaction parameter for blends containing polycarbonates. Pt 1: Tetramethyl bisphenol-A-polycarbonate/polystyrene, *Polymer*, 33, 8, 1630–1639, 1992.

20. L. Zeman and D. Patterson, Effect of the solvent on polymer incompatibility in solution, *Macromolecules*, 5, 513, 1972.

# 4 Keesom Forces and Group Solubility Parameter Approach

## LEARNING OBJECTIVES

- Hildebrand solubility parameter
- Hansen three-dimensional solubility parameter
- Hydrogen bonding in polymer blends
- Association model
- Combinatorial entropy
- Chemical and physical forces
- Equilibrium rate constant
- Estimation of solubility parameter for EOS theories
- Measurements of Keesom forces using FTIR

As discussed in Chapter 1, the Nobel laureate P. Flory alluded to possible favorable interactions among "polar substituents" leading to polymer–polymer miscibility. Polar interactions can be referred to as *Keesom* forces. When the polar interactions are between dipoles and induced dipoles they are called *Debye* forces. When both interaction forces are due to induced dipoles they are called *London* forces. Keesom is when the interactions are due to dipoles and dipole *ab initio*. When Keesom forces become strong enough they lead to the formation of hydrogen bonds.

## 4.1 HILDEBRANDT SOLUBILITY PARAMETER

Hildebrandt and Scott [1] introduced a solubility parameter that can be used to better predict solubility in nonelectrolytes:

$$\delta_v = \sqrt{\frac{\Delta H_v}{v_L}}, \qquad (4.1)$$

where $\delta_v$ is the Hildebrandt solubility parameter, $v_L$ is the molar volume in the saturated liquid phase, and $\Delta H_v$ is the enthalpy change of vaporization. The Hildebrandt solubility parameter can be related to the Flory–Huggins interaction parameter by use of regular solution theory. Thus,

$$\chi = \frac{v_m}{RT}(\delta_1 - \delta_2)^2, \tag{4.2}$$

where $v_m$ is the molar volume of the liquid, $\delta_1$ and $\delta_2$ are the Hildebrandt solubility parameters of the polymers 1 and 2 in a binary blend, $T$ is the temperature of the blend, and $R$, the universal molar gas constant.

The Hildebrandt solubility parameter can be calculated from a considered EOS of polymer as given in Chapter 2 as follows:

$$dH = dU + PdV - VdP = TdS + VdP, \tag{4.3}$$

$$\left[\frac{\partial \Delta H}{\partial V}\right]_P = T\left(\frac{\partial S}{\partial V}\right)_P. \tag{4.4}$$

From Maxwell's relation, Equation (A.16):

$$\left(\frac{\partial T}{\partial P}\right)_S = \left(\frac{\partial V}{\partial S}\right)_P. \tag{4.5}$$

Combining Equation (4.5) with Equation (4.4):

$$\left[\frac{\partial \Delta H}{\partial V}\right]_P = T\left(\frac{\partial P}{\partial T}\right)_S. \tag{4.6}$$

From Equations (2.18) and (2.21), Equation (4.6) can be rewritten as

$$\left[\frac{\partial \Delta H}{\partial V}\right]_P = T\left(\frac{\beta}{\kappa}\right) \cong \frac{\Delta H_v}{v_L} = \delta_v^2. \tag{4.7}$$

The Hildebrandt solubility parameter can be calculated from the EOS parameters of isothermal compressibility, $\kappa$, and volume expansivity, $\beta$. Kumar [2] gave a similar expression from an analysis using internal energy change, $\Delta U$. He defines another solubility parameter, $\delta_{ip}$, for internal pressure. The rate of change of internal energy change with volume is the internal pressure of the fluid.

The EOS developed by Sanchez and Lacombe using the lattice fluid theory can be used to obtain estimates of Hildebrandt solubility parameter, $\delta_v$. Equation (2.61) for isothermal compressibility, $\beta$, and volume expansivity, $\kappa$, from the lattice fluid EOS for polymers is used to obtain an expression for $\delta_v$ as

$$\delta_v^2 = \left(\frac{P^* V^{*2}}{v^2} + P\right). \tag{4.8}$$

The solubility parameter of the polymer is a function of its molecular weight. Molecular weight effects are taken into account through the molar volume, $v$.

## Example 4.1

For PMMA, poly(methyl methacrylate), with a degree of polymerization of 1000, calculate the Hildebrand solubility parameter using the EOS from lattice fluid theory.
From Table 2.4:

Characteristic pressure for PMMA, $P^* = 5169$ bar
Characteristic volume for PMMA, $V^* = 1000*100/1.2812 = 78,052$ cc/mole
Molar volume of PMMA $= 10^5$ cc/mole
From Equation (4.8), $\delta_v = 55.4$ atm$^{1/2} = 17.6$ MPa$^{1/2}$

## 4.2   HANSEN THREE-DIMENSIONAL SOLUBILITY PARAMETER

Hansen [3] separated the total Hildebrandt solubility parameter, $\delta_v$, into three components: (1) dispersive, $\delta_D$; (2) polar, $\delta_P$, and (3) hydrogen bonding, $\delta_H$. Thus,

$$\delta_v^2 = \delta_D^2 + \delta_P^2 + \delta_H^2. \tag{4.9}$$

The Hansen parameters are additive. The numerical values for the component solubility parameters are determined in a stepwise fashion. The homograph method can be used to obtain the dispersive component. The homomorph of a polar molecule is the nonpolar molecular closely resembling it in size and structure. The Hildebrandt value for the nonpolar homograph due to dispersive forces is assigned to the polar molecule as its dispersion component value. The square of the dispersion component is subtracted from the Hildebrandt value squared. The remainder represents the polar interaction between the molecules. By trial and error, and by use of numerous solvents and polymers, Hansen separated the polar value into polar and hydrogen bonding component parameters from the best fit of experimental data. Further, he derived polymer solubilities. The spherical volume of solubility was formed for each polymer by doubling the dispersion parameter axis. An interaction radius was defined. The solubility parameter values for some polymers are provided in Table 4.1.

The fraction of the net solubility parameter that is dispersive, polar, or hydrogen bonding is called the *fractional solubility parameter*. Graphical representations of these fractional parameters on triangular graph paper are referred to as TEAS graphs. Solvent blends can be selected to dissolve a polymer by matching the solubility parameter values.

## 4.3   SPECIFIC INTERACTIONS

Coleman and Painter [4–6] consider hydrogen bonding as the central strong interaction in polymers that cause the observed phase behavior and miscibility. They have used Fourier transform infrared (FTIR) spectroscopy to study the hydrogen bonding in polymer blends in systems such as polyamides and polyurethanes. A large class

**TABLE 4.1**

**Hansen Solubility Parameters for Some Polymers**

| Polymer | $\delta_D$ (MPa$^{1/2}$) | $\delta_P$ (MPa$^{1/2}$) | $\delta_H$ (MPa$^{1/2}$) |
|---|---|---|---|
| Cellulose acetate (CA) | 18.6 | 12.7 | 11.0 |
| Chlorinated polypropylene (ClPP) | 20.3 | 6.3 | 5.4 |
| Epoxy | 20.4 | 12.0 | 11.5 |
| Isoprene elastomer | 16.6 | 1.4 | 0.0 |
| Cellulose nitrate (CN) | 15.4 | 14.7 | 8.8 |
| Polyamide (PA) | 17.4 |  | 14.9 |
| Polyisobutylene (PIB) | 14.5 | 2.5 | 4.7 |
| Polyethlyl methacrylate (PEMA) | 17.6 | 9.7 | 4.0 |
| Polymethyl methacrylate (PMMA) | 18.6 | 10.5 | 7.5 |
| Polystryrene (PS) | 21.3 | 5.8 | 4.3 |
| Polyvinyl acetate (PVA) | 20.9 | 11.3 | 9.6 |
| Polyvinyl butyral (PVB) | 18.6 | 4.4 | 13.0 |
| Polyvinyl chloride (PVC) | 18.2 | 7.5 | 8.3 |
| Polyester (saturated) | 21.5 | 14.9 | 12.3 |

of polymer blends form miscible blends and exhibit interesting phase behaviors with appropriate partners. This class of polymer is defined by the presence of functional groups that are capable of forming hydrogen bonds. These associations are polar in nature and hence consistent with Flory's predictions as discussed in Chapter 1. In Chapter 3, a number of miscible polymer blend systems were discussed and observations explained using the binary interaction model and the intramolecular repulsion among the polymers. Exothermic enthalpy of mixing was causative in driving the free energy of mixing as favorable for miscibility.

Early attempts to model polymer miscibility started with the Flory–Huggins theory. Modifications to the theory to account for the free volume effects have been proposed in the literature. The Flory interaction parameter, $\chi$, was allowed to assume negative values to explain observed polymer–polymer miscibility. This approach was not found to be satisfactory for four reasons:

1. The number of hydrogen bonding contacts are not random. The interaction term has to take a more complex composition dependent form than the $\varphi_1 \varphi_2 \chi_{12}$.
2. Formation of hydrogen bonds results in loss of rotational degree of freedom in the molecules.
3. A specific and nonspecific interactions cannot be lumped into a single interaction parameter, $\chi$.
4. The number of hydrogen bonds can be counted by experimental techniques such as FTIR spectroscopy.

Coleman and Painter [4–6] have developed a model that uses solubility parameters to account for the nonspecific interactions. The solubility parameters are estimated from chosen group contributions that exclude the association effects. The associative contribution was measured using infrared methodologies for determination of parameters that describe the stoichiometry of hydrogen bonding, and this must be taken into account when considering their contribution to the free energy change of blending. The authors have initiated light scattering and neutron scattering experiments on certain polymer systems.

Consider a blend with one of the components consisting of a halogenated system such as chlorinated PVC and another component containing a phenol group. Self-assembly or hydrogen bonding can be expected between the bonded chlorine atom on one component and the –OH group in the other component. Due to the bulkiness of the electron cloud in the chlorine atom, it assumes a fractional negative charge, and the hydrogen in the OH group assumes a fractional positive charge. The permanent dipole formed can be viewed as a Keesom interaction. When this Keesom interaction becomes strong, the interaction can lead to a hydrogen bond. If the repeat unit in the polymer where the hydrogen bonding occurs is denoted by $A$, then

$$A_j + A_1 \leftrightarrow A_{j+1}. \tag{4.10}$$

Flory wrote an equilibrium constant to describe $j$ mers that participate in hydrogen bonding by use of a volume fraction:

$$K_A = \frac{\phi_{A(j-1)}}{\phi_{A(j)}\phi_{A1}} \frac{jr}{(j+r)}, \tag{4.11}$$

where $r$ is the ratio of the molar volumes of the two components in the blend. Free energy of the blend can be defined in terms of the equilibrium distribution of self-associating species. Thus,

$$\frac{\Delta G_H}{RT} = \sum_j n_{Aj} \ln\left(\frac{\phi_{Aj}}{j}\right) + \sum_j n_{Aj1} \ln\left(\frac{\phi_{Aj}}{j+r}\right)$$
$$+ n_1 \ln\phi_1 + \sum_j n_{Aj}(j-1) + \sum_j n_{Aj1}j + n_1(r-1). \tag{4.12}$$

Another term in the right-hand side of Equation (4.12) can be added in terms of $z$, the coordination number of the lattice, and $\sigma$ a symmetry number. Coleman and Painter [4–6] assumed that the association model to describe hydrogen bonding interactions can be used and added to the contribution to the Flory–Huggins equation for the free energy of mixing:

$$\frac{\Delta G_m}{RT} = N_A \ln\phi_A + N_1 \ln\phi_1 + n_A\phi_1\chi + \frac{\Delta G_H}{RT}. \tag{4.13}$$

where $N_A$ and $N_1$ are the number of polymer molecules of A and 1 present and $n_A$ is the total number of A segments. They also recognize a combinatorial term. The excess combinatorial entropy of mixing can be realized as

$$-\frac{\Delta S_c}{R} = \sum_j n_{Aj}^0 \ln\phi_A + n_1 \ln\phi_1 = \frac{n_A}{j^0}\ln\phi_A + n_1 \ln\phi_1. \tag{4.14}$$

where $n_{Aj}^0$ is the number of molecules of $j$-mer present in pure A and $j^0$ is the number average hydrogen bond chain length in pure A. Equation (4.14) depicts a blend in a hypothetical situation where the distribution of hydrogen bonds remains the same as in pure A. A regular solution type combinatorial entropic contribution may be subtracted:

$$-\frac{\Delta S_c}{R} = \sum_j n_A \ln\phi_A + n_1 \ln\phi_1. \tag{4.15}$$

Coleman and Painter [4–6] have established a theoretical basis for using a Flory-type lattice approach for describing hydrogen bonding in polymer blends. They consider randomly mixed, non-hydrogen-bonded polymer chains, and then impose constraints due to hydrogen bonding by obtaining the probability that the appropriate segments would be situated next to one another to form an equilibrium distribution of hydrogen bonds. The free energy of mixing can be written as

$$\frac{\Delta G_m}{RT} = n_A \ln\left(\frac{\phi_A}{\phi_A^0}\right) + n_1 \ln\phi_1 + n_A K_A(\phi_A - \phi_A^0)$$

$$+n_A(1 - K_A\phi_A)\left(\frac{K_1\phi_1}{r + K_1\phi_1}\right) + n_A\phi_1\chi. \tag{4.16}$$

Equation (4.15) gives the combinatorial entropy change in the form of regular solution combinatorial entropy. Thus, three main components can be recognized in the free energy term for mixing in polymer blends. These are

1. Combinatorial entropy:

$$\frac{\phi_1}{M_1}\ln\phi_1 + \frac{\phi_A}{M_A}\ln\phi_A. \tag{4.17}$$

The contribution from the combinatorial entropy term is favorable for miscible system formation, but the absolute values of the contribution are very small, especially for high molecular weight blends.

2. "Physical" forces: Unfavorable for miscible system formation and provide a positive contribution:

$$\chi \phi_1 \phi_A, \tag{4.18}$$

where $\chi = V_r \, / \, RT(\delta_1 - \delta_A)^2$ in terms of the solubility parameters of polymers A and 1.

3. "Chemical" forces: Favorable for miscible system formation and provide a negative contribution:

$$\frac{\Delta G_H}{RT}. \tag{4.19}$$

The equilibrium constants $K_A$, $K_1$, and enthalpies of formation of A and 1 can be measured using FTIR spectroscopy methods. The fraction of free and hydrogen-bonded groups can be used to calculate these quantities.

The investigators [4–6] delineated the contributions to the free energy of mixing from "strong" and "weak" forces. The weak forces are calculated from solubility parameter values. The strong forces are estimated by use of an association model. The contributions to the free energy of mixing from strong and weak forces have very different temperature and compositional dependencies. These affect the shape and form of phase diagrams. Miscibility is driven by a balance between the contributions from the $\chi \phi_1 \phi_A$ and $\Delta G_H/RT$. When there is no contribution to the free energy of mixing from chemical forces, the equation reverts to the case of Flory–Huggins. The critical value of the interaction parameter, $\chi$, for a miscible single phase, $\chi_{crit}$, $< 0.002$ for high molar mass. In other words, the solubility parameters of the two polymers must lie within 0.1 $(cal.cm^{-3})^{0.5}$ to ensure miscibility. Miscible polymer systems cannot be predicted when only dispersive forces are considered in this approach. The contribution from the $\Delta G_H/RT$ term can indicate favorable conditions for miscible systems to form. This translates to a non-hydrogen-bonded solubility parameter difference of about 2.4 $(cal.cm^{-3})^{0.5}$.

The magnitude of the favorable contribution to the free energy of mixing from the distribution of hydrogen–bonded species depends on two factors: degree of self-association and degree of interassociation. This is the salient factor that can be used to determine the miscibility window of polymer–polymer systems. The degree of interassociation can be expressed in terms of equilibrium constant in the reversible hydrogen bonding reaction as given by Equation (4.10). A second factor that determines the magnitude of the $\Delta G_H/RT$ term is the number of specific interacting species per unit volume in the blend. The magnitude of this term will reduce when the number of specific interacting sites per unit volume decreases. The free-volume effects have been neglected. Free-volume differences between polymers have been found to be small and hence not expected to make an appreciable difference in miscibility predictions.

### 4.3.1 EXPERIMENTAL DETERMINATION OF EQUILIBRIUM RATE CONSTANTS

Coleman and Painter [5] have used FTIR spectroscopy to determine the number of fraction of functional groups that are hydrogen bonded or non-hydrogen bonded. This information is used to determine the equilibrium rate constants of interassociation and self-association. Polymers that contain OH functional groups such as phenol and propanol have been found to self-associate. Hydrogen bonds are dynamic in nature. They form in a continuous manner, break up, and reform under the influence of thermal motion. As a result, a distribution of hydrogen bonds is present. They may be hydrogen-bonded dimmers and hydrogen-bonded multimers. This distribution can be expected to change with temperature and concentration. The hydroxyl stretching region of the infrared spectrum at different concentrations was obtained. The spectral information is used in the determination of equilibrium rate constants and enthalpies of hydrogen bond formation in interassociation and self-association. Alcohols and phenols form miscible solutions with inert solvents because of the large contribution of combinatorial entropy to the free energy of mixing.

Quantitation of equilibrium constants of self-association needs the fraction of free monomers that are present in dilute solutions of known concentration. Suitable assumptions were made such as the intensity of the free hydroxyl band as a measure of the free monomers. The intensity from absorbance of the isolated hydroxyl band, I is related to the absorptivity coefficient, $\varepsilon$, the concentration, $c$, and the path length $l$ by the Beer–Lambert law:

$$I = (\varepsilon l c). \tag{4.20}$$

Two dimensionless equilibrium constants were found necessary to adequately describe the self-association of OH groups. The formation of dimers with an equilibrium rate constant of $K_2$ and higher repeat multimer complex with an equilibrium rate constant of $K_A$:

$$A_1 + A_1 \overset{K_2}{\Longleftrightarrow} A_2$$
$$A_j + A_1 \overset{K_A}{\Longleftrightarrow} A_{j+1}. \tag{4.21}$$

Coleman and Painter [5] have developed an iterative least squares fitting procedure to obtain the best fit of experimental data. The fraction of OH groups that participate in hydrogen bonding is related to the dimensionless equilibrium rate constants $K_2$ and $K_A$ in Equation (4.22):

$$f_m^{OH} = \frac{\phi_{A1}}{\phi_A} = \left\{ \left( 1 - \frac{K_2}{K_A} \right) + \frac{K_2}{K_A} \left( \frac{1}{(1 - K_A \phi_{A1})^2} \right) \right\}^{-1}. \tag{4.22}$$

The difficulty experienced using this method is when systems such as polyvinyl phenol (PVPh) and ethylenevinyl alcohol (EVOH) need to be measured. The self-association equilibrium constants measured in such cases will be confounded by association of OH groups from both polymers, making the interpretation of the experimental information tedious.

Intermolecular association between two functional groups located in different polymers can be accounted for as follows:

$$A_j + B \overset{K_B}{\Longleftrightarrow} A_jB. \tag{4.23}$$

Infrared methods can be used to obtain the equilibrium constant of interassociation Keesom interactions. Equilibrium constants representing self-association and interassociation cannot be obtained independently from experimental data. Values of a self-association have to be available a priori to calculation of interassociation equilibrium rate constants. The relative magnitude of the equilibrium constants representing self-association and interassociation is a critical factor in determining their contributions to the free energy of mixing. The approaches vary from functional group to functional group. For example, when dealing with a carbonyl functional group, the equilibrium constant of an interassociation can be obtained without dilution of inert solvent. Infrared spectra of ethyl phenol/ethyl isobutyrate mixture at room temperature reveal that carbonyl stretching frequency is split into two bands at about 1736 and 1707 cm$^{-1}$. These are assigned to non-hydrogen-bonded and hydrogen-bonded carbonyl groups. The fraction of free carbonyl groups $f_F^{C=0}$, can be estimated from the relative intensities of the two bands. By stoichiometry of the system it can be shown that

$$f_F^{C=0} = \frac{\phi_{B1}}{\phi_B} = \left(1 + K_B\phi_{A1}\left(\left(1 - \frac{K_2}{K_A}\right) + \left(\frac{K_2}{K_A}\right)(1 - K_A\phi_{A1})^{-1}\right)\right)^{-1}, \tag{4.24}$$

where $\phi_B$ is the total volume fraction of $B$, and $\phi_{B1}$ and $\phi_{A1}$ are the volume fractions of the non-Keesom $B$ and $A$ species in the mixture, respectively. The value of $K_B$ is determined by the least-squares fit of Equation (4.24) from mixtures of varying composition. The equilibrium constants that represent interassociation between two copolymers are determined directly from experimental infrared studies of single-phase polymer blend samples in the solid state. Model compounds or low molecular weight analogs of the polymers may also be used when necessary.

### 4.3.2 Phase Behavior of Miscible Blends with Keesom Interactions

Coleman and Painter [5] studied blends of styrenic copolymers containing hexafluoro-2-hydroxy-2-propyl styrene. The OH group was found to weakly self-associate but strongly interassociate with oxygen atoms of aliphatic ester, ester, and acetoxy groups. They looked at blends containing the alternating copolymer of polytetrafluoroethylene/vinyl alcohol. This copolymer has a solubility parameter of 6.2 (cal.cm$^{-3}$)$^{0.5}$. The physical forces are unfavorable but the Keesom forces are favorable. The balance between the competing forces is the key to uncovering more miscible polymer–polymer systems.

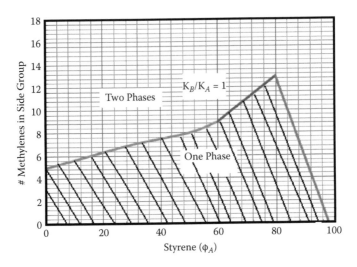

**FIGURE 4.1** Compositional window of miscibility for blends of polyalkyl methacrylates with a styrene copolymer and hydroxyl group.

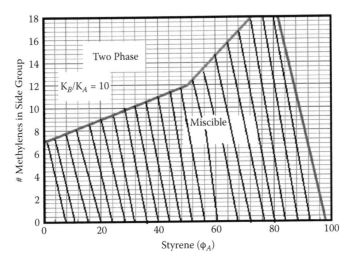

**FIGURE 4.2** Compositional window of miscibility for blends of poly(alkylmethacrylates) with a styrene copolymer and hydroxyl group when interassociation is favored over self-association.

Hydrogen-bonded polymer blends may exhibit lower critical solution temperature (LCST) and closed-loop two-phase regions in the phase diagram. Coleman and Painter [5] have developed a phase calculator computer program that can be used for simulating miscibility composition windows. The compositional window of miscibility for blends of poly(alkylmethacrylates) with a styrene copolymer containing a comonomer with a hydroxyl group is shown in Figures 4.1 and 4.2. These phase diagrams were successfully compared with experimental data. From Figure 4.2, it

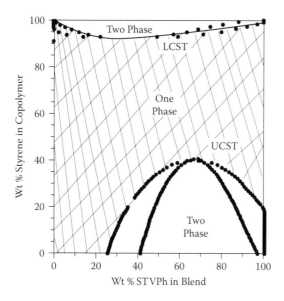

**FIGURE 4.3** Miscibility window for poly(hexylmethacrylate) and styrene–vinyl phenol copolymer.

can be seen that for $K_B/K_A$ values of 10.0, the interassociation is favored over self-association. The case shown in Figure 4.1 is when the self-association and interasso-ciation contributions are equal to each other. When the self-association is not lower than interassociation contributions, the miscibility window is narrowed.

The balance of forces is key to the phase behavior of the Keesom-forces-driven miscible polymer blends. This balance is dominated by the physi-cal and chemical contributions to the free energy of mixing. The miscibility window for poly(hexylmethacrylate) and styrene–vinyl phenol copolymer is shown in Figure 4.3. The phase diagram shown in Figure 4.3 has been verified by experiment. Polystyrene (PS) homopolymer is found to be immiscible with poly(hexylmethacrylate) (PHMA). The non-hydrogen-bonded solubility param-eters for PS and PHMA are $\delta_{PHMA} = 8.5$ (cal.cm$^{-3}$)$^{0.5}$ and $\delta_{PS} = 9.5$ (cal.cm$^{-3}$)$^{0.5}$, respectively. This implies a blend interaction parameter, $\chi$, of about 0.2. This is two orders of magnitude greater than that required for miscibility in poly-mer systems that exhibit no specific interactions. There is no appreciable favor-able contribution from chemical forces with which to overcome an unfavorable $\chi$ contribution to the free energy of mixing. However, when vinyl phenol repeat units are introduced as comonomer in the copolymer, both the blend interaction parameter, $\chi$, and $\Delta G_H$ contributions change. The non-hydrogen-bonded solubil-ity parameter of the styrene–vinyl phenol copolymer will increase over polysty-rene homopolymer as $\delta_{PVPh} = 10.6$ (cal.cm$^{-3}$)$^{0.5}$, making the physical contribution more unfavorable. From Figure 4.3, it can be seen that when there is 10–56% of vinyl phenol the chemical forces are greater than the physical forces, and hence a miscible system results. Beyond 56% vinyl phenol, the solubility parameter dif-ference between PHMA and styrene–vinyl phenol copolymer becomes greater

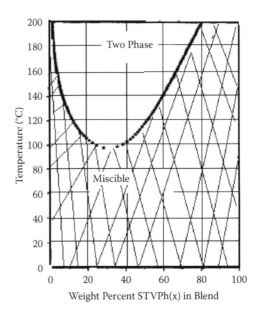

**FIGURE 4.4** LCST Behavior for PHMA and styrene/vinylphenol system.

than the contribution from the blend interaction parameter, $\chi$, term. The compositional window of miscibility is determined primarily by the rate of change of the solubility parameter of the styrene–vinyl phenol copolymer relative to the $\Delta G_H$ contribution from hydrogen bonding as a function of copolymer composition. LCST behavior was found [5] for the PHMA and styrene–vinyl phenol copolymer blends. The theoretical prediction of LCST is shown in Figure 4.4. The boundary between the miscible and immiscible region is called the *edge of miscibility*. Here, the phase separation is a pronounced function of temperature and composition of the blend. Experimental data using light scattering confirms the theoretical predictions [5]. The skewness of the phase diagram with changes in copolymer composition can also be predicted by theory. The skewness of the curves is found to be independent of the molecular weight of the copolymers. This is due to the balance between interassociation and self-association forces.

Small molecule and polymer blends can be predicted using the solubility parameter approach. The polymer additives used can be explained using this theory.

## 4.4 SUMMARY

Stronger Keesom forces promote hydrogen bonding among two different polymers, leading to miscible polymer–polymer systems. Hildebrandt and Scott [1] introduced a solubility parameter

$$\delta_v = \sqrt{\frac{\Delta H_v}{v_L}}. \tag{4.25}$$

The Hildebrandt solubility parameter can be related to the Flory–Huggins interaction parameter by use of regular solution theory. The Hildebrandt solubility parameter can be calculated from one of the EOS theories discussed in Chapter 2 for polymeric systems. It was shown that

$$\delta_v^2 = T\left(\frac{\beta}{\kappa}\right). \tag{4.26}$$

Hansen [3] represented the total Hildebrandt solubility parameter, $\delta_v$, by introducing three components: dispersive, polar, and hydrogen bonding:

$$\delta_v^2 = \delta_D^2 + \delta_P^2 + \delta_H^2. \tag{4.27}$$

Representative values of fractional solubility parameter values of $\delta_D$, $\delta_H$, $\delta_P$ for 14 commonly used polymer systems are given in Table 4.1.

Coleman and Painter [4–6] consider hydrogen bonding as the central strong interaction in polymers that causes the observed phase behavior and miscibility. They have measured this bonding using FTIR spectroscopy. Their developed expressions for contributions to the free energy of mixing of the blend form are (1) the combinatorial entropy of mixing, (2) the physical forces, and (3) the chemical forces. The $\Delta G$ can be calculated from $K_{eq}$, the equilibrium rate constant for Keesom interactions. By use of FTIR spectroscopy, the number of fractions of functional groups that are hydrogen bonded are estimated. This information is used to determine the equilibrium rate constants of interassociations and self-associations.

Hydrogen-bonded polymer blends may exhibit LCSTs and a closed-loop two-phase region in the phase diagram of a binary blend. A phase computer program was developed for simulating compositional window of miscibility. The miscibility window for poly(hexylmethacrylate) PHMA and styrene–vinyl phenol copolymer is shown in Figure 4.3. Theoretical predictions of LCST behavior found for this system are shown in Figure 4.4. The skewness of LCST curves was found to be independent of molecular weight. This is due to the balance between interassociation and self-association forces.

## EXERCISES

1. Estimate the Hildebrandt solubility parameter using the Sanchez–Lacombe lattice fluid theory described in Chapter 2 and given as Equation (4.8) for low density polyethylene (LDPE) with a degree of polymerization of 2000.

2. Estimate the Hildebrandt solubility parameter using the Sanchez–Lacombe lattice fluid theory described in Chapter 2 and given as Equation (4.8) for polystyrene with a degree of polymerization of 1000.

3. Estimate the Hildebrandt solubility parameter using the Sanchez–Lacombe lattice fluid theory described in Chapter 2 and given as Equation (4.8) for poly(dimethyl siloxane) (PDMS) with a degree of polymerization of 5000.

4. Estimate the Hildebrandt solubility parameter using the Sanchez–Lacombe lattice fluid theory described in Chapter 2 and given as Equation (4.8) for polytetrafluoroethylene (PTFE) with a degree of polymerization of 4000.
5. Estimate the Hildebrandt solubility parameter using the Sanchez–Lacombe lattice fluid theory described in Chapter 2 and given as Equation (4.8) for poly(butadiene) with a degree of polymerization of 2500.
6. Estimate the Hildebrandt solubility parameter using the Sanchez–Lacombe lattice fluid theory described in Chapter 2 and given as Equation (4.8) for poly(ethylene oxide) (PEO) with a degree of polymerization of 6000.
7. Estimate the Hildebrandt solubility parameter using the Sanchez–Lacombe lattice fluid theory described in Chapter 2 and given as Equation (4.8) for istotactic polypropylene with a degree of polymerization of 2000.
8. Estimate the Hildebrandt solubility parameter using the Sanchez–Lacombe lattice fluid theory described in Chapter 2 and given as Equation (4.8) for poly(ethylene terepthalate) (PET) with a degree of polymerization of 1000.
9. Estimate the Hildebrandt solubility parameter using the Sanchez–Lacombe lattice fluid theory described in Chapter 2 and given as Equation (4.8) for poly(phenylene oxide) (PPO) with a degree of polymerization of 5000.
10. Estimate the Hildebrandt solubility parameter using the Sanchez–Lacombe lattice fluid theory described in Chapter 2 and given as Equation (4.8) for polycarbonate with a degree of polymerization of 10,000.
11. Estimate the Hildebrandt solubility parameter using the Sanchez–Lacombe lattice fluid theory described in Chapter 2 and given as Equation (4.8) for poly(ether ether ketone) (PEEK) with a degree of polymerization of 2500.
12. Estimate the Hildebrandt solubility parameter using the Sanchez–Lacombe lattice fluid theory described in Chapter 2 and given as Equation (4.8) for polyamide 6 with a degree of polymerization of 6000.
13. Derive an expression for Hildebrandt solubility parameter as given by Equation (4.7) for the Flory– Orwoll–Vrij EOS summarized as

$$\frac{P_r V_r}{T_r} = \frac{V_r^{1/3}}{\left(V_r^{1/3} - 1\right)} - \frac{1}{T_r V_r}. \tag{4.28}$$

14. Derive an expression for Hildebrandt solubility parameter as given by Equation (4.7) for the Prigogine square-well cell equation of state (EOS) summarized as

$$\frac{P_r V_r}{T_r} = \frac{V_r^{1/3}}{\left(V_r^{1/3} - 0.8909\right)} - \frac{2}{T_r}\left(\frac{1.2045}{V_r^2} - \frac{1.011}{V_r^4}\right). \tag{4.29}$$

15. Derive an expression for Hildebrandt solubility parameter as given by Equation (4.7) for the model proposed by Dee and Walsh. The EOS as given by Equation (2.41) is summarized as

$$\frac{P_r V_r}{T_r} = \frac{V_r^{1/3}}{\left(V_r^{1/3} - 0.8909q\right)} - \frac{2}{T_r}\left(\frac{1.2045}{V_r^2} - \frac{1.011}{V_r^4}\right).$$

(4.30)

16. Derive an expression for Hildebrandt solubility parameter as given by Equation (4.7) for the Tait EOS summarized as

$$V = V_0\left(1 - C\ln\left(1 + \frac{P}{B}\right)\right).$$

(4.31)

17. Derive an expression for Hildebrandt solubility parameter as given by Equation (4.7) for the perturbed-hard chain theory (PCHT) EOS summarized as

$$\frac{PV}{RT} = Z = 1 + cZ^{CS} + c\sum \frac{A_{nm}}{V_r^m T_r^{n-1}},$$

(4.32)

where $Z^{CS}$ can be obtained from the Carnahan–Starling equation.

18. Derive an expression for Hildebrandt solubility parameter as given by Equation (4.7) for the Born–Green–Yvon (BGY) integral EOS summarized as

$$P_r + T_r\left[\ln\left(1 - \phi\right) - \frac{z\ln(1 - \phi)}{2(1 - \xi)} + \left(1 - \frac{1}{r}\right)\phi\delta_{kr} + \frac{zJ\xi^2}{2}\right] = 0,$$

(4.33)

where, $\delta_{kr}$ is the Kronecker delta and $J$ is given by

$$J = \frac{e^{-\frac{2}{zT_r}} - 1}{1 - \xi\left(1 - e^{-\frac{2}{zT_r}}\right)}.$$

(4.34)

19. Derive an expression for Hildebrandt solubility parameter as given by Equation (4.7) for the lattice cluster (LC) EOS summarized as

$$P_r + T_r\left[\begin{array}{c}\ln\left(1 - \phi\right) + \phi + \phi^2\left(\dfrac{1}{T_r} - \dfrac{4}{zT_r} + \dfrac{1}{zT_r^2} + \dfrac{1}{z} - \dfrac{3}{z^2}\right) \\ + \phi^3\left(\dfrac{4}{zT_r} - \dfrac{4}{zT_r^2} - \dfrac{20}{3z^2}\right) + +\phi^4\left(\dfrac{3}{zT_r^2} + \dfrac{6}{z^2}\right)\end{array}\right] = 0.$$

(4.35)

20. Derive an expression for Hildebrandt solubility parameter as given by Equation (4.7) for the Panayiotou and Vera (PV) EOS summarized as

$$P_r + T_r \left[ \ln(1-\phi) - \frac{z\ln(1-\phi)}{2(1-\xi)} \right] + \xi^2 = 0. \tag{4.36}$$

21. Derive an expression for Hildebrandt solubility parameter as given by Equation (4.7) for the Simha–Somcynsky (SS) EOS summarized as

$$\frac{P_r V_r}{T_r} = \left[ 1 - 2^{-\frac{1}{6}} y(yV_r)^{-\frac{1}{3}} \right]^{-1} + \frac{2y}{T_r}(yV_r)^{-2} \left[ 1.011(yV_r)^{-2} - 1.2045 \right]. \tag{4.37}$$

The reduced variables are different for the SS equation of state: $y$ is the fraction of occupied sites depending here on $r$ and on an additional parameter defined such that there are $3c$ degrees of freedom per chain. The reduced variables are given by

$$V_r = \frac{V}{Nrv^*}; \ T_r = \frac{c}{qz\beta\varepsilon}; \ P_r = \frac{Pr\,v^*}{qz\varepsilon} \tag{4.38}$$

where $v^*$ is the characteristic volume per unit segment. Calculate the EOS parameters for the SS equation.

22. Derive an expression for Hildebrandt solubility parameter as given by Equation (4.7) for the van der Waals cubic EOS summarized as

$$\left( P + \frac{a}{V^2} \right)(V - b) = RT. \tag{4.39}$$

23. Derive an expression for Hildebrandt solubility parameter as given by Equation (4.7) for the Virial EOS summarized as

$$Z = \frac{PV}{RT} = 1 + \frac{B}{V} + \frac{C}{V^2} + \frac{D}{V^3} + \dots\dots$$
$$Z = 1 + B'P + C'P^2 + D'P^3 + \dots \tag{4.40}$$

24. Derive an expression for Hildebrandt solubility parameter as given by Equation (4.7) for the Redlich and Kwong EOS summarized as

$$P = \frac{RT}{V - b} - \frac{a}{V(V + b)}. \tag{4.41}$$

Note that $a$ and $b$ are given as functions of temperature and other parameters as follows:

$$Z = \frac{PV}{RT} = 1 + \frac{B}{V} + \frac{C}{V^2} + \frac{D}{V^3} + \dots\dots$$
$$Z = 1 + B'P + C'P^2 + D'P^3 + \dots \tag{4.42}$$

where $T_c$ and $P_c$ are critical temperature and critical pressure, $\psi$ and $\Omega$ are pure numbers independent of substance and determined for a particular equation of state. The parameter $\alpha$ is a function of the reduced temperature $T_r = T/T_c$

## REVIEW QUESTIONS

1. Calculate the window of miscibility in terms of molecular weights of homopolymer A and homopolymer A binary blend using the solubility parameter approach.
2. Are Debye and London forces considered in the estimates of fractional solubility parameter values?
3. Will the Keesom forces between polymer–polymer chains end up crosslinking them?
4. What is the difference in the solubility parameter estimate of a copolymer with alternating sequence distribution and random sequence distribution with the same composition?
5. Two solvents were combined to get polycarbonate into solution. How would you estimate the solubility parameter of the solvents when compared with the polymer?
6. Does the boiling point of the solvent have a bearing on the solubility parameter value of a polymer?
7. What is the effect of the glass transition temperature on the Hildebrandt solubility parameter estimate of the polymer?
8. Can the end groups in a polymer hydrogen bond with the end groups of another polymer chain?
9. Some investigators report novel polymer composite materials that possess a "negative coefficient of expansion." This means that the volume expansivity is negative. Can $\delta_v^2$ be negative for such materials, as predicted in Equation (4.7).
10. Can enthalpy of vaporization take on "negative" values, meaning energy is released by the process as opposed to absorbed?
11. Is the occurrence of LCST in binary systems because of the reversible nature of the hydrogen bonding or because of EOS effects as derived using mathematical models?
12. What is the significance of UCST in hydrogen bonded systems?
13. Can the dispersive and polar effects oppose each other for some systems?
14. Can the hydrogen bonding and dispersive forces oppose each other for some systems?
15. How is the solubility parameter of a copolymer expected to vary with composition of the copolymer?
16. Where is the volume fraction of the blend accounted for in predicting blend miscibility using the solubility parameter approach?
17. How is the solubility parameter approach applicable to a ternary blend?
18. How is the solubility parameter approach applied to terpolymer–terpolymer miscibility?
19. How are binary blends of copolymer and terpolymer with one common monomer treated using the solubility parameter approach?
20. How are binary blends of copolymer and terpolymer with two common monomers treated using the solubility parameter approach?

# REFERENCES

1. J. H. Hildebrandt and R. L. Scott, *The Solubility of Non-Electrolytes*, Dover, New York, 1964.
2. S. K. Kumar, Thermodynamics of random copolymer miscibility, PMSE, *Proc. of the 214th American Chemical Society National Meeting*, Dallas, TX, Vol. 78, 120, March/April, 1998..
3. C. M. Hansen, The three-dimensional solubility parameter—Key to paint component affinities: I. Solvents plasticizers, polymers and resins, *J. Paint Technol.*, Vol. 39, 505 (1967).
4. M. M. Coleman, J. F. Graf and P. C. Painter, *Specific Interactions and the Miscibility of Polymer Blends,* Technomic Publishing, Lancaster, PA, 1991.
5. M. M. Coleman and P. C. Painter, Hydrogen bonded polymer blends, *Prog. Polym. Sci.*, Vol. 20, 1–59, 1995.
6. M. M. Coleman, D. J. Skrovanek, J. Hu, and P. C. Painter, Hydrogen bonding in polymer blends. 1. FTIR studies of urethane-ether blends, *Macromolecules*, Vol. 21, 1, 59–65, 1988.

# 5 Phase Behavior

## LEARNING OBJECTIVES

- Construct phase diagrams
- Binodal and spinodal curves
- Appearance of first new phase, disappearance of phase boundary
- Temperature, molecular weight, glass transition effects
- Hydrogen bonded systems
- Partially miscible blends

## 5.1 INTRODUCTION

The phase diagrams of polymer–polymer blend systems and polymer–solvent mixtures can be constructed as a function of temperature versus composition. The mixture will try to reach a state where the free energy of mixing of the blend will be minimized. From this point, when the temperature is increased the free energy will change. Often, when the free energy of the blend is less than or equal to zero, the blend would be a miscible one-phase mixture. As the temperature is increased, the free energy of the blend changes in a monotonic and continuous manner. At some point, the free energy of the blend will be greater than zero. This is the onset of phase separation. The loci of all such points can be seen to be the lower critical solution temperature (LCST). Further increase in temperature again may change the free energy in such a fashion that it may become negative at a third temperature. The loci of such points can be seen to be the upper critical solution temperature (UCST).

There are two important factors that need to be considered when constructing phase diagrams:

1. Equilibrium
2. Stability

Consider a closed system containing $k$ polymer species and $\pi$ phases in which the temperature and pressure are uniform. The system is initially at a nonequilibrium state with respect to the mass transfer between phases. Changes that occur in the system are irreversible in nature, and they take the system closer to the equilibrium state. Let the system be placed in surroundings such that the system and surroundings are at thermal and mechanical equilibrium. Heat transfer and expansion work are accomplished in a reversible manner. The entropy change of the surroundings can be written as

$$dS_{sur} = \frac{dQ_{sur}}{T_{sur}} = -\frac{dQ_{sys}}{T_{sys}}. \tag{5.1}$$

The heat transfer for the system is $dQ_{sys}$ and has a sign opposite to that of the surroundings, $dQ_{sur}$; the temperature of the system $T_{sys}$ replaces $T_{sur}$. Both temperatures must have the same value for reversible heat exchange. The second law of thermodynamics (Appendix B) requires that

$$dS^{sys} + dS^{sur} \geq 0, \tag{5.2}$$

where $S^{sys}$ is the total entropy of the system. Combining Equations (5.1) and (5.2),

$$dQ^{sys} \leq TdS^{sur}. \tag{5.3}$$

The first law of thermodynamics for closed systems (Appendix B) can be written as

$$dU^{sys} = dQ + dW = dQ - PdV^{sys}. \tag{5.4}$$

Combining Equation (5.4) with Equation (5.3),

$$dU^{sys} + PdV^{sys} - TdS^{sys} \leq 0. \tag{5.5}$$

Equation (5.5) deals with properties only. It must be satisfied for changes in state of any closed system of uniform temperature, $T$, and pressure, $P$. No restrictions on the mechanical and thermal reversibilities. Equation (5.5) is applicable for incremental changes. It governs the direction of change that leads toward equilibrium. Equation (5.5) may be used for changes between equilibrium states. A corollary to Equation (5.5) at constant entropy and constant volume is

$$(dU^{sys})_{S,V} \leq 0. \tag{5.6}$$

In a similar token, at constant internal energy and constant volume,

$$(dS^{sys})_{U^{sys}, V^{sys}} \geq 0. \tag{5.7}$$

An isolated system is necessarily at constant internal energy and constant volume. For such systems a direct corollary from the second law of thermodynamics is Equation (5.7). Let the blending process be restricted to constant temperature, $T$, and pressure, $P$. Then Equation (5.5) can be written as

$$dU^{sys} + d(PV^{sys}) - d(TS^{sys}) \leq 0, \tag{5.8}$$

or

$$d(U^{sys} + PV^{sys} - TS^{sys}) \le 0, \tag{5.9}$$

or

$$d(H^{sys} - TS^{sys}) = dG^{sys} \le 0. \tag{5.10}$$

$$\text{Thus, } (dG^{sys})T,P \le 0. \tag{5.11}$$

Equation (5.11) is an interesting variation of Equation (5.5). This is because in prac-tice, temperature and pressure can be readily measured and controlled. It can be seen from Equation (5.11) that all irreversible processes that occur at constant tempera-ture, $T$, and constant pressure, $P$, proceed in a direction as to cause a decrease in the Gibbs free energy, $G$, of the system. Thus [1], the equilibrium state of a closed system is that state at which the total Gibbs free energy is a minimum with respect to all possible changes at the given temperature, $T$, and pressure, $P$.

This criterion of equilibrium provides a general method of determination of equi-librium states. The expression for free energy, $G^{sys}$, can be expressed in terms of the number of moles of the polymer species in several phases. The values of mole fractions that minimize the free energy of the system can be determined. This can form the procedure for determining phase equilibrium in polymer–polymer blends and polymer–solvent systems. At the equilibrium state, differential variations can occur in the system at constant temperature, $T$, and pressure, $P$, without affecting any changes in the expression for free energy of the system, $G^{sys}$. Thus, a criterion for equilibrium can be written as

$$(dG^{sys})_{T,P} = 0. \tag{5.12}$$

In order to apply Equation (5.12), an expression for $dG^{sys}$ is obtained in terms of the volume fraction of the component in the blend, copolymer composition, temperature, and pressure. Phase equilibrium is established. Equation (5.11) must be satisfied by any single phase that is stable with respect to the alternative of splitting into two phases. It requires that the Gibbs free energy for an equilibrium state be the mini-mum value with respect to all possible changes at the given temperature, T, and pres-sure, P. Two cases are shown in Figure 5.1:

1. Curve A—Complete miscibility
2. Curve B—Partially miscible; two phases, $\alpha$ and $\beta$

When different polymer components are blended together, the free energy of the system changes. $\Delta G^{sys}$ or the free energy changes by blending needs to be negative for obtaining a *miscible blend*. The free energy expression needs to be monotonic and negative. Curve A in Figure 5.1 is for a system that is miscible and stable. In the case of curve B in Figure 5.1, the system has achieved a lower free energy state by formation of two phases from a single phase. This is represented by $\alpha$ and $\beta$ in curve B in Figure 5.1. The construction line that joins $\alpha$ and $\beta$ in Figure 5.1 represents Gibbs free energy, G, that would result in a range of states consisting of two phases of compositions, $\varphi\alpha$ and $\varphi\beta$. The stable region is the curve between $\alpha$ and $\beta$ that

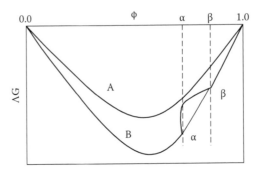

**FIGURE 5.1** Complete miscibility and partial miscibility,

denotes two phases. This leads to a criterion of stability. At constant temperature and pressure, $\Delta G$ and its first and second derivatives must be monotonic and continuous functions of the volume fraction $\varphi_1$ of the first component in the blend. The second derivative of change in $\Delta G$ energy with respect to the volume fractions must also be positive everywhere:

$$\frac{\partial^2 \Delta G}{\partial \phi_1^2} > 0 \quad \text{(constant T, P).} \tag{5.13}$$

Metastable regions are different from stable and unstable regions. When the two criteria outlined in this chapter—phase behavior on equilibrium and stability—are both met as necessary and sufficient conditions, stable regions in the phase diagram can be expected. When both conditions are not met, unstable regions arise. When one of the conditions, the stability condition is met and the equilibrium condition not met this gives rise to what is called "metastable region."

In Chapters 2–4, the mathematical models described, such as the binary interaction model, group solubility parameter approach, and entropic difference model, can be used to obtain an expression for free energy of mixing of the blend. The molar volume in the entropic term can be expressed in terms of temperature and pressure using the different EOS discussed in Chapter 2. Then Equations (5.12) and (5.13) can be used to obtain the phase boundaries in the phase diagrams. Several cases are possible. Six cases are discussed below.

## 5.2  LCST AND UCST

### 5.2.1  CASE I: LCST

Figure 5.2 of the phase diagram ($T$ versus $\varphi_1$) shows one LCST.

### 5.2.2  CASE II: UCST

Figure 5.3 phase diagram ($T$ versus $\varphi_1$) shows one UCST.

One method of calculating the different cases of phase diagrams listed here is by expressing the polymer blend interaction parameter, $\chi$, as a function of temperature

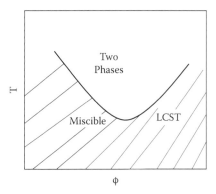

**FIGURE 5.2**   Phase diagram (T versus $\varphi_1$)–I LCST (lower critical solution temperature).

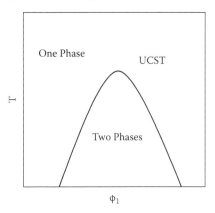

**FIGURE 5.3**   Phase diagram (T versus $\varphi_1$)–II UCST (upper critical solution temperature).

and composition. Mumby et al. [2] gave the following expression for the interaction parameter:

$$\chi = \left(1 + b_1\phi + b_2\phi^2\right)\left(d_0 + \frac{d_1}{T} + d_2\ln(T)\right). \tag{5.14a}$$

Where $b_1$, $b_2$, $d_0$, $d_1$, and $d_2$ are adjustable parameters. It was assumed that the excess partial molar heat capacity of blend is independent of temperature. A quadratic dependence on composition was chosen from observations made in polymer–solvent systems. When the coefficients of $d_1$ and $d_2$ in Equation (5.14a) are both positive, the resulting phase diagram is of the combined LCST/UCST or hourglass type, as discussed for different cases in this section. The system acetone and polystyrene has been found to undergo a transition from a combined LCST/UCST to the hourglass behavior as the molecular weight of the polystyrene is increased.

A combined Case I LCST and Case II UCST can be illustrated for a polymer–polymer system with the interaction parameter varying as follows:

$$\chi = \left(-0.6366 + \frac{60}{T} + 0.18\ln(T)\right). \tag{5.14b}$$

## 5.3   CIRCULAR ENVELOPE IN PHASE DIAGRAM

### 5.3.1   Case III: Circular Phase Envelope UCST and LCST

In the Figure 5.4 phase diagram, $T$ versus $\varphi_1$ shows the circular phase boundary with UCST convex downward and LCST concave upward.

### 5.3.2   Case IV: Circular Phase Envelope LCST and UCST

In Figure 5.6 showing the phase diagram ($T$ versus $\varphi_1$), the circular phase boundary appears with UCST convex upward and LCST concave downward.

The free energy change happens by blending two polymers 1 and 2, where polymer 1 is a copolymer with monomers A and B as repeat units, and polymer 2 is a homopolymer of monomer repeat unit C, which can be written as follows:

$$\Delta G_m = \Delta H_m - T\Delta S_m. \tag{5.15}$$

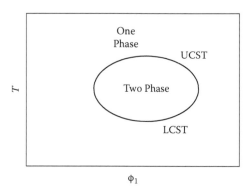

**FIGURE 5.4**   Phase diagram ($T$ versus $\varphi_1$)–circular phase boundary with UCST convex downward and LCST concave upward.

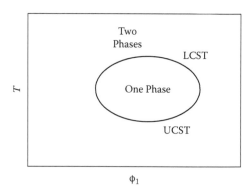

**FIGURE 5.5A**   Phase diagram ($T$ versus $\varphi_1$)–circular phase boundary with UCST convex upward and LCST concave downward.

The enthalpic contribution can be obtained from the binary interaction model as explained in Chapter 3. This is restated as follows for the considered blend of polymers 1 and 2.

$$\frac{\Delta H_m}{V} = B_{AC}x(1-\phi)\phi + B_{BC}(1-x)(1-\phi)\phi - B_{AB}x(1-x)(1-\phi)^2, \quad (5.16)$$

where x is the mole fraction of monomer repeat unit A in the copolymer 1, with repeat units A and B, and $\phi$ is the volume fraction of polymer 1 in the blend.

The entropy change upon blending the polymers can be calculated as shown in Chapter 6.

$$\Delta H_{blend} = \Delta(U_{blend} + (PV)_{blend})$$

$$C_p dT = T_g \Delta S_{blend}(-PdV + PdV + VdP)_{blend}$$

$$0 = T_g \Delta S_{blend} + (VdP)_{blend}$$

$$\Delta S_{blend} = -\frac{V_{blend}}{T_g}\Delta P_{blend} \quad (5.17)$$

where $T_g$ is the glass transition temperature of the blend. Combining Equations (5.17), (5.15), and (5.14):

$$\frac{\Delta G_m}{V_{blend}} = B_{13}x(1-\phi)\phi + B_{23}(1-x)(1-\phi)\phi - B_{12}x(1-x)(1-\phi)^2 + \frac{T\Delta P_{blend}}{T_g}. \quad (5.18)$$

$$\Delta P_{blend} \cong \phi P_{AB} + (1-\phi)P_C - \phi(1-\phi)\Delta P_{AB/C}. \quad (5.19)$$

The equation of state for the polymers can be selected from Chapter 2, depending on the application. Examining an EOS that is developed from the lattice fluid theory, it can be seen that pressure, P, is directly proportional to temperature, T. Thus:

$$\Delta P_{blend} = \xi T + \zeta, \quad (5.20)$$

where, $\xi$ and $\zeta$ are constants that depend on the polymer used in the blend.

Combining Equation (5.20) and Equation (5.18):

$$\frac{\Delta G_m}{V_{blend}} = B_{13}x(1-\phi)\phi + B_{23}(1-x)(1-\phi)\phi - B_{12}x(1-x)(1-\phi)^2 + \frac{\xi T^2 + \zeta T}{T_g}.$$

$$(5.21)$$

Applying Equation (5.12), the condition for equilibrium, Equation (5.21) becomes at equilibrium

$$0 = ((B_{13} - B_{23})x + B_{23})(\phi - \phi^2) - B_{12}x(1 - x)(1 + \phi^2 - 2\phi) + \frac{\xi T^2 + \zeta T}{T_g}, \quad (5.22)$$

or

$$\phi^2 - \phi \frac{(x(B_{13} - B_{23}) + 2B_{12} + B_{23})}{(x(B_{13} + B_{12} - B_{23}) + B_{23})} + \frac{B_{12}x}{x(B_{13} + B_{12} - B_{23}) + B_{23}}$$

$$+ T^2\left(\frac{\xi}{T_g}\right) + T\left(\frac{\zeta}{T_g}\right) = 0. \quad (5.23)$$

The equation of a circle in a phase diagram of T versus $\phi$ can be written as

$$\left(\frac{T}{T_g} - \beta\right)^2 + (\phi - \alpha)^2 = R^2, \quad (5.24)$$

where, $\alpha$ and $\beta$ are the coordinates of the center of the circular envelope in the phase diagram, and $R$ is the radius of the circle. A circle is a simple shape of Euclidean geometry. This consists of all those points in a plane that lie equidistant from a point called the center of the circle. The common distance of the points of a circle from its center is called the radius of the circle, $R$. Circles are simple closed curves which divide the plane into two regions an interior and one exterior. This can be used to obtain the phase envelope or delineation between the single phase and two phase regions in the temperature versus composition phase diagram. A circle is a special case of ellipse when the two foci are coincidental.

The phase diagram of the considered polymer system in the T versus $\phi$ diagram becomes circular when Equation (5.23) takes on the form of Equation (5.24). When this happens,

$$\alpha = \sqrt{\frac{B_{12}x}{x(B_{13} + B_{12} - B_{23})}} = \frac{(x(B_{13} - B_{23}) + 2B_{12} + B_{23})}{(2x(B_{13} + B_{12} - B_{23}) + B_{23})} \quad (5.25)$$

$$\xi = \frac{1}{T_g} \quad (5.26)$$

$$R = \beta = -\frac{\zeta}{2}.$$

In Figure 5.6, Equation (5.24) is plotted for $\alpha = 0.4$, $\beta = 0.35$, $T_g = 50°C$. This has the UCST convex upward and LCST concave downward. The phase diagram is circular in nature from the standpoint of equilibrium. For stability,

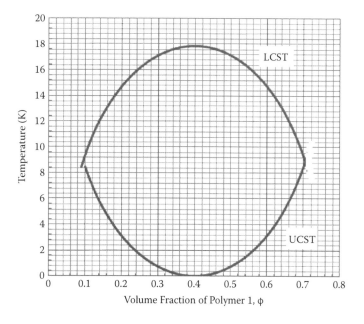

**FIGURE 5.6** Circular phase envelope when $\alpha = 0.4$ and $\beta = 0.35$ and $T_g = 50°C$.

$$\frac{\partial^2 \Delta G}{\partial \phi^2} > 0. \tag{5.27}$$

Equation (5.24) can also be written in the parametric form as

$$\frac{T}{T_g} = \beta + RCost = R(1 + Cost) \tag{5.28}$$

$$\phi = \alpha + RS\,in(t),$$

where

$$Tant = \frac{\phi - \alpha}{\dfrac{T}{T_g} - \beta}, \tag{5.29}$$

where t is a parametric variable. It can be interpreted geometrically as the angle that the ray from the origin to $(\phi, T)$ makes with the x-axis.

## 5.4   HOURGLASS BEHAVIOR IN PHASE DIAGRAMS

### 5.4.1   CASE V: LCST AND UCST HOURGLASS BEHAVIOR

Figure 5.7 of the phase diagram ($T$ versus $\varphi_1$) shows the hourglass behavior. The phase diagram can take the form of a *hyperbola* when Equation (5.21), the criteria for the equilibrium state, assumes the form

$$\frac{\phi^2}{\alpha^2} - \frac{T^2}{\beta^2 T_g^2} = 1. \tag{5.30}$$

The hyperbola defined by Equation (5.30) is about the origin at the (0,0) location. A hyperbola is a smooth planar curve having two connected components or branches, each a mirror image of the other. They resemble infinite bows. The hyperbola is a kind of conic section or intersection of a plane and a cone when the plane makes a smaller angle with the axis of the cone than does the cone itself. Hyperbolic shape can be expected when temperature, $T$, varies with the composition of the blend, $\varphi$, in the reciprocal manner $1/\varphi$. The hyperbola consists of two disconnected curves, referred to as arms or branches. A hyperbola has two focal points. The line connecting these foci is called the *transverse axis* and the midpoint between the foci is known as the *hyperbola's center*. The line through the center that is normal to the transverse axis is referred to as the *conjugate axis*. The transverse axis and conjugate axis are called the two principal axes of the hyperbola. The hyperbola exhibits mirror symmetry about its principal axes. It is also symmetrical about a $180°$ turn about the hyperbola's center. The transverse and conjugate axes are sometimes called the *semimajor* and *semiminor axis,* respectively. The points at which the hyperbola crosses the transverse axis are known as the *vertices* of the hyperbola, and are located $2\alpha$ apart, where $\alpha$ is the distance from the center to each vertex. At large distances from the center the hyperbola approaches its asymptotes two straight lines. These straight lines intersect at the hyperbola's center. A degenerate hyperbola consists only of its

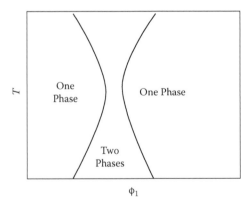

**FIGURE 5.7**   Phase diagram ($T$ versus $\phi_1$)–hourglass behavior.

asymptotes. A hyperbola can also be defined as a second degree polynomial equation in the Cartesian coordinate system:

$$A_{\phi\phi}\phi^2 + 2A_{\phi T}T\phi + A_{TT}T^2 + 2B_{\phi}\phi + 2B_T T + C = 0 \tag{5.31}$$

Comparing Equation (5.31) and Equation (5.23):

$$A_{\phi\phi} = 1; \; A_{\phi T} = 0; \; A_{TT} = \xi/T_g \tag{5.32}$$

$$B_{\phi} = -\frac{1}{2}\left(\frac{x(B_{13} - B_{23}) + 2B_{12} + B_{23}}{x(B_{13} + B_{12} - B_{23}) - B_{23}}\right); \; B_T = \frac{1\xi}{2T_g} \tag{5.33}$$

$$C = \frac{B_{12}x}{x(B_{13} + B_{12} - B_{23}) - B_{23}}. \tag{5.34}$$

Equation (5.23) will be a hyperbola, provided the determinant $D$ is less than zero:

$$D = \begin{vmatrix} A_{\phi\phi} & A_{\phi T} \\ A_{\phi T} & A_{TT} \end{vmatrix} < 0. \tag{5.35}$$

This happens when

$$\frac{\xi}{T_g} - 0 < 0, \tag{5.36}$$

and $\xi$ is negative. This depends on the EOS and how the pressure, temperature, and density relations of the particular polymer considered behave in the domain of interest.

In a special case of a hyperbola, the degenerate hyperbola consisting of two intersecting lines occurs when another determinant is zero:

$$\Delta := \begin{vmatrix} A_{\phi\phi} & A_{\phi T} & B_{\phi} \\ A_{\phi T} & A_{TT} & B_T \\ B_{\phi} & B_T & C \end{vmatrix} = 0. \tag{5.37}$$

The center of the hyperbola $(\phi_c, T_c)$ can be calculated from

$$\phi_c = -\frac{\begin{vmatrix} T_g & B_{\phi} & 0 \\ \xi & \zeta & \xi \\ 2T_g & T_g \end{vmatrix}} = -B_{\phi} = \frac{1(x(B_{13} - B_{23}) + 2B_{12} + B_{23})}{2(x(B_{13} + B_{12} - B_{23}) - B_{23})}. \tag{5.38}$$

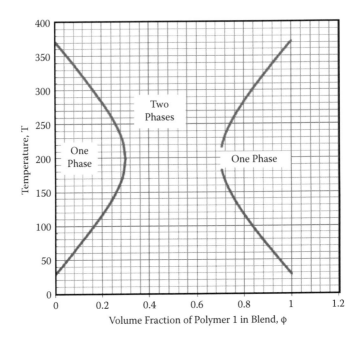

**FIGURE 5.8** Phase envelope assumes the form of hyperbola $(\varphi_c, T_c) = (0.5, 200)$ and $(\alpha, \beta T_g) = (0.2, 75)$.

$$
T_c = -\frac{T_g}{\xi}
\begin{vmatrix}
1 & B_\varphi \\
\dfrac{\zeta}{2T_g} & \dfrac{\xi}{T_g}
\end{vmatrix}.
\tag{5.38}
$$

An example of the phase envelope taking on the form a hyperbola is shown in Figure 5.8. This holds for $(\varphi_c, T_c) = (0.5, 200)$ and $(\alpha, \beta T_g) = (0.2, 75)$ in Equation (5.30).

## 5.5 MOLECULAR ARCHITECTURE

Bates [3] discussed the role of molecular architecture in polymer–polymer phase behavior. There are a number of molecular configurations available to a pair of chemically distinct polymer species. *Star polymers* with a specific arm number with predetermined molecular weights and with narrow molecular weight distribution can be synthesized using anionic polymerization. Diblock and multiblock arrangements are possible. Different polymers can be combined into a single material in several different ways that can lead to a variety of phase behaviors. Four factors control the phase behavior of polymer–polymer systems [3]. These are

1. The choice of monomers
2. Molecular architecture

3. Composition of the polymer
4. Molecular size

The most significant factor in determining block copolymer phase behavior is the covalent bond that restricts macroscopic separation of chemically dissimilar polymer blocks. The formation of microscopic heterogeneities in composition at length scales comparable to the molecular dimensions around 50–1000 Å contrasts sharply with the macroscopic phase separation associated with binary mixtures. Let φ be the overall volume fraction of a component assigned to a star-block copolymer composition, $AN$ is the number of segments per polymer molecular, ρ the polymer density, $M$ the molecular weight, and $AN$ the Avagadro number. Then,

$$N = \frac{\rho VAN}{M}.$$  (5.40)

The Flory–Huggins segment–segment interaction parameter can be written as

$$\chi = \frac{1}{k_B T}\left(\varepsilon_{AB} - \frac{1}{2}(\varepsilon_{AA} + \varepsilon_{BB})\right),$$  (5.41)

where $\varepsilon_{ij}$ is the contact energy between $i$ and $j$ segments, and $k_B$ is the Boltzmann constant.

The contact energy between $i$ and $j$ segments for nonpolar polymer such as polyethylene, polystyrene, and polyisoprene that are characterized by dispersive van der Waal interactions can be represented by

$$\varepsilon_{ij} = -\sum_{i,j} \frac{3}{4}\frac{I_i I_j}{(I_i + I_j)}\frac{\alpha_i \alpha_j}{r_{ij}^6},$$  (5.42)

where $r_{ij}$ is the segment–segment separation distance, α and $I$ are the segment polarizability and ionization potential, respectively. If there is no volume change ($\Delta V_m = 0$) or preferential segment orientation upon mixing. Equation (5.41) and Equation (5.42) can be combined to yield an interaction parameter:

$$\chi = \frac{3}{16}\frac{I}{k_B T}\frac{z}{V^2}(\alpha_A - \alpha_B)^2.$$  (5.43)

When cubic lattice is assumed with $I_i = I_j = I$,

$$\chi = \frac{\alpha}{T} + \beta.$$  (5.44)

In general, $\alpha$ and $\beta$ represent experimentally determined enthalpy and excess entropy coefficients for a particular composition; $\alpha$ and $\beta$ may be a function of

1. Volume fraction, $\varphi$
2. Function of fraction, $f = N_A/N$
3. Number of segments per polymer molecule, $N$
4. Temperature, $T$
5. Molecular architecture

Specific interactions, preferential segment orientation, and EOS effects play an important role, although not well understood, in dictating phase behavior. Phase state is governed by a balance between enthalpic and entropic contributions. Thus,

$$\frac{\Delta G_m}{k_B T} = \frac{\phi_A}{N_A} \ln(\phi_A) + \frac{(1-\phi_A)}{N_B} \ln(1-\phi_A) + \phi_A(1-\phi_A)\chi. \qquad (5.45)$$

The expression for entropy comes from the mean field theory. Spatial fluctuations in composition are neglected. For $N = 1$, Equation (5.44) reduces to the expression predicted from regular solution theory frequently encountered in solution thermodynamics. As $N$ is increased, the nonmean field region decreases to a small area around the critical point. In contrast, for monomers $N = 1$, composition fluctuations influence the entire phase diagram. Bates [3] verified the Ginsburg criterion by studying the LCST behavior of polystyrene (PS) and poly(vinyl methyl ether) (PVME), and the UCST of poly(ethylene propylene) and polyisoprene above a molecular weight of $1.8*10^{-5}$ g/mol. They observed that symmetric mixtures of dueterated and protonated polybutadienes undergo phase separation. Bates [3] studied mixing and demixing and band polarizability, noting some deficiencies in the Flory–Huggins combinatorial entropy of mixing. An example of a homogeneous alloy is Noryl, the commercial plastic composted of poly phenylene oxide and polystyrene.

The phase separation process can be divided into two categories: (1) nucleation and growth and (2) spinodal decompositions. Spinodal decomposition results in a disordered bicontinuous two-phase structure. The path in the phase diagram traverses the critical point. The morphology is discontinuous. Interfacial tension is a salient consideration. Uniform particle size distribution is possible during spinodal decomposition.

The order and disorder in diblock copolymer systems were discussed in [3]. The ensemble of molecular configurations that produces the minimum overall free energy, G, can be estimated. In some cases periodic mesophases can be formed. Order-to-disorder transition (ODT) criteria can be estimated. The theories that deal with block copolymer phase behavior can be divided into two categories:

1. Strong segregation limit
2. Weak segregation limit

Well-developed microdomain structures are assumed to occur with relatively sharp interfaces, and chain stretching is explicitly accounted for. Lower amplitude sinusoidal composition profiles and unperturbed Gaussian coils can be seen in some cases. Chain stretching is neglected. One example in this regard is the poly(ethylene propylene)/poly(ethylene ethylene) diblock copolymer where three distinct, ordered phases aligning with a disordered phase can be accessed by varying temperature. Thus, polymer–polymer phase behavior is a complex interdisciplinary subject.

## 5.6   SUMMARY

Phase diagrams of polymer–polymer blend systems and polymer-solvent mixtures can be constructed as a function of temperature vs. composition. Two important factors need to be considered when constructing phase diagrams: (1) equilibrium and; (2) stability. It can be shown from the laws of thermodynamics and reversibility that the equilibrium sate of a closed system is that state at which the total Gibbs free energy is a minimum with respect to all possible changes at the given temperature and pressure. At equilibrium:

$$\left(dG^{sys}\right)_{T,P} = 0.$$

For stability:

$$\left(dG^{sys}\right)_{T,P} \le 0.$$

Curve A in Figure 5.1 is for a system that is miscible and stable and curve B is for a system that has achieved a lower free energy state by formation of two phases from a single phase. A criteria for stability can be derived as

$$\left(\frac{\partial^2(\Delta G)}{\partial \phi_1^2}\right)_{T,P} > 0.$$

Five possible types of phase diagrams were discussed. These are

Case I—LCST
Case II—UCST
Case III—Circular envelope; merged LCST/UCST
Case IV—Circular envelope; merged UCST/LCST
Case V—Merged LCST/UCST; hourglass behavior; merged LCST/UCST; degenerate hyperbola

The polymer blend interaction parameter $\chi$ is expressed as a function of temperature and composition in Equation (5.14a). When the coefficients $d_1$ and $d_2$ are both positive, the resulting phase diagram is a combined LCST/UCST hourglass type. When interaction parameter, $\chi$, varies inversely proportional with temperature and directly proportional with ln(T), the UCST or LCST-type behavior can be expected. The equation of circle in a phase diagram of temperature versus composition, $\varphi$, is given by Equation (5.24). Equation (5.23) is the condition for equilibrium with the entropy of mixing expressed in terms of $\Delta P_{blend}$ and $\Delta P_{blend}$

written as a linear function of temperature in Equation (5.20). When Equation (5.23) takes the form of Equation (5.24), the phase envelope for the considered system becomes circular.

The phase diagram can assume the form of a hyperbola when Equation (5.21) assumes the form

$$\frac{\phi^2}{\alpha^2} - \frac{T^2}{\beta^2 T_g^2} = 1.$$

Bates [3] discussed the role of molecular architecture in polymer–polymer phase behavior.

## EXERCISES

1. When the temperature is increased, the free energy of the polymer blend
   a. Decreases
   b. Increases
   c. Stays the same
   d. Depends on the polymer system
   e. Becomes negative
2. What is the difference between metastable and stable states of equilibrium?
3. What is the difference between binodal and spinodal curves?
4. What is the difference between LCST and UCST curves?
5. What is meant by merged LCST/UCST?
6. What is meant by a "degenerate" hyperbola?
7. What is meant by hourglass behavior?
8. What is meant by circular phase behavior?
9. Can you apply the phase rule to polymer blends?
10. Can a pressure–composition diagram be constructed along similar lines as a temperature–composition diagram?
11. Can the discussions on binary polymer blends be extended to ternary polymer blends?
12. What is the difference between a ternary blend and a blend of terpolymer and homopolymer?
13. Show that at equilibrium:

$$\left( \frac{\partial U}{\partial V} \right)_S^{sys} \leq -P.$$

14. Show that at equilibrium:

$$\left( \frac{\partial U}{\partial S} \right)_V^{sys} \leq T.$$

15. What are the differences between complete miscibility and partial miscibility?
16. Rewrite Equation (5.14b) with T as a function of interaction parameter $\chi$.
17. Use the solubility parameter approach and rewrite Equation (5.16). What are the implications with the circular envelope in phase diagrams?

18. Use the entropic difference model and rewrite Equation (5.16). What are the implications with the circular envelope in phase diagrams?
19. Does Equation (5.17) imply that the polymer blending process is isochoric?
20. Rewrite Equation (5.20) using the FOV equation of state. What are the implications in the phase diagram type?
21. Rewrite Equation (5.20) with Prigogine square well model and discuss the ramifications of it in the phase diagram type.
22. Rewrite Equation (5.20) with the virial equation of state and discuss the ramifications of it on the phase diagram type.
23. What is the significance in polymer science of parameters $\alpha$ and $\beta$?
24. Can the phase diagram type become elliptical in nature?
25. Can the phase diagram type become parabolic in nature?

## PROBLEMS

1. Construct the phase diagram for a poly(methylmethacrylate) (PMMA) and styrene acrylonitrile (SAN) blend. Use the binary interaction parameter values tabulated in Chapter 3, and use the Sanchez and Lacombe free volume theory as the equation of state.
2. Construct the phase diagram for a PMMA and AMSN–AN (alphamethyl–styrene acrylonitrile blend). Use the binary interaction parameter values tabulated in Chapter 3, and use the Sanchez–Lacombe free volume theory as the equation of state.
3. Construct the phase diagram for the styrene methylmethacrylate and styrene acrylonitrile (MMA–S) blend. Use the binary interaction parameter values tabulated in Chapter 3, and use the Sanchez–Lacombe free volume theory as the equation of state.
4. Construct the phase diagram for the alphamethyl–styrene methylmethacrylate and alpha(methyl–styrene acrylonitrile) (MMA–AMS) blend. Use the binary interaction parameter values tabulated in Chapter 3, and use the Sanchez–Lacombe free volume theory as the equation of state.
5. Construct the phase diagram for the PMMA–SAN blend. Use the binary interaction parameter values tabulated in Chapter 3, and use the FOV theory as the equation of state.
6. Construct the phase diagram for the PMMA and AMS–AN blend. Use the binary interaction parameter values tabulated in Chapter 3, and use the FOV theory as the equation of state.
7. Construct the phase diagram for the MMA-S–SAN blend. Use the binary interaction parameter values tabulated in Chapter 3, and use the FOV theory as the equation of state.
8. Construct the phase diagram for the MMA–AMS and AMS–AN blend. Use the binary interaction parameter values tabulated in Chapter 3, and use the FOV as the equation of state.
9. Construct the phase diagram for the PMMA–SAN blend. Use the binary interaction parameter values tabulated in Chapter 3, and use the Prigogine square-well cell theory as the equation of state.
10. Construct the phase diagram for the PMMA and AMS–AN blend. Use the binary interaction parameter values tabulated in Chapter 3, and the Prigogine square-well cell theory as the equation of state.

11. Construct the phase diagram for the MMA–S and SAN blend. Use the binary interaction parameter values tabulated in Chapter 3, and use the Prigogine square-well cell theory as the equation of state.

12. Construct the phase diagram for the MMA–AMS and AMS–AN blend. Use the binary interaction parameter values tabulated in Chapter 3, and the Prigogine square-well cell theory as the equation of state.

13. From the slopes of degenerate hyperbola in phase diagram, calculate the EOS parameters for the lattice fluid theory.

14. Develop a procedure to calculate the binary interaction parameter values from cloud-point curves.

15. Develop a procedure to calculate the solubility parameter values from the cloud-point curves.

## REFERENCES

1. J. M. Smith, H. C. van Ness, and M. M. Abbott, *Introduction to Chemical Engineering Thermodynamics*, 7th edition, McGraw Hill, New York, 2005.

2. S. J. Mumby, C. Qian, and B. E. Eichinger, Phase diagrams of quasi-binary polymer systems with LCST/UCST spinodals and hourglass cloud-point curves, *Polymer*, 33, 23, 5105–5105, 1992.

3. F. S. Bates, Polymer-polymer phase behavior, *Science*, 251, 898–905, 1991.

# 6 Partially Miscible Blends

## LEARNING OBJECTIVES

- Identify partially miscible polymer systems
- Develop mathematical framework to predict two-glass transition temperatures
- Explore physical significance of entropic difference model
- Apply the quadratic expression for $T_g$ to different systems
- Derive expression for change in change in entropy of mixing
- Cubic equation for prediction of glass transition temperature of blend
- Effect of chain sequence distribution of copolymer
- Entropy of copolymerization
- Dyads and triads

The change in entropy of mixing at the glass transition stage of a polymer blend in relation to the changes in heat capacity of the blend at the glass transition can be used to account for the observations of two distinct glass transition temperatures in partially miscible polymer blends. A quadratic expression for the mixed system glass transition temperature is derived:

$$A\, T_g^{\,2} + B\, T_g + C = 0, \tag{6.1}$$

where

$$A = 2\, \varphi_1\, (\, \Delta\Delta C_{p12}{}^{gl}) + 2\, \Delta C_{p2}{}^{gl} + \Delta\Delta S_m{}^{gl}, \tag{6.2}$$

$$B = 2\, \varphi_1\, (\, \Delta\Delta C_{p12}{}^{gl}) + 2\, \varphi_1\, \Delta T_{g21}\, (\Delta C_{p1}{}^{gl} + \Delta C_{p2}{}^{gl}) + 2\Delta C_{p2}{}^{gl}\, \Delta T_{g12} + \Delta\Delta S_m{}^{gl}, \tag{6.3}$$

$$C = 2\, \varphi_1\, (\, \Delta\Delta C_{p21}{}^{gl})\, T_{g1}T_{g2} - 2\Delta C_{p2}{}^{gl}\, T_{g1}T_{g2} + T_{g1}T_{g2}\, \Delta\Delta S_m{}^{gl}. \tag{6.4}$$

It can be seen that when $\alpha = \beta = 0$, the sum of the roots is $(T_{g1} + T_{g2})$ and the product of the roots is $T_{g1} T_{g2}$. This is the same as the homopolymer glass transition temperatures, and hence it can denote the immiscible blends. This is the case when $\Delta\Delta S_m{}^{gl}$ is extremely large compared with the $\Delta C_{p12}{}^{gl}$ or when the heat capacity changes at the glass transition temperature. The case when the change in entropy of mixing is zero results in a miscible blend with a single glass transition temperature. For a symmetric blend when $x_1 = \frac{1}{2}$, $\Delta\Delta S_m{}^{gl} = 0$, and $\Delta\Delta C_{p12}{}^{gl} = 0$, $A = 2\, \Delta C_p{}^{gl}$, $B = 0$, and $C = -2\, \Delta C_p{}^{gl}\, T_{g1}T_{g2}$. The roots of the quadratic equation are then $\pm\mathrm{sqrt}(T_{g1}T_{g2})$. The partially miscible systems lie intermediate to these two extremes with two real positive roots to the quadratic equation, where

$$\alpha = \frac{\left(\Delta C_{p1}^{gl} + \Delta C_{p2}^{gl}\right)}{\Delta \Delta S_m^{gl}}. \tag{6.5}$$

$$\beta = \frac{\Delta C_{p12}^{gl}}{\Delta \Delta S_m^{gl}}. \tag{6.6}$$

## 6.1   COMMERCIAL BLENDS THAT ARE PARTIALLY MISCIBLE

Most polymer blends that are commercial products in the industry are partially miscible. Partially miscible polymer blends are those that exhibit some shift from their pure component glass transition temperatures. Thus, a binary miscible blend will exhibit one glass transition temperature [1], and a partially miscible blend may exhibit two distinct glass transition temperatures other than their pure component values [2,3]. Some experimental systems such as polyethylene terepthalate (PET) and poly-hydroxybenzoic (PHB), polycarbonate (PC), and styrene acrylonitrile (SAN) have been reported [4]. Very little mathematical description of partially miscible systems is available in the literature.

Mixed system glass transition temperature, $T_g$, is defined by the requirement that entropy for the glassy state is identical to that for the rubbery state [5]. A revisit of the Couchman equation without neglecting the differences of entropy of mixing in the glassy and rubbery state is used to develop a mathematical framework to represent multiple glass transition temperatures in partially miscible copolymers in blends.

## 6.2   ENTROPY DIFFERENCE MODEL ($\Delta \Delta S_M$)

For blends, at a molecular level during phase transition the molar volume and bulk density changes along with the heat capacity. This implies that $\Delta S_m^g$, the entropy of mixing in the glassy state, is different from $\Delta S_m^l$, the entropy of mixing in the rubbery phase. The mixed system entropy is given by

$$S = \varphi_1 S_1 + (1 - \varphi_1) S_2 + \Delta S_m \tag{6.7}$$

where $\Delta S_m$ is the excess entropy of mixing. Consider a binary blend with the composition of the blend $x_1$ and $1 - x_1$. The entropy of the components at the glass transition temperatures is

$$S_1 \text{ at } T_{g1} = S_1^0 \tag{6.8}$$

$$S_2 \text{ at } T_{g2} = S_2^0. \tag{6.9}$$

The entropy at the glassy and rubbery states is then calculated as

$$S^g = \varphi_1 \left( S_1^0 + \int_{T_{g1}}^{T_g} C_{p1}^g d(\ln T) \right) + (1 - \varphi_1) \left( S_2^0 + \int_{T_{g2}}^{T_g} C_{p2}^g d(\ln T) \right) + \Delta S_m^g, \tag{6.10}$$

$$S^l = \phi_1 \left( S_1^0 + \int_{T_{gl}}^{T_g} C_{pl}^l d(\ln T) \right) + (1 - \phi_1) \left( S_2^0 + \int_{T_{g2}}^{T_g} C_{p2}^l d(\ln T) \right) + \Delta S_m^l. \quad (6.11)$$

From the continuity of entropy considerations at the glass transition the Equations (6.10) and (6.11) are equated, i.e., $S^g = S^l$. Then,

$$\phi_1 \Delta C_{p1}^{gl} \ln\left(\frac{T_g}{T_{g1}}\right) + (1 - \phi_1) \Delta C_{p2}^{gl} \ln\left(\frac{T_g}{T_{g2}}\right) + \Delta\Delta S_m^{gl}. \quad (6.12)$$

By the Taylor approximation,

$$\ln\left(\frac{T_g}{T_{g1}}\right) \cong \frac{2\left(\dfrac{T_g}{T_{g1}} - 1\right)}{\left(\dfrac{T_g}{T_{g1}} + 1\right)}, \quad (6.13)$$

$$\ln\left(\frac{T_g}{T_{g2}}\right) \cong \frac{2\left(\dfrac{T_g}{T_{g2}} - 1\right)}{\left(\dfrac{T_g}{T_{g2}} + 1\right)}. \quad (6.14)$$

Now, substituting Equations (6.13) and (6.14) into Equation (6.12) and rearranging the terms,

$$A\, T_g^2 + B\, T_g + C = 0 \quad (6.15)$$

where,

$$A = 2\phi_1 (\Delta\Delta C_{p12}^{gl}) + 2 \Delta C_{p2}^{gl} + \Delta\Delta S_m^{gl}, \quad (6.16)$$

$$B = 2\,\phi_1 (\Delta\Delta C_{p12}^{gl}) + 2\,\phi_1 \Delta T_{g21} (\Delta C_{p1}^{gl} + \Delta C_{p2}^{gl}) + 2\Delta C_{p2}^{gl} \Delta T_{g12} + \Delta\Delta S_m^{gl}, \quad (6.17)$$

and

$$C = 2\,\phi_1 (\Delta\Delta C_{p21}^{gl}) T_{g1} T_{g2} - 2\Delta C_{p2}^{gl} T_{g1} T_{g2} + T_{g1} T_{g2} \Delta\Delta S_m^{gl}. \quad (6.18)$$

Special cases for Equation (6.15) can be identified.

For a symmetric blend, when $\phi_1 = \frac{1}{2}$, $\Delta\Delta S_m^{gl} = 0$, and $\Delta \Delta C_{p12}^{gl} = 0$,

$$A = 2 \Delta C_p^{gl}, \quad (6.19)$$

$$B = 0, \text{ and} \quad (6.20)$$

$$C = -2 \Delta C_p^{gl} T_{g1} T_{g2}. \quad (6.21)$$

The roots of the Equation (6.15) are then

$$\pm\text{sqrt}(T_{g1} T_{g2}). \quad (6.22)$$

Thus, only one possible glass transition results for the mixed system, $T_g$. This may denote the miscible blends.

The sum and product of the roots of Equation (6.15) for the general case may be written as

$$-\frac{B}{A} = \frac{(T_{g1} + T_{g2})(T_{g2} - T_{g1})}{(1+\alpha)((T_{g1} + T_{g2})\beta - 1)},$$  (6.23)

$$\frac{C}{A} = T_{g1}T_{g2}\frac{(1-\alpha)}{(1+\alpha)},$$  (6.24)

$$\alpha = \frac{\left(\Delta C_{p1}^{gl} + \Delta C_{p2}^{gl}\right)}{\Delta\Delta S_m^{gl}}, \text{ and}$$  (6.25)

$$\beta = \frac{\Delta C_{p12}^{gl}}{\Delta\Delta S_m^{gl}}.$$  (6.26)

It can be seen that when $\alpha = \beta = 0$, the sum of the roots is $(T_{g1} + T_{g2})$ and the product of the roots is $T_{g1}T_{g2}$. This is the same as the homopolymer glass transition temperatures, and hence it can denote the *immiscible blends*. This case when $\Delta\Delta S_m^{gl}$ is extremely large compared with the $\Delta C_{p12}^{gl}$ or the heat capacity changes at the glass transition temperature. The partially miscible systems lie intermediate to these two extremes with two real positive roots to the quadratic equation (Sharma [2, 3, 5–14]). The quadratic function of the blend glass transition temperature $T_g$, $f(T_g)$ given by Equation (6.27) is shown in Figure 6.1 for the PC/SAN system. It can be seen that there are three possibilities: miscible blend, immiscible blend, and partially miscible blend and these are shown in the figure. When the system is miscible the blend exhibits one single glass transition temperature. When the system is immiscible the two roots of the quadratic result in the homopolymer glass transition temperatures, $T_{g1}$ and $T_{g2}$, respectively. For the case of the partially miscible blends as discussed in Example 6.1 the system exhibits two glass transition temperatures.

To recap:

1. A mathematical framework is provided to account for the multiple glass transition temperatures found in partially miscible polymer blends.

$$A\,T_g^2 + B\,T_g + C = 0$$  (6.27)

where

$$A = 2\,\varphi_1\,(\Delta\,\Delta C_{p12}^{gl}) + 2\,\Delta C_{p2}^{gl} + \Delta\Delta S_m^{gl},$$  (6.28)

$$B = 2\,\varphi_1\,(\Delta\,\Delta C_{p12}^{gl}) + 2\,x_1\,\Delta T_{g31}\,(\Delta C_{p1}^{gl} + \Delta C_{p2}^{gl}) + 2\Delta C_{p2}^{gl}\,\Delta T_{g13} + \Delta\Delta S_m^{gl}, \text{ and}$$  (6.29)

$$C = 2\varphi_1\,(\Delta\,\Delta C_{p21}^{gl})\,T_{g1}T_{g2} - 2\Delta C_{p2}^{gl}\,T_{g1}T_{g2} + T_{g1}T_{g2}\,\Delta\Delta S_m^{gl}$$  (6.30)

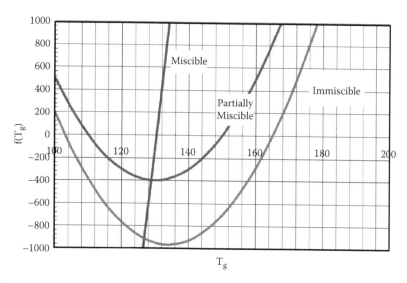

**FIGURE 6.1** Blend entropy change at the glass transition as a function of glass transition temperature(s) of the blend.

2. The change in entropy of mixing at the glass transition temperature of the blend can be used to account for the observations.

3. A quadratic expression for the mixed glass transition temperature is developed from the analysis.

4. For a symmetric blend, the conditions when the solution to the quadratic equation is one root is derived. This may denote the miscible blends. At this point the entropy of mixing can be seen to be zero. For a symmetric blend when $x_1$ = ½, $\Delta\Delta S_m^{gl} = 0$, and $\Delta \Delta C_{p12}^{gl} = 0$, A = 2 $\Delta C_p^{gl}$, B = 0; and C = −2 $\Delta C_p^{gl} T_{g1}T_{g2}$. The roots of the equation are then ±sqrt $(T_{g1}T_{g2})$. Thus, only one possible glass transition results for the mixed system $T_g$. This may denote the miscible blends and the point where the entropy of mixing may be taken as zero.

5. A special case of the quadratic equation is when the solution values are identical with that of the glass transition temperatures of the constituent polymers of the blend as they were prior to blending. In this case, the entropy of mixing is much larger compared with the changes in heat capacity at glass transition of the blend. It can be seen that when α = β = 0, the sum of the roots is $(T_{g1} + T_{g2})$, and the product of the roots is $T_{g1} T_{g2}$. This is the same as the homopolymer glass transition temperatures, and hence it can denote the immiscible blends, where

$$\alpha = \frac{\left(\Delta C_{p1}^{gl} + \Delta C_{p2}^{gl}\right)}{\Delta\Delta S_m^{gl}},$$ (6.31)

$$\beta = \frac{\Delta C_{p12}^{gl}}{\Delta\Delta S_m^{gl}}.$$ (6.32)

6. The intermediate cases with two positive real roots to the quadratic Equation (6.27) may denote the partially miscible blends [6].

## Example 6.1

Consider the partially miscible blend of polycarbonate and styrene acrylonitrile copolymer (SAN). The two glass transition temperatures were found to be 110°C and 150°C. Discuss the change in entropy of mixing at the glass transition of the blend. The homopolymer glass transition temperature of PC is 165°C and SAN at azeotropic composition of 25% AN is 103°C.

Equation (6.27) for this system can be written as

$$T_g^2 - 260T_g + 16,500 = 0, \text{ and}$$

$$C/A = 110*150 = 16,500; \quad -B/A = (110 + 150) = 260.$$

From Equation (6.23) and Equation (6.24) using an MS Excel spreadsheet, $\alpha$ and $\beta$ are calculated as

$$\alpha = \frac{\left(1 - \dfrac{C}{AT_{g1}T_{g2}}\right)}{\left(1 + \dfrac{C}{AT_{g1}T_{g2}}\right)} = 0.0148 \text{ and}$$

$$\beta = 1 + \frac{\left(T_{g2} - T_{g1}\right)}{(1-\alpha)\left(-\dfrac{B}{A}\right)} = 1.242.$$

The $\Delta C_p$ increase at glass transition for some polymers are provided by van Krevelen [15]. This is given below in Table 6.1.

### TABLE 6.1
### $\Delta C_p$ Values from van Krevelen and Nijenhius [15]

| Polymer | $T_g(K)$ | $\Delta C_p$ (J mol K$^{-1}$) |
|---|---|---|
| Polycarbonate (PC) | 424 | 56.4 |
| Polyethylene terepthalate (PET) | 342 | 77.8 |
| Polyether ether ketone (PEEK) | 419 | 78.1 |
| Polyolefin (PO) | 358 | 25.7 |
| Polyphenylene sulfide (PPS) | 363 | 33.0 |
| Polyphenylene oxide (PPO) | 483 | 32.2 |
| Polyethylene napthalate (PEN) | 391 | 81.3 |
| Polybutylene terepthatlate (PBT) | 248 | 107 |

## 6.3 ESTIMATES OF CHANGE IN ENTROPY OF MIXING AT GLASS TRANSITION, $\Delta\Delta S_M$

In order to obtain better estimates of the $\alpha$ and $\beta$ as shown in Equations (6.31) and (6.32), and hence better estimates of the sum and product of the roots of the quadratic expression for blend $T_g$ as given in Equation (6.15), the following analysis may be applicable. Consider a reversible blending of two polymers as considered in the above section. Assume that the process is reversible.

$$\Delta H_{blend} = \Delta(U_{blend} + (PV)_{blend})$$

$$C_p dT = T_g \Delta S_{blend}(-PdV + PdV + VdP)_{blend}$$

$$0 = T_g \Delta S_{blend} + (VdP)_{blend} \tag{6.33}$$

$$\Delta S_{blend} = -\frac{V_{blend}}{T_g}\Delta P_{blend}.$$

Thus, at the glass transition, the change in entropy of mixing of the blend can be expressed in terms of the change in change of pressure of the blend. This can be estimated from the EOS theories as discussed in Chapter 2. Thus, from Equation (6.33), assuming isochoric mixing at the glass transition,

$$\Delta\Delta S^{gl} = -\frac{V}{T_g}\Delta\Delta P^{gl}_{blend}. \tag{6.34}$$

Equation (6.34) can be combined with Equation (6.12) to yield

$$\phi_1 \Delta C^{gl}_{p1}\ln\left(\frac{T_g}{T_{g1}}\right) + (1-\phi_1)\Delta C^{gl}_{p2}\ln\left(\frac{T_g}{T_{g2}}\right) - \frac{V}{T_g}\Delta\Delta P^{gl}_{blend} = 0. \tag{6.35}$$

$$\text{Let } \frac{T_g}{T_{g1}} = z; \Delta\Delta C^{gl}_{p12} = \Delta C^{gl}_{p1} - \Delta C^{gl}_{p2}; \gamma = \frac{T_{g1}}{T_{g2}}; < \Delta C^{gl}_{p12} >= \frac{\Delta C^{gl}_{p1} + \Delta C^{gl}_{p2}}{2}, \tag{6.36}$$

then Equation (6.35) becomes

$$\phi_1 \Delta C^{gl}_{p1}\ln(z) + (1-\phi_1)\Delta C^{gl}_{p2}\ln(\gamma z) - \frac{V}{zT_{g1}}\Delta\Delta P^{gl}_{blend} = 0, \tag{6.37}$$

and by Taylor approximation,

$$\ln(z) = \frac{2(z-1)}{z+1}; \quad \ln(\gamma z) = \frac{2(\gamma z - 1)}{(\gamma z + 1)}. \tag{6.38}$$

Equation (6.37) then becomes

$$\frac{(z)(\gamma z+1)(z-1)2\phi_1\Delta C_{p1}^{gl}+2(1-\phi_1)(z)(z+1)(\gamma z-1)\Delta C_{p2}^{gl}-V\dfrac{\Delta\Delta P_{blend}^{gl}}{T_{g1}}(z+1)(\gamma z+1)}{z(z+1)(\gamma z+1)}=0.$$

(6.39)

A cubic equation for $T_g$ can be recognized from Equation (6.39). For values of $z$ where the denominator in Equation (6.39) is not zero, Equation (6.39) becomes

$$z^3\left(2\phi_1\gamma\Delta\Delta C_{p12}^{gl}+2\gamma\Delta C_{p2}^{gl}\right)+z^2((4\phi_1)<\Delta C_{p12}^{gl}>(1-\gamma)-2(1-\gamma)\Delta C_{p2}^{gl}-V\gamma\frac{\Delta\Delta P_{blend}^{gl}}{T_{g1}})$$

(6.40)

$$-2z\left(\phi_1\Delta C_{p12}^{gl}+\Delta C_{p2}^{gl}-V((1+\gamma)\frac{\Delta\Delta P_{blend}^{gl}}{2T_{g1}}\right)-V\frac{\Delta\Delta P_{blend}^{gl}}{T_{g1}}=0.$$

The equation for the glass transition temperature of the blend obtained from relating the change in the entropy of mixing to the change in pressure of the blend at the glass transition turns out to be cubic in nature. Equation (6.40) can be written as

$$az^3+bz^2-cz-d=0.$$

where

$$a=\left(2\phi_1\gamma\Delta\Delta C_{p12}^{gl}+2\gamma\Delta C_{p2}^{gl}\right),$$

$$b=(4\phi_1)<\Delta C_{p12}^{gl}>(1-\gamma)-2(1-\gamma)\Delta C_{p2}^{gl}-V\gamma\frac{\Delta\Delta P_{blend}^{gl}}{T_{g1}},$$

$$c=2\left(\phi_1\Delta C_{p12}^{gl}+\Delta C_{p2}^{gl}-V((1+\gamma)\frac{\Delta\Delta P_{blend}^{gl}}{2T_{g1}}\right),$$

(6.41)

$$d=V\frac{\Delta\Delta P_{blend}^{gl}}{T_{g1}}.$$

The glass transition temperature of the blend can be obtained from Equation (6.41). If it is miscible, there will be one glass transition temperature. If it is partially miscible, multiple glass transition temperatures can be expected. These values are different from the homopolymer values. For the case of immiscible blends, the glass transition temperature would be equal to the homopolymer values. Equation (6.41) can be solved using computer numerical solutions such as the Newton–Raphson method. For the special case when $b=0$, the depressed cubic equation can be solved for by the method of Vieta's substitution. A procedure to obtain closed form analytical solutions to Equation (6.41) is as follows:

$$z^3+\frac{b}{a}z^2-\frac{c}{a}z-\frac{d}{a}=0.$$

(6.42)

Consider the substitution, $z = y + \zeta$. Equation (6.42) becomes

$$(y+\zeta)^3 + \frac{b}{a}(y+\zeta)^2 - \frac{c}{a}(y+\zeta) - \frac{d}{a} = 0$$

$$y^3 + \left(3\zeta + \frac{b}{a}\right)y^2 + (3\zeta^2 + 2\zeta\frac{b}{a} + \frac{c}{a})y + \zeta^2 + \zeta^2\frac{b}{a} + \zeta\frac{c}{a} + \frac{d}{a} = 0$$

(6.43)

Choose $\zeta = -b/3a$; Equation (6.43) becomes

$$y^3 + (3\zeta^2 + 2\zeta\frac{b}{a} + \frac{c}{a})y + \zeta^2 + \zeta^2\frac{b}{a} + \zeta\frac{c}{a} + \frac{d}{a} = 0$$

(6.44)

or

$$y^3 + ey + f = 0,$$

(6.45)

where $e = 3\zeta^2 + 2\zeta\frac{b}{a} + \frac{c}{a}$; $f = \zeta^2 + \zeta^2\frac{b}{a} + \zeta\frac{c}{a} + \frac{d}{a}$.

Consider Vieta's substitution:

$$y = w + \frac{s}{w}$$

(6.46)

$$\left(w + \frac{s}{w}\right)^3 + e\left(w + \frac{s}{w}\right) + f = 0,$$

(6.47)

Expanding the cubic term in Equation (6.47) and multiplying throughout by $w^3$. Then,

$$w^6 + (3s + e)w^4 + fw^3 + s(3s + e)w^2 + s^3 = 0).$$

(6.48)

Let $s = -e/3$, and then the equation becomes

$$w^6 + fw^3 - \frac{e^3}{27} = 0.$$

(6.49)

Let $v = w^3$. Then

$$v^2 + fv - \frac{e^3}{27} = 0.$$

(6.50)

Equation (6.50) is quadratic in $v^2$. Thus, the general form of cubic equation was *quadratisized*.

## 6.4 COPOLYMER AND HOMOPOLYMER BLEND

Consider a copolymer (polymer 1) with two comonomers with composition of the monomer A in the copolymer given by $x_A$. Consider a homopolymer for the second polymer 2. Consider a binary blend of copolymer A and homopolymer B. The entropic difference model can be applied to this system as follows. Equation (6.7) can be written for this system as

$$S = \phi_1 S_1 + (1-\phi_1)\, S_2 + \Delta S_m. \qquad (6.51)$$

The entropy of the copolymer 1 can be calculated from the copolymerization reactions that occur during the formation of polymer 1. There are 4 propagation steps (this is discussed in detail in later sections of the textbook) that occur during the formation of polymer 1:

$$A + A* \xrightarrow{k_{AA}} [AA]$$

$$B + A* \xrightarrow{k_{BA}} [BA]. \qquad (6.52)$$

$$B + B* \xrightarrow{k_{BB}} [BB]$$

$$A + B* \xrightarrow{k_{AB}} [AB].$$

The entropy of copolymer 1 can be written in terms of the *random sequence distribution* of the copolymer. Although this is discussed in greater detail in later sections of the textbook, here the entropy of copolymerization is written in terms of the probabilities of sequences of a repeat unit, A, in the copolymer. Thus:

$$S_1 = -(S_{AA}x_A P_{AA} + S_{BB}(1 - x_A)P_{BB} + S_{AB}x_A P_{AB} + S_{BA}(1 - x_A)P_{BA}). \qquad (6.53)$$

Equation (6.35) can be rewritten in terms of a interaction parameter, $\chi$, where

$$\chi = (S_{AB} + S_{BA}) - (S_{AA} + S_{BB}). \qquad (6.54)$$

Other contributions to the entropy expression are (1) residual entropy at 0 K in the glassy state, (2) entropy due to chain configuration, and (3) entropy due to copolymerization randomness. The residual entropy can be neglected, as its contribution is small. The configurational entropic contribution is also small. The entropy of *copolymerization randomness* can be calculated. It can be viewed as an entropy of mixing. The probability of a sequence length of L repeat units of comonomer A in the copolymer 1 can be given by geometric distribution as shown by Sharma [17–19]:

$$P_L = P_{AA}^{L-1}(1 - P_{AA}). \qquad (6.55)$$

The number of sequences per chain is bounded by the ceiling O(2n). Roughly half of them are sequences with comonomer A repeat units, and half of them are sequences with comonomer B repeat units in the extreme case of a alternating copolymer. The number of A sequences of length L is then

$$nP_{AA}^{L-1}(1 - P_{AA}). \qquad (6.56)$$

Let there be $a_1$ sequences of A monads, $a_2$ sequences of AA dyads, $a_3$ sequences of AAA triads, ... and, $a_r$ sequences of AAA ... A of length $r$. Thus:

$$n = a_1 + a_2 + a_3 + ... + a_r. \tag{6.57}$$

The number of permutations possible to these AAA sequences can be written as

$$W = \frac{n_A!}{a_1! a_2! a_3! ... a_r!}. \tag{6.58}$$

In entropy analysis, Stirling's approximation can be used to simplify large numbers in the argument of logarithm term

$$\ln(n!) = n \ln(n) - n. \tag{6.59}$$

Thus, for large values of $a$,

$$\ln(W) = n \ln(n) - a_1 \ln(a_1) - a_2 \ln(a_2) - .... - a_r \ln(a_r). \tag{6.60}$$

The entropy of mixing $\Delta S_{mix}$ per mole can be written as

$$\Delta S_{mix} = R \ln(W) = \frac{n \ln(n) - a_1 \ln(a_1) - a_2 \ln(a_2) - ... - a_r \ln(a_r)}{a_1 + 2a_2 + 3a_3 + ... + ra_r}. \tag{6.61}$$

When $r$ is large, Equation (6.61) can be simplified by expansion of the terms $P_{AA}$ and $P_{BB}$. It can be shown that

$$a_1 + 2a_2 + 3a_3 + ... + ra_r \cong \frac{1}{1 - P_{AA}} \tag{6.62}$$

and

$$\sum_{i=1}^{r} a_i \ln(a_i) \cong n\left(\ln(n) + \ln(1 - P_{AA})\right) + \frac{P_{AA}}{(1 - P_{AA})} \ln(P_{AA}). \tag{6.63}$$

The inequalities in Equation (6.63) become equalities when $r$ is infinite. Substituting Equation (6.63) into Equation (6.61);

$$\Delta S_{mix} = -R\left(P_{AA} \ln(P_{AA}) + (1 - P_{AA}) \ln(1 - P_{AA})\right) \tag{6.64}$$

For block copolymers, $P_{11} = P_{22} = 1$ and from Equation (6.64), $\Delta S_{mix} = 0$. For completely alternating copolymers, $P_{11} = P_{22} = 0$ and again from Equation (6.64), $\Delta S_{mix} = 0$. The entropy of copolymer is the sum of the entropy of copolymerization reactions plus the entropy of randomness. Thus,

$$\Delta S = -R(x_A(1 - P_{AA}) + P_{AA} \ln(P_{AA})) + (1 - x_A)(P_{BB} \ln(P_{BB})) + (1 - P_{BB})(\ln(1 - P_{BB})). \tag{6.65}$$

## 6.5  SEQUENCE DISTRIBUTION EFFECTS ON MISCIBILITY

It can be noted that the observed miscible polymer systems may be due to intra-molecular repulsion rather than intermolecular attraction as originally hypothesized. The miscible blends of random copolymer (AB) and a homopolymer (C) in AB/C type blends were found miscible over a range of compositions of comonomer 1 in copolymer AB. This indicates a possible effect of the chain sequence distribution on the miscibility. Even in cases where the homopolymers were not miscible, the copolymer and homopolymer were found to be miscible. The mean-field theory of random copolymer blends is an important development in the thermodynamics of polymer miscibility. The compositional window of miscibility has been found for systems where A, B, and C were found to be immiscible as homopolymers, but were found to be miscible as A–B copolymer and C homopolymer.

Karasz et al. [20] considered AB–AB type blends. Per the first-order random copolymer blend theory as described in Section 3.9 the interaction parameter of the blend can be written as

$$\chi_{blend} = \chi_{AB}(x_{A1} - x'_{A2})^2,  \tag{6.66}$$

where $\chi_{AB}$ is the segment pair interaction parameter, and the $x_{A1}$ and $x_{A2}$ are the compositions of monomer A in copolymer 1 and copolymer 2. By this approximation, the isothermal boundary between the miscible and immiscible regions in the phase diagram are represented by a pair of straight lines. These straight lines are symmetrically placed about a diagonal in the composition phase diagram of copolymer/copolymer system. The lines are defined by the equation

$$\chi_{crit} = \chi_{blend}.  \tag{6.67}$$

The critical value of the blend interaction is derived from the equation for the combinatorial entropy of the blend:

$$\chi_{crit} = \frac{1}{2}\left(\frac{1}{\sqrt{n_1}} + \frac{1}{\sqrt{n_2}}\right)^2,  \tag{6.68}$$

where $n_1$ and $n_2$ are the number-average degrees of polymerization for components 1 and 2. Blends for which $\chi_{blend} < \chi_{crit}$ are in the one phase region. Blends for which $\chi_{crit} < \chi_{blend}$ are in the two phase region. The miscibility boundaries for chlorinated polyethylene (CPE) and CPE blends prepared in the laboratory were found not to be parallel straight lines. The miscible region was found to be enlarged by symmetrical bulges convex to the diagonal in the copolymer–copolymer phase diagram plot. This was attempted to be explained in terms of the chain sequence distribution effect for both AB/S and AB/AB polymer systems. The interactions between two given monomer units are dependent on the nature of the respective intramolecular nearest neighbors. The interactions between different *triads* can be used to better account for the experimental observations on miscibility. For the copolymer with the two monomers A and S, there can be 6 possible triads. These are

   1. AAA
   2. AAS
   3. ASA
   4. SSS
   5. SAS
   6. SSA

For a given composition, there is an optimal range of sequence distributions in which AB/C blends are miscible. This may be the reason for the lines in the phase diagram to intersect. It was assumed that

$$\overline{\chi}_{AC} = \chi_{BAB,C} = \chi_{AAB,C} = \chi_{BAA,C} \neq \chi_{AAA,C} \tag{6.69}$$

$$\overline{\chi}_{BC} = \chi_{ABA,C} = \chi_{ABB,C} = \chi_{BBA,C} \neq \chi_{BBB,C}$$

Difference parameters were defined as

$$\Delta\chi_A = \chi_{AAA,C} - \overline{\chi}_{AC}$$

$$\Delta\chi_B = \chi_{BBB,C} - \overline{\chi}_{BC}. \tag{6.70}$$

where BAB, AAB, etc. are the triad sequences in the copolymer of the monomer repeat units, and $\chi_{triad,C}$ is the interaction parameter of the particular triad with the monomer unit of C of the homopolymer C. Different investigators such as Cantow and Shulz [21] made different assumptions about the triad interaction parameters. These are given by

$$\chi_{AAA,C} - \chi_{AAB,C} = \chi_{AAB,C} - \chi_{BAB,C}$$

$$\chi_{BBB,C} - \chi_{BBA,C} = \chi_{BBA,C} - \chi_{ABA,C}. \tag{6.71}$$

Zhikuan [20] studied three AB/AB systems: (1) CPVC/CPVC, (2) MMA/EMA-MMA-EMA, and (3) CPE/CPE systems and derived interaction parameters between AB/C and AB/AB blends using Cantow and Schulz theory.

For AB/C blends of copolymer and homopolymer the effective blend interaction parameter based on the assumptions given by Equations (6.69) and (6.70):

$$\chi_{blend}\phi_1 = \chi_{BAB,C}x_A\phi_1\left(\frac{P_{AB}}{x_A}\right)^2 + 2\chi_{BAA,C}x_A\phi_1\left(\frac{P_{AA}P_{AB}}{x_A^2}\right) + \chi_{AAA,C}x_A\phi_1\left(\frac{P_{AA}}{x_A}\right)^2$$

$$+ \chi_{ABA,C}(1-x_A)\phi_1\left(\frac{P_{AB}}{1-x_A}\right)^2 + 2\chi_{ABB,C}(1-x_A)\phi_1\left(\frac{P_{BB}P_{AB}}{(1-x_A)^2}\right)$$

$$+ \chi_{BBB,C}(1-x_A)\phi_1\left(\frac{P_{BB}}{1-x_A}\right)^2 \tag{6.72}$$

$$+ \overline{\chi}_{AB}\frac{\phi_1^2}{(1-\phi_1)}x_A(1-x_A) - \overline{\chi}_{AB}\frac{\phi_1^2}{(1-\phi_1)}x_A(1-x_A),$$

where $\varphi_1$ is the volume fraction of the copolymer AB in the blend and $(1-\varphi_1)$ is the volume fraction of the homopolymer C in the blend; $P_{AA}$, $P_{BB}$, and $P_{AB}$ are the probabilities of finding the dyads AA, BB, and AB, respectively; $x_A$ is the copolymer AB volume fraction of monomer A; and $(1-x_A)$ is the copolymer AB volume fraction of monomer B. The first terms in the right-hand side of Equation (6.72) represents the triad interactions in the copolymer and homopolymer. The seventh and eighth terms in the RHS of Equation (6.72) are interaction of the monomer units A and B in the mixture and interaction of monomer units A and B before the mixing process, respectively.

Using the Cantow and Schulz assumption [21], the effective blend interaction parameter [22] can be written as follows:

$$\chi_{blend} = x_A \chi_{AAA,C} + (1-x_A)\chi_{BBB,C} - x_A(1-x_A)\chi_{AB} - 4(\Delta\chi_A + \Delta\chi_B)\theta x_A(1-x_A) \quad (6.73)$$

$\theta$ was introduced by Balasz et al. [22] as

$$P_{AB} = 2\theta x_A(1-x_A). \tag{6.74}$$

The $\theta$ parameter denotes the binary sequence distribution of the monomer repeat units in a copolymer chain. The dyad probabilities can be calculated from the geometric distribution for random copolymers synthesized by the free radical method of polymerization as explained in Chapter 8. The interaction parameter is a linear function of $\theta$. Further, it can be seen that

$$\frac{\partial\chi_{blend}}{\partial\theta} = -4(\Delta\chi_A + \Delta\chi_B)x_A(1-x_A). \tag{6.75}$$

It can be verified that either a more blocky or a more alternating structure will favor mixing in an AB/C system. There is no extrema in the relation between $\chi_{blend}$ and $\theta$.

For AB/AB blends 15 triad interaction parameters are required to completely describe the system. Six of the fifteen interaction parameters can be set to zero because the interacting pairs have identical center monomers, that is, $\chi_{12} = \chi_{13} = \chi_{23} = \chi_{45} = \chi_{46} = \chi_{56} = 0$. Using the Cantow and Schulz assumption as given in Equation (6.73), five independent equations can be derived. Five interaction parameters can be solved for from the system of five equations and five unknowns.

$$\Delta_1 = \chi_{36} - \chi_{26} = \chi_{26} - \chi_{16}$$

$$\Delta_2 = \chi_{35} - \chi_{25} = \chi_{25} - \chi_{15}$$

$$\Delta_3 = \chi_{34} - \chi_{24} = \chi_{24} - \chi_{14}$$

$$\Delta_4 = \chi_{36} - \chi_{35} = \chi_{35} - \chi_{34} \tag{6.76}$$

$$\Delta_5 = \chi_{26} - \chi_{25} = \chi_{25} - \chi_{24}$$

$$\Delta_6 = \chi_{16} - \chi_{15} = \chi_{15} - \chi_{14}.$$

It can be verified that

$$\Delta_6 - \Delta_5 = \Delta_5 - \Delta_4 = \Delta_3 - \Delta_2 = \Delta_2 - \Delta_1. \tag{6.77}$$

Similar to $\varphi_1$ and P for the first chain, $\varphi_1'$ and P′ for the second copolymer chain are introduced. Thus, the effective interaction parameter of the blend of the copolymer–copolymer system can be written as follows:

$$\chi_{blend} = \chi_{25}(x_A - x_A')^2 + \Delta_2(X_A - X_A') - \Delta_5(X_B - X_B')(x_A - x_A') \\ + (\Delta_6 - \Delta_5)(X_A - X_A')(X_B - X_B') \tag{6.78}$$

where $X_A = P_{AA} - P_{AB}$; $X_B = P_{BB} - P_{AB}$; $X_A' = P'_{AA} - P_{AB}'$; $X_B' = P_{BB}' - P_{AB}'$. Equation (6.78) reverts to Equation (6.66) when there is no sequence distribution effects. This happens when all the $\Delta_i = 0$. For copolymer 1 and 2 with the same composition, Equation (6.78) reverts to

$$\chi_{blend} = 4(\Delta_6 - \Delta_5)(P_{AB} - P_{AB}')^2. \tag{6.79}$$

Miscibility is then determined by the sign magnitude of $(\Delta_6 - \Delta_5)$. When this is positive, the blend can be expected to be immiscible. An extremum was found in the expression for the interaction parameter of the blend as a function of the sequence distribution parameter, $\theta$.

$$\frac{\partial^2 \chi_{blend}}{\partial \theta^2} = 32(\Delta_6 - \Delta_5)(x_A - x_A^2 - x_A' + x_A'^2)^2 \tag{6.80}$$

The interaction parameter of the blend, $\chi_{blend}$ exhibits a *minima* when $\Delta_6 > \Delta_5$. When $\Delta_6 < \Delta_5$, either blocky or alternating copolymer blends can be expected to be more *miscible*.

## 6.6 SUMMARY

A mathematical framework is provided to account for the multiple glass transition temperatures found in partially miscible polymer blends

$$A\,T_g^2 + B\,T_g + C = 0,$$

where

$$A = 2\,\varphi_1\,(\Delta\Delta C_{p12}{}^{gl}) + 2\,\Delta C_{p2}{}^{gl} + \Delta\Delta S_m{}^{gl}$$

$$B = 2\,\varphi_1\,(\Delta\Delta C_{p12}{}^{gl}) + 2\,x_1\,\Delta T_{g31}\,(\Delta C_{p1}{}^{gl} + \Delta C_{p2}{}^{gl}) + 2\Delta C_{p2}{}^{gl}\,\Delta T_{g13} + \Delta\Delta S_m{}^{gl}$$

$$C = 2\varphi_1\,(\Delta\Delta C_{p21}{}^{gl})\,T_{g1}T_{g2} - 2\Delta C_{p2}{}^{gl}\,T_{g1}T_{g2} + T_{g1}T_{g2}\,\Delta\Delta S_m{}^{gl}$$

The change in entropy of mixing at the glass transition temperature of the blend can be used to account for the observations. A quadratic expression for the mixed glass transition temperature is developed from the analysis. For a symmetric blend the conditions when the solution is one root denoting the miscible blends is derived. For a symmetric blend when $x_1 = \frac{1}{2}$, $\Delta\Delta S_m^{gl} = 0$, and $\Delta\,\Delta C_{p12}^{gl} = 0$; $A = 2\,\Delta C_p^{gl}$; $B = 0$; and $C = -2\,\Delta C_p^{gl}$ $T_{g1}T_{g2}$. The roots of the equation are then $\pm$sqrt $(T_{g1}T_{g2})$. Thus only one glass transition result is possible for the mixed system, $T_g$. This may denote the miscible blends. This is when the entropy of mixing may be taken as zero. When the roots of the quadratic expression $T_g$ takes on the homopolymer values, the entropy of mixing is very large compared with the changes in heat capacity at the glass transition. It can be seen that when $\alpha = \beta = 0$, the sum of the roots is $(T_{g1} + T_{g2})$ and the product of the roots is $T_{g1}\,T_{g2}$. This is the same as the homopolymer glass transition temperatures, and hence it can denote the immiscible blends. This is the case when $\Delta\Delta S_m^{gl} \gg \Delta C_p^{gl}$, where

$$\alpha = \frac{\left(\Delta C_{p1}^{gl} + \Delta C_{p2}^{gl}\right)}{\Delta\Delta S_m^{gl}}$$

$$\beta = \frac{\Delta C_{p12}^{gl}}{\Delta\Delta S_m^{gl}}$$

The intermediate cases with two positive real roots to the quadratic equation for blend $T_g$ may denote the partially miscible blends.

Many polymer blends that are used in industrial practice have been found to be partially miscible. Examples are PVC/SAN, PC/SAN at certain AN compositions, PET/PHB, etc. The entropic difference model was developed by taking into account the change in entropy of mixing at glass transition in the Couchman model. Equation (6.15) quadratic expression for blend $T_g$s is obtained upon Taylor approximation of the relation obtained by equating the entropy of the blend in glassy phase with the entropy of blend in rubbery phase at glass transition of the blend.

Estimates of entropy change of mixing can be made from $\Delta P_{blend}$. Information about $\Delta P_{blend}$ is available from EOS theories discussed in Chapter 2. A cubic equation for $T_g$s can be recognized from Equation (6.39).

Entropy of random copolymer can be calculated from the reactions between comonomers that were involved in the formation of the copolymer. The chain sequence distribution information can be used in estimating the entropy of the copolymer. The sequence distribution of random copolymer can be used to interpret the observations of polymer blend miscibility. The Flory–Huggins interaction parameter can be expressed as a function of dyad and triad probabilities.

## NOMENCLATURE

| | |
|---|---|
| A | $2\,\varphi_1\,(\Delta\Delta C_{p12}^{gl}) + 2\,\Delta C_{p2}^{gl} + \Delta\Delta S_m^{gl}$ |
| B | $2\,\varphi_1\,(\Delta\Delta C_{p12}^{gl}) + 2\,\varphi_1\,\Delta T_{g31}\,(\Delta C_{p1}^{gl} + \Delta C_{p2}^{gl}) + 2\Delta C_{p2}^{gl}\,\Delta T_{g13} + \Delta\Delta S_m^{gl}$ |
| C | $2\,\varphi_1\,(\Delta\Delta C_{p21}^{gl})\,T_{g1}T_{g2} - 2\Delta C_{p2}^{gl}\,T_{g1}T_{g2} + T_{g1}T_{g2}\,\Delta\Delta S_m^{gl}$ |
| $C_p$ | Heat capacity (J/mole/K) |
| $\Delta\Delta C_{p12}^{gl}$ | Change after glass transition in the change in heat capacity between components 1 and 2 (J/mole/K) |

| $\Delta C_{p1}{}^{gl}$ | Change after glass transition in heat capacity in component 1 (J/mole/K) |
| $\Delta C_{p2}{}^{gl}$ | Change after glass transition in heat capacity in component 1 (J/mole/K) |
| $S$ | Entropy (J/mole/K) |
| $\Delta S_m$ | Entropy of mixing of blend (J/mole/K) |
| $\Delta \Delta S_m{}^{gl}$ | Change after glass transition in the entropy of mixing between components 1 and 2 (J/mole/K) |
| $T_g$ | Glass transition temperature (K) |
| $\Delta T_{g31}$ | Difference in glass transition temperature between components 2 and 1 (K) |
| $X$ | Mole fraction |

## SUBSCRIPTS

| 1 | Component 1 |
| 2 | Component 2 |
| g | Glassy |
| l | Liquidy |
| m | Mixing |

## SUPERSCRIPTS

| 0 | Reference state |
| g | Glassy state |
| L | Liquidy state |
| Gl | Change from glassy to liquid |

## GREEK

$$\alpha = \frac{\left( \Delta C_{pl}^{gl} + \Delta C_{p2}^{gl} \right)}{\Delta \Delta S_m^{gl}}$$

$$\beta = \frac{\Delta C_{p12}^{gl}}{\Delta \Delta S_m^{gl}}$$

$$\Delta = \qquad \text{Difference}$$

# EXERCISES

1. When heat capacity variation with temperature is taken into account, what would happen to Equation (6.15)?
2. The two blend glass transition temperatures of a partially miscible blend of PC/SAN $T_g^{blend}$ are given as 110 and 150, respectively. The homopolymer $T_g$ values of SAN and PC are 100°C and 165°C. What can be inferred

about the change in entropy of mixing and change in heat capacity at glass transition?

3. When the quadratic equation derived for blend glass transition temperature(s) predicts as a solution the two homopolymer glass transition temperature values, what does it mean?

4. What does it mean when blend glass transition temperature results are monovalued and when found equal to be $\sqrt{T_{g1}T_{g2}}$ ?

5. What does the cubic nature of Equation (6.41) tell you?

6. What is the physical significance of the terms A, B, and C in the expression to predict the blend glass transition temperatures in Equation (6.15)?

7. What is the physical significance of the coefficients of the third degree equation for blend glass transition temperature derived in Equation (6.41)?

8. Is $S_{AB} = S_{BA}$ in Equation (6.54)?

9. Is there a maxima in Equation (6.64) for the entropy of mixing for a random copolymer as a function of dyad fractions?

10. What is the significance of the mean-field theory of random copolymer blends in the thermodynamics of polymer miscibility?

11. Why were the miscibility boundaries of chlorinated polyethylene, CPE and CPE blends not parallel straight lines?

12. How many triad interactions are needed to completely describe AB/AB blends?

13. Why can 6 of the 15 triad interaction parameters be set to zero?

14. How many triad interactions are needed to completely describe ABC/EFG blends?

15. How many triad interactions are needed to completely describe ABC/ABC blends?

16. What is the significance of Cantow and Schulz assumptions about the triad interaction parameters as given in Equation (6.71)?

17. What do the seventh and eighth terms in the RHS of Equation (6.72) signify?

18. What is the significance of the $\theta$ parameter introduced by Balasz et al. [22]?

19. Why is geometric distribution better suited to describe sequence distribution information?

20. When does the expression for $\chi_{blend}$ exhibit a minima in Equation (6.80)? What does this mean?

## REFERENCES

1. D. R. Paul and J. W. Barlow, A binary interaction model for miscibility of copolymers in blends, *Polymer*, Vol. 25, 487–494, 1984.

2. K. R. Sharma, Mathematical Modeling of Glass Transition Temperature of Partially Miscible Polymer Blends and Copolymers, 214th ACS National Meeting, Dallas, TX, 1998.

3. K. R. Sharma, Mesoscopic Simulation and Entropic Difference Model for Glass Transition Temperature of Partially Miscible Copolymers in Blends, 1999, ANTEC, New York.

4. P. R. Couchman, *Macromolecules*, 1156–1161, 1978.

5. K. R. Sharma, Change in Entropy of Mixing and Two Glass Transitions for Partially Miscible Blends, 228th ACS National Meeting, Philadelphia, PA, August 2004.

6. K. R. Sharma, Dimensionless Ratio of Entropy of Mixing and Change in Heat Capacity and Prediction of Two Glass Transitions for Partially Miscible Blends, CHEMCON 2004, Mumbai, India, December 2004.

7. K. R. Sharma, Dimensionless Ratio of Heat Capacity Change to that of Entropy of Mixing for Partially Miscible Polymer Blends, 95th AIChE Annual Meeting, San Francisco, November 2003.

8. K. R. Sharma, Quantitation of Partially Miscible Systems in Polymer Blends—Entropy of Mixing Change at Glass Transition, 37th Annual Convention of Chemists, Haridwar, India, November 2000.

9. K. R. Sharma, Entropic Difference Model during Glass Transition for Partially Miscible Blends with Styrenic Systems, 218th ACS National Meeting, New Orleans, LA, August 1999.

10. K. R. Sharma, Mesoscopic Simulation of Glass Transition and Comparison with Entropic Difference Model, Annual Technical Conference for Society of Plastics Engineers, ANTEC 1999, New York, May 1999.

11. K. R. Sharma, Spreadsheets and Simulation in Polymer-Polymer Phase Diagrams, Annual Technical Conference for Society of Plastics Engineers, ANTEC 1999, New York, May 1999.

12. K. R. Sharma, Dissimilar Entropy of Mixing Effects on Glass Transition Temperature of Polymer Blends, 30th ACS Central Regional Meeting, Cleveland, OH, May 1998.

13. K. R. Sharma, Mathematical Modeling of Partially Miscible Copolymers in Blends, 215th ACS National Meeting, Dallas, TX, March 1998.

14. K. R. Sharma, Effect of Chain Sequence Distribution on Contamination Formation in Alphamethylstyrene Acrylonitrile Copolymers during Its Manufacture in Continuous Mass Polymerization, 5th World Congress of Chemical Engineering, San Diego, CA, July, 1996.

15. D. W. van Krevelen and K. te Nijenhius, *Properties of Polymers: Their Correlation with Chemical Structure; Their Numerical Estimation and Prediction from Additive Group Contributions*, 4th edition, Elsevier, Amsterdam, Netherlands, 2009.

16. K. Renganathan, Sequence Distribution Effects on Contamination Formation in Alphamethyl Styrene Acrylonitrile Copolymers, presentation to Vice President J. Tuley, Plastics Division, Monsanto, February 1991.

17. K. R. Sharma, Terpolymer Sequences and Geometric Distribution, Annual Technical Conference of Society of Plastics Engineers, ANTEC 02, San Francisco, CA, May 2002.

18. K. R. Sharma, First Order Markov Model Representation of Chain Sequence Distribution of Alphamethylstyrene Acrylonitrile Prepared by Reversible Free Radical Polymerization, 229th ACS National Meeting, San Diego, CA, March 2005.

19. K. R. Sharma, Geometric Distribution Effects in Quality of Continuous Copolymerization of Alphamethylstyrene Acrylonitrile, 91st AIChE Annual Meeting, Dallas, TX, October/ November 1999.

20. C. Zhikuan, S. Ruona, and F. E. Karasz, Miscibility in blends of copolymers: Sequence distribution effects, *Macromolecules*, Vol. 25, 6113–6118, 1992.

21. H. J. Cantow and O. Schulz, Effect of chemical and configurational sequence distribution on the miscibility of polymer blends - 1. blends of monotactic homopolymers, *Polym. Bull.*, Vol. 15, 449, 1986.

22. A. C. Balasz, I. C. Sanchez, I. R. Epstein, F. E. Karasz, and W. J. Macknight, Effect of sequence distribution on the miscibility of polymer/copolymer blends, *Macromolecules*, Vol. 18, 2188, 1985.

# 7 Polymer Nanocomposites

## LEARNING OBJECTIVES

- Intercalated and exfoliated nanocomposites
- 20 precursor morphologies in block copolymer systems
- Commercial products
- Carbon nanotubes (CNTs), fullerenes
- Morphologies of nanotubes

## 7.1 INTRODUCTION

The term *polymer nanocomposites* pertains to the synthesis, characterization, and applications of polymeric materials with at least one or more dimensions less than 100 nm. Materials made with dimensions on the nanoscale offer unique and different properties compared to those in the macro scale developed with conventional technology. The field holds a lot of promise not only to the practitioner of the technology but also for inventors and scientists. The benefits to *miniaturization,* as realized in the microprocessor and microelectronics industry as a result or as part of the computer revolution, are being tapped into by the emergence of nanotechnology.

The maximum *resolution* or the minimum detectable size achievable is half the wavelength of light, that is, 200 nm. This is the Rayleigh criterion. The Prescott criterion, introduced by Intel in the Pentium IV chip, has dimensions of about 90 nm, 40 nm lower from the previous minimum size of 130 nm. As integrated circuits (ICs) with dimensions of the images smaller than the wavelength of light, the Rayleigh criterion is reinterpreted in terms of x-ray and helium ion characterization techniques such as small angle x-ray scattering (SAXS), wide angle x-ray scattering (WAXS), helium ion microscope (HeIM). HeIM boasts the world record of minimum achievable size of surface resolution to 0.24 Angstroms.

## 7.2 COMMERCIAL PRODUCTS

The first *clay nanocomposite* was commercialized by Toyota. A nylon-6 matrix filled with 5% clay was used for heat-resistant timing-belt covers. The material possesses 40% higher tensile strength, 60% higher flexural strength, 12.6% higher flexural modulus, and a higher heat distortion temperature, ranging from 65–152°C.

General Motors commercialized the first exterior trim application of *nanocomposite* in their 2002 mid-sized vans. The part was stiffer, lighter, and less brittle in cold temperatures and more recyclable. Molded of a thermoplastic olefin clay nanocomposite, it offers weight savings and improved surface quality. About 540,000 lbs/yr of nanocomposite material is used by GM. The GM research and development team worked with Basell, the supplier of polyolefins, and southern clay products, the supplier of nanoclay. The technology for dispersing the clay came from Basell and the southern clay products supplied the Impala project with cloisited nanoclay, that is, organically modified montmorillonite.

Organically modified nanometer scale layered magnesium aluminum silicate platelets measuring 1 nm thick and 70–150 nm across were used as additives. Foster Corp. manufactures selectively enhanced polymer (SEP), a line of nanocomposite nylon. Demand for nanocomposite nylon has grown in medical device applications, especially in the catheter tube area. Foster president, Larry Aquorlo, introduced the product in 2001. Made from nylon-12 and nanoclay particles, one of the dimensions is less than 1 nm and the other is 1500 times longer. It offers 65% increase in flexural modulus, 135% elongation of the material with increased stiffness, and rigidity without brittleness.

Altairnano was the first company to replace traditional graphite materials used in conventional lithium-ion batteries with a proprietary, nanostructured lithium titanate. The company has overcome the limitations with conventional lithium-ion technologies such as cycle and calendar life, safety, recharge time, power delivery and the ability to operate in extreme temperatures. The breakthrough in technology was announced by Altairnano in 2005.

## 7.3   THERMODYNAMIC STABILITY

The solvation thermodynamics define the stability of a system. The system can be stable, metastable, or unstable. UCST and LCST define the critical consolute temperatures of two-phase systems. When a supersaturated system is disturbed, the particles begin to nucleate, grow, and form into stable structures. The process can be arrested sufficiently early to form nanospheres. But the free energy of formation of the structure and the surface energy association with the solid embryo particle would be equal to each other at equilibrium. For stability the free energy has to be negative or equal to zero. From this criterion, a minimum stable size of the solid particle formed can be calculated. This depends on the solid–liquid surface tension values and other parameters of the system. In one system, for example, the engineering thermoplastic, acrylonitrile–butadiene–styrene (ABS), the smallest stable butadiene particle size, can be calculated from Gibbs free energy as 200 nm. When attempted to be made any smaller, the rubber phase particles agglomerated with each other to a size larger than 200 nm. It was reported by a number of investigators that making the rubber particles smaller and smaller was difficult. Maybe if they were made into tubes, they could be made into nanotubes. The free energy and surface energy analysis will still hold well. The shape is different in the derivation.

The four thermodynamically stable forms of carbon are diamond, graphite, $C_{60}$, Buckminster fullerene, and carbon nanotubes. It would be a challenge to extend the experience gained in CNT to nanotubes made of other material than carbon. It would also be interesting to form stable spherical structures in the nanoscale dimensions without agglomeration. At what scale would the quantum analysis for atoms be applicable when compared with the Newtonian mechanics used to describe macro systems? Nanostructures of all the different shapes, and Bravais lattices in several materials, need to be established. Nanostructures that are known today and successfully used in the industry are in the form of tubular morphology, gate patterns of oxidation, and packing transistors that leave some features at the nanoscale dimensions. Why is tubular morphology favored over spherical morphology during the formation of carbon nanotubes? The layered materials of the nanoscale dimension made using atomic layer deposition techniques break no known laws of thermodynamics. But one issue is the layer rearrangement due to Marangoni instability.

## 7.4   VISION AND REALITIES

Richard Feynman's "after-dinner talk" in 1959 provided the vision for the storage of an encyclopedia in the area of a pinhead, a method of writing small using ions, the focus of electrons on a small photoelectric screen, etc. He called for the design and development of better electron microscopes with the capability to view atoms, better $f$-value lenses, and taking computers that filled several rooms and reducing their function to the submicroscopic. He alluded to a process of evaporation and formation of layered materials much like the atomic layer deposition (ALD) methods used currently. He called for drilling holes, cutting things, soldering things, stamping out things, and molding different shapes at an infinitesimal level. A pantograph can make a smaller pantograph that can make a smaller pantograph and so on and so forth. He mused whether atoms can be rearranged at will. He offered $1000 to the first person who could take the information on the page of a book and put on an area 1/25,000 smaller in linear scale in such a fashion that it could be read using an electron microscope.

Drexler and Smalley debated whether "molecular assemblers," devices capable of positioning atoms and molecules for precisely defined reactions in any environment, are possible or not. Smalley objected to the immortality implied by Drexler's vision of a molecular assembler in the form of the "fat fingers problem." *Chemical and Engineering News* carried the exchange between Drexler and Smalley; Smalley writing three letters that Drexler countered, and Smalley concluding the exchange with the "sticky fingers problem." Drexler alluded to the long-term goal of molecular manufacturing. Smalley sought agreement that precision picking and placing of individual atoms through the use of "Smalley-fingers" are impossible. Computer-controlled fingers would be too fat and too sticky for providing the control needed. Enzyme and ribosome were necessary to complete the reactions. He quarreled with the vision of a self-replicating nanobot. Water is needed to provide the nutrients. Drexler applauded Smalley's goal of debunking nonsense in nanotechnology. Direct positional control of reactants are revolutionary and achievable.

## 7.5   FULLERENES

Fullerenes ($C_{60}$) are the third allotropic form of carbon. The Nobel Prize for their discovery was awarded in 1996 to Curl, Smalley, and Kroto. Soccer-ball-structured $C_{60}$, with a surface filled with hexagons and pentagons, satisfies Euler's law. Euler's law states that no sheet of hexagons will close. Pentagons have to be introduced for hexagon sheets to close. The stability of $C_{60}$ requires Euler's 12 pentagon closure principle and the chemical stability conferred by pentagon nonadjacency. $C_{240}$, $C_{540}$, $C_{960}$, and $C_{1500}$ can be built with icosahedra symmetry.

Howard patented the first-generation combustion synthesis method for fullerene production, an advance over the carbon arc method. In the second-generation combustion, the synthesis method optimizes the conditions for fullerene formation. A continuous high flow of hydrocarbon is burned at low pressure in a three-dimensional chamber. Manufacturing plants have been constructed in Japan and the United States with production capacities of fullerenes at 40 metric tons/year. Purity levels are greater than 98%. The reaction chamber consists of a primary zone where the initial phase of combustion synthesis is conducted, and a secondary zone where combustion products with higher exit age distribution do not mix with those with lower exit age distribution. Flame control and flame stability are critical in achieving higher throughputs of fullerenes. Typical operating parameters include residence time in the primary zone of 2–500 ms, residence time in the secondary zone from 5 ms–10 s, a total equivalence ratio in the range of 1.8–4.0, pressure in the range of 10–400 torr, and temperature in the range of 1500–2500 K.

Fullerene crystals can be produced at high yield. By counter diffusion from fullerene solution to pure isopropyl alcohol solvent, fullerene single crystal fibers with needle shape were formed. Needle diameters were found to be 2–100 μm and their lengths were 0.15 = 5 mm. Buckyball-based sintered carbon materials can be transformed into polycrystalline diamonds at less severe conditions using powder metallurgy methods.

A chemical route has been developed by Scott to synthesize $C_{60}$. Corannulene is synthesized from naphthalene structure. As the rings fuse and the sheet forms, it is then rolled into a soccer ball structure. The challenge is how to stitch up the seams between the arms to make the ball. Oligoarenes are transformed into highly strained, curved π surfaces. The molecule needs to bend to effect ring closure on a "soccer ball" structure at 1000°C. A 60-carbon-ring system can be built by acid catalyzed aldol trimerization of ketone. Oligoarene "zips-up" to the soccer ball structure affected by cyclodehydrogenations.

In order to generate higher yield, supercritical ethanol is used to react with naphthalene with ferric chloride as a catalyst for 6 hours. The reaction products are subjected to extraction with toluene. The reactor temperature range is 31–500°C, with a pressure range of 3.8–60 MPa. Smalley patented a process to make fullerenes by tapping into solar energy. The carbon is vaporized by applying focus of solar arrays and conducting the carbon vapor to a dark zone for fullerene growth and annealing. Fullerene content of soot deposits collected on the inside of the Pyrex tube is analyzed by extraction with toluene.

In the electric arc process for fullerene production, carbon material is heated using an electric arc between two electrodes to form carbon vapor. Fullerene molecules are condensed later and collected as soot. Fullerenes are later purified by extraction of soot using a suitable solvent, followed by evaporation of the solvent to yield the solid fullerene molecules.

Applications of fullerenes include high temperature superconductors. Polymerized fullerene molecules have been found to have interesting properties. They possess a superconducting transition temperature of 100–150 K; improved catalytic properties as "bucky-onions," which can be used to prepare styrene from ethylbenzene with increased selectivity; improved mechanical strength as nanocomposites; improved electromechanical systems; and with an improved property for use as synthetic diamonds.

## 7.6 CARBON NANOTUBES (CNTs)

CNTs are made up of graphene sheets of atoms; each graphene sheet is rolled about an axis of rotation that is coplanar with the graphene sheet of atoms, called "its needle axis." They are 0.7–100 nm diameter and a few microns in length. Carbon hexagons are arranged in a concentric manner with both ends of the tube capped by pentagons containing a Buckminster fullerene-type structure. They possess excellent electrical, thermal, and toughness properties. Young's modulus of CNTs has been estimated at 1 TPa with a yield strength of 120 GPa. S. Ijima verified fullerenes in 1991, and observed multiwalled CNT formed from carbon arc discharge.

Five methods of synthesis of CNTs are discussed: (1) arc discharge, (2) laser ablation, (3) chemical vapor deposition (CVD), (4) high pressure carbon monoxide (HIPCO) process, and (5) surface-mediated growth of vertically aligned. The arc discharge process was developed by NEC in 1992. Two graphite rods are connected to a power supply spaced a few mm apart. At 100 amps carbon vaporizes and forms a hot plasma. Typical yields are 30–90%. The single-walled nanotube (SWNT), and multi-walled nanotube (MWNT), are short tubes with diameters 0.6–1.4 nm in diameter. They can be synthesized in open air, but the product needs purification. The CVD chemical vapor deposition, process was invented by Nagano (Japan). The substrate is placed in an oven, heated to 600°C, and a carbon-bearing gas such as methane is slowly added. As the gas decomposes, it frees up the carbon atoms, which recombine as a nanotube. Yield range is 20–100%. Long tubes with diameters ranging from 0.6–4 nm are formed. The CVD process can be easily scaled up to industrial production. The SWNT diameter is controllable. The tubes are usually multiwalled and riddled with defects. The laser vaporization process was developed by Smalley in 1996. Graphite is blasted with intense laser pulses to generate carbon gas. Prodigious amount of SWNTs are formed, and a yield of up to 70% is found. Long bundles of tubes, 5–20 μm with diameters in the range of 1–2 nm are formed. The product formation is primarily SWNTs. Good diameter control is possible, and few defects are found in the product. The reaction product is pure, but the process is expensive.

The HIPCO high pressure carbon monoxide process was also developed by Smalley in 1998. A gaseous catalyst precursor is rapidly mixed with carbon monoxide (CO) in a chamber at high pressure and temperature. The catalyst precursor decomposes and nanoscale metal particles form the decomposition product. CO reacts on the catalyst surface and form solid carbon and gaseous carbon dioxide ($CO_2$). The carbon atoms roll up into CNTs. The entire product is SWNT, and the process is highly selective. Samsung patented a method for vertically aligning CNTs on a substrate. A CNT support layer is stacked on the substrate filled with pores. A self-assembled monolayer (SAM) is arranged on the surface of the substrate. On end of each of the CNTs are attached portions of the SAM exposed through the pores formed between the colloid particles present in the support layer. CNTS can be vertically aligned on the substrate, having the SAM on it with the help of pores formed between the colloid particles.

CNTs possess interesting physical properties. Thermal conductivity of CNTs is in excess of 2000 w/m/K. They have unique electronic properties. Applications include electromagnetic shielding, electron field emission displays for computers and other high-tech devices, photovoltaics, super capacitors, batteries, fuel cells, computer memory, carbon electrodes, carbon foams, actuators, material for hydrogen storage, and adsorbents.

CNTs can be produced with different morphologies. Processes are developed to prepare CNTs with desired morphology. Phase-separated copolymers and stabilized blends of polymers can be pyrolyzed along with sacrificial materials to form the desired morphology. The sacrificial material is changed to control the morphology of the product. Self-assembly of block copolymers can lead to 20 different complex-phase separated morphologies. As is the precursor, so is the product. Therefore, even a greater variety of CNT morphologies can be synthesized.

## 7.7  MORPHOLOGY OF CNTs

Carbon nanostructures can be produced with different morphologies. Some examples of different morphologies of CNTs are

1. Single-walled nanotube (SWNT)
2. Double-walled nanotube (DWNT)
3. Multiwalled nanotube (MWNT)
4. Nanoribbon
5. Nanosheet
6. Nanopeapods
7. Linear and branched CNTs
8. Conically overlapping bamboo-like tubules
9. Branched Y-shaped tubules
10. Nanorope
11. Nanowire
12. Nanofiber

Inorganic nanopeapods were grown at the Max Planck Institute of Microstructure Physics in Halle, Germany. Facile control of both the size and separation of platinum nanoparticles within a $CoAl_2O_4$ nanoshell was possible. Nanowires with alternating layers of cobalt and platinum are electrodeposited within a nanoporous anodic aluminum oxide membrane. Annealing the membrane at 700°C allows for the reaction between cobalt and alumina, forming continuous $CoAl_2O_4$ pods. The heating also causes the platinum to agglomerate into spherical "peas" within these shells. The lengths of the cobalt and platinum segments determine the diameter and distance between each of the platinum peas. Long cobalt segments lead to large spaces between the particles, whereas short cobalt segments form more tightly packed peas.

Strong van der Waals attraction forces allow for spontaneous roping of nano-structures leading to the formation of extended carbon structures. A single-walled nanotube has only one atomic layer of carbon atoms. A double-walled nanotube has two atomic layers of carbon atoms, and a multiwalled nanotube comprises up to a thousand cylindrical graphene layers concentrically folded about the needle axis. MWNTs have excellent strength, small diameter (less than 200 nm), and near metallic electrical conductivity and other interesting properties. They can be used as additives to enhance structural properties of carbon–carbon compos-ites, carbon–epoxy composites, metal–carbon composites, and carbon–concrete composites. The properties of these materials depend on the topology, morphol-ogy, and quality of the nanotubes. In a flat panel display application, the align-ment of the tubes is important. There is sufficient interest in developing processes that can generate nanotubes with predetermined morphology in a economic and environmentally safe manner. There is an identified need for a nanostructure having a high surface area layer containing uniform pores with a high effective surface area. This would increase the number of potential chemical reactions or catalyst sites on the nanostructure. These sites may also be functionalized to enhance chemical activity. Nanofibers are needed with large surface areas filled with additional pores, as well as more functionality with an increased number of reaction/catalyst sites.

Matyjaszewski et al. [2] patented a novel and flexible method for the preparation of CNTs with predetermined morphology. Phase-separated copolymers/stabilized blends of polymers can be pyrolyzed to form the carbon tubular morphology. These materials are referred to as *precursor materials*. One of the comonomers that form the copolymers can be acrylonitrile, for example. Another material added along with the precursor material is called the *sacrificial material*. The sacrificial material is used to control the morphology, self-assembly, and distribution of the precursor phase. The primary source of carbon in the product is the precursor. The polymer blocks in the copolymers are immiscible at the micro scale. Free energy and entropic considerations can be used to derive the conditions for phase separation. Lower critical solution temperatures and upper critical solution temperatures (LCST and UCST) are also important considerations in the phase separation of polymers. But the polymers are covalently attached, thus preventing separation at the macro scale. *Phase separation* is limited to the nanoscale. The nanoscale dimensions typical of these structures range from 5–100 nm. The precursor phase pyrolyzes to form car-bon nanostructures. The sacrificial phase is removed after pyrolysis.

When the phase-separated copolymers undergo pyrolysis, they form two differ-ent carbon-based structures such as a pure carbon phase and a doped carbon phase. The topology of the product depends on the morphology of the precursor. Due to the phase separation of the copolymer on the nanoscale, the phase-separated copo-lymers self-assemble on the molecular level into the phase-separated morphologies. ABC block copolymer may self-assemble into over 20 different complex phase sep-arated morphologies (Figure 7.1). Typical are spherical, cylindrical, and lamellar morphologies. Phase-separated domain may also include gyroid morphologies with two interpenetrating continuous phases. The morphologies are dependent on many factors, such as volume ratio of segments, chemical composition of the segments,

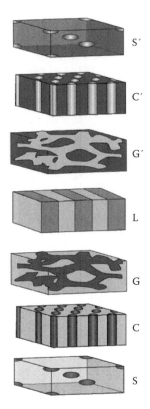

**FIGURE 7.1**   Different precursor morphologies.

their connectivity, Flory–Huggins interaction parameters between the segments, and processing conditions.

The nanoscale morphologies formed by spontaneous or induced self-assembly of segmented copolymers exist below a certain critical temperature. The morphologies are reversibly formed. When a morphology with a cylindrical domain is selected as precursor, the product is made of carbon nanotubes. Oriented nanostructures can be formed. The thickness of the film used to affect the process affects the length of the cylinder. Spinning or extrusion of the polymer at the surface prior to carbonization leads to nanowires used in nanoelectronics after pyrolysis. Nanotubes, nanowires, and nanofibers can be formed in this manner.

Polyacrylonitrile (PAN) has been used in the industry to form carbon fibers. Cross-linking of microscale-phase-separated domains in the polymer blend that forms the precursor can lead to the formation of nanoclusters in the product. $Tr$ clusters have a lot of applications in semiconductors.

The precursor polymer blend can be made out of (1) linear block copolymers, (2) AB block copolymers, (3) ABA block copolymers, (4) ABC block copolymers, (5) multiblock copolymers, (6) symmetrical and asymmetrical star block copolymers, (7) blends of polymers, (8) graft copolymers, (9) multibranched copolymers, (10) hyperbranched or dendritic block copolymers, and (11) novel brush copolymers.

The blocks of each of the polymers may be comprised of (1) homopolymer blocks, (2) statistical polymer blocks, (3) random copolymer blocks, (4) tapered copolymer blocks, (5) gradient copolymer blocks, and (6) other forms of block copolymers. Any copolymer topology that allows immiscible segments present in the system to phase separate inherently or upon annealing, either alone or in the presence of a solvent, is a suitable precursor material.

## 7.8 NANOSTRUCTURING OPERATIONS

Nanostructures can be of several kinds: nanowires, nanorods, nanotetrapods, nano-crystals, quantum dots, nanosheets, nanocylinders, nanocubes, nanograins, nano-filaments, nanolamella, nanopores, nanotrenches, nanotunnels, nanovoids, and nanoparticles. Nanostructuring methods encompass a wide range of technologies. Nanostructures can be generated either by building up from atoms using methods classified as bottom-up strategy or by diminishing of size from nanoparticles using methods grouped under top-down strategies. Bottom-up strategies use self-assembly concepts, are cheap, more scalable, more flexible, and lead to molecular-level engi-neering. Top-down strategies are expensive, less scalable, and inflexible.

Vacuum synthesis method of nanostructuring used sputtering of molecular ions under ultrahigh vacuum. The sputtering process is followed by annealing process. Crystalline silicon is made to form into isolated quantum wires. The gas evapora-tion technique is a dry process to make ultra fine metallic magnetic powders. Metal is evaporated onto a thin film under vacuum conditions. Metal atoms are allowed to impinge on the surface of the dispersing medium. Condensation of metal atoms can also be accomplished using a cooling nozzle. Nanoparticles in the size range of 50 nm–4 μm can be prepared using this technique.

Triangular nanoprisms can be generated by exposure to light at different wave-lengths of 400–700 nm. Ostwald ripening concepts are used. Edge lengths ranging of 31 to 134 nm can be prepared. Nanorods may be produced using the con-densed phase synthesis method. The starting material is heated until the material vaporizes. Later the vapor is condensed. Aggregates of nanoparticles are formed. Particles are delivered by boundary layer delivery and thermophoresis-assisted deposition to form the epitaxial deposit. CNTs can also be made using this method.

Subtractive and additive fabrication methods can be used for nanostructuring operations. Lithography, etching, and galvanic fabrication processes are subtractive. A series of chemical reactions effects removal of the layer in the apertures of the mask and transfer of material into a gas phase. Lift-off processes are employed in the fabrication of circuit integrated. Nanotips and nanorods can be efficiently formed using conventional CMOS processes. Patterning iridium oxide nanostructures con-sists of (1) forming a substrate with first and second regions adjacent to each other, (2) growing $IrO_x$ structures from a continuous oxide film overlying the first region, (3) simultaneously growing $IrO_x$ nanostructures from a co-continuous oxide film overlying the second region, (4) selectively etching an area of the second region, (5) lifting off overlying $IrO_x$ structures, and (6) forming a substrate with nanostructures overlying the first region. The second material can be $SiO_x$.

Dip-pen lithography, SAM, hot embossing, nanoimprint lithography, electron beam lithography, dry etching, and reactive ion etching are techniques that can be used to prepare nanostructures with 50–70 nm dimensions. Nanomechanical techniques include processes that involve local transfer of material from a tool onto a substrate when either the tool or the substrate is prestructured.

Quantum dots (QDs) are structures where quantum confinement effects are significant. Reproducibility of organized arrays of quantum dots is an identified problem. Techniques such as sol gel, solid state precipitation, molecular beam epitaxy, chemical vapor deposition, and lithography were developed to enhance uniformity, control morphology, and determine spatial distribution of quantum dots in thick and thin films. QDs can be prepared at low temperatures by precipitation from solution by sol–gel methods. Uniformly sized QDs are affected by control of nucleation and growth of particles within a lithographically or electrochemically designed template. Face centered cubic (FCC) packing of silica balls can be used as a template for melt/infusion of the semiconductor InSb.

Nanostructures of metal oxides can be prepared by sol–gel processing methods. Chemical reactions are conducted in solution to produce nanosized particles called *sols*. The sols are connected into a three-dimensional network called a gel. Controlled evaporation of the liquid phase leads to dense porous solids called *xerogel*. Surface instabilities and pattern formation in polymer thin films can lead to formation of nanostructures. The control of morphology of phase separated polymer blends can result in nanostructures. Kinetic and thermodynamic effects during phase separation can be used in preparation of nanostructures. Nanostructures can be synthesized by quenching of a partially miscible polymer blend below the critical temperature of demixing. Spin coating can be used to prepare polymer film. Pattern formation from polymer solvent systems is stage wise. A stage of layered morphology, followed by destabilization of layers by capillary instability and surface instability, leads to nanostructure formation.

Cryogenic milling is a top-down approach to prepare nanoscale titanium of 100–300 nm size. Several mechanical deformations of large grains into ultra fine powder and degassing lead to nanopowder with improved characteristics.

Atomic lithography is the method of choice to generate structures with less than 50 nm dimensions. A laser beam forms a high intensity optical spot allowing formation of the 2-dimensional pattern of atoms on the surface of the substrate. Nanostructured thin film nanocomposites can be manufactured using the electrodeposition method. Electrodeposition can be used to form nanostructured films within the pores of mesoporous silica. Silica is then removed from the nanocomposite by dissolution in a suitable etching solvent such as hydrogen fluoride (HF). Plasma compaction techniques can be used to form nanoparticles of semiconductor compounds resulting in improvements in the figure of merit. Nanoparticles can be expected as reaction products after the reactant mixture is subjected for a sufficient time, under a prescribed temperature and pressure. A plasma compaction apparatus may comprise two high strength pistons capable of compressive pressure in the range of 100–1000 MPa to a sample of nanoparticles that is disposed within a high strength cylinder. The desired level of compaction is achieved by varying the applied pressure, applied current, and time duration of the process.

In direct write lithography, a tip can be used to pattern a surface and prepare polymeric nanostructures. Polymerization is initiated by the tip and a pattern is formed. Polymer brush nanostructures can be synthesized using ring opening metathesis polymerization (ROMP). Edge distances of less than 100 nm are possible, and control over feature size, shape, and interfeature distance is achievable.

Nanofluids are composed of nanoparticles dispersed in a suitable solvent. The surface-to-volume ratio is increased by 1000 times. Nanofluids are expected to have enhanced thermal conductivity comparable to that of copper. Materials with predefined morphology can be made using a self-assembly with the block copolymer principle. Nanospheres, nanolamellae, and nanopores both cylindrical and spherical are possible using this approach. Thickness of the interfacial layer can be 2–30 nm. Using pulse-layered deposition (PLD), laser pulses evaporate the starting material and then deposit it onto a substrate to produce thin films with profound nanotechnological significance. Typical temperature of the evaporation surface is 5000 K, average velocity of atoms in vapor flow is 2000 m/s, and the expansion front moves at 6000 m/s. Laser intensity is optimized for more efficient evaporation of the target.

## 7.9 POLYMER THIN FILMS

In applications in the semiconductor industry, polymer structures are required on length scales down to individual molecules. A bottom-up approach is better than a top-down approach in order to achieve this. A lateral resolution less than 100 nm can be created by surface instabilities and pattern formation in polymer films. Steiner [6] discussed demixing of polymer blends and pattern formation by capillary instabilities for nanostructure formation.

As a matter of generality, two different polymers are immiscible with each other. A compatible blend is one when two polymers are mixed and the property of the blend shows some improvement with commercial applications. Different from this is the requirement of molecular mixing. When two or more polymers mix with each other at a molecular level, they are considered to be miscible blends. Control of the morphology of phase separated polymer blends can result in nanostructures. Polymer blends can also be partially miscible. Miscible polymer blends at a single glass transition temperature, much like a homopolymer made out of a single monomer. Immiscible polymer blends will exhibit the same two-glass transition temperatures as their homopolymers that went into the make of the blend. The partially miscible blends can be expected to exhibit glass transition temperatures distinct from the homopolymer glass transition temperatures.

When polymers undergo phase separation in thin films, the kinetic and thermodynamic effects are expected to be pronounced. The phase separation process can be controlled to effect desired morphologies. Under suitable conditions a film deposition process can lead to pattern replication. Demixing of polymer blends can lead to structure formation. The phase separation process can be characterized by the binodal and spinodal curves. UCST is the upper critical solution temperature, which is the temperature above which the blend constituents are completely miscible in each other in all proportions. LCST behavior is not found as often in systems other than among polymers. LCST is the lower critical solution temperature. This is the temperature above which the polymers that were miscible below this temperature now exhibit immiscibility. The binodal and spinodal curves can be calculated by the stability criteria of Gibbs free energy:

$$\Delta G = \Delta H - T\Delta S,$$

$$\Delta G \leq 0,$$

$$\frac{\partial^2 G}{\partial \phi^2} > 0,$$

where $G$ is the Gibbs free energy, $H$ is the enthalpy, $S$ is the entropy, and $\phi$ is the phase composition. When the free energy change is negative, the polymer blend can be expected to be miscible. The immiscible and miscible regions are separated by the binodal curve. The spinodal curve can be found within the binodal where the curvature of the free energy curve becomes negative. When the material falls on the spinodal region of the phase diagram, spinodal decomposition can be expected to occur.

Phase *morphology* with a *single-characteristic-length scale* can be synthesized by quenching a partially miscible polymer blend below the critical temperature of demixing. A well-defined spinodal pattern is formed and becomes larger with passage of time. Polymer films can be made by the spin-coating solvent-casting method. The polymers and solvent form a homogeneous mixture at first. Solvent evaporation during spin coating causes an increase in polymer phase volume, resulting in traversal to the spinodal region of the phase diagram. This can be expected to lead to polymer–polymer demixing. Characteristic phase morphology can be found in the polymer film as shown in Figures 7.2–7.5. The polystyrene and poly(vinylpyridine) are mixed in the tetrahydrafuran (THF) solvent.

Different phase separated morphologies can be found in different polymer solvent systems. The pattern formation consists of several stages. In the initial stage, phase separation results in a layered morphology of the two solvent swollen phases. As more solvent evaporates, this double layer is destabilized in two ways: (1) capillary instability of the interface, and (2) surface instability. Each of the mechanisms results in different morphological length scales. Core shell spherical domains in phase-separated ternary systems have also been found. The shell thickness can be a few nanometers.

## 7.10  NANOSTRUCTURING FROM SELF-ASSEMBLY OF BLOCK COPOLYMERS

Materials with nanoscopic features have several interesting applications in the areas of low dielectric materials, catalysis, membrane separation, molecular engineering, photonics, biosubstrates, and various other fields. Synthesis methods include use of polymers as templates, phase separation processes, and multistep templating chemical reactions. Deficiencies in these methods include complexity in execution; economically prohibitive for scale-up; mechanically weak, nonuniform in structure materials with little use in application.

IBM patented a process for making materials with predefined morphology. Table 7.1 provides a list of experiments and theoretical predictions for forming materials with predetermined morphology. Self-assembly of block copolymer principles are used in this approach. Let the block copolymers be characterized by volume fractions, $\varphi_1$ and $\varphi_2$ respectively. A cross-linkable polymer is also added to the mixture that is miscible with the second block copolymer. The volume fraction of the cross-linkable polymer is $\varphi_3$. The first and second block species have volume fractions of $\varphi_{1A}$ and $\varphi_{2A}$, respectively, in the two components of the mixture. The mixture is applied to a substrate. The assembly consists of a substrate, a mixture coating/structural layer having nanostructures, and a interfacial layer without nanostructures. The thickness of the layer is 0.5–50 nm.

One example of a self-assembling block copolymer is polystyrene-$b$-polyethylene oxide (PS-$b$-PEO). Typically, the volume fraction of the PS block and the PEO block present in the copolymer is in a range of 0.9–1. The cross-linkable polymer used is polymethylsilsesquioxane (PMSSQ). The PMSSQ is miscible with the PEO block of the copolymer. The PS-$b$-PEO and PMSSQ are dissolved in toluene and 1-propoxy-2-propanol, respectively. A thin film of the mixture is deposited onto the substrate and is spin cast. A typical spin speed was 3000 rpm and falls in the range of 50–5000 rpm. The mixture is spin cast and annealed at a temperature of about 100°C for about 10 hrs. After annealing, the mixture is further heated. The PMSSQ crosslinks at a temperature in a range from about 150–200°C. The result of the spin casting process is the material having a predefined morphology. The spin casting process may be replaced with *photochemical irradiation, thermolysis, spraying, dip coating,* and *doctor-blading* processes. The predetermined morphologies and block volume fractions of copolymers and the cross-linkable polymer are given in Table 7.1. Morphologies that can be tailored are nanolamellae, cylindrical nanopores, spherical nanopores, and nanospheres.

A cross-sectional transmission electron microscopy (TEM) image of a material with predetermined morphology of spherical pores was examined. The structure consists of an interfacial layer, structural layer, and a substrate. The substrate is in direct mechanical contact with the interfacial layer. The structural layer is composed of spherical nanopores nanostructure, and essentially consists of the cross-linkable polymer. The interfacial layer lacks the spherical nanopores. The thickness of the interfacial layer is 2–30 nm. The structural layer thickness is of the range 50–300 nm.

## 7.11  INTERCALATED AND EXFOLIATED NANOCOMPOSITES

In the area of polymer technology prior to the emergence of nanotechnology, nanoscale dimensions were recognized in some areas:

1. Phase-separated polymer blends often feature nanoscale phase dimensions;
2. Block copolymer domain morphology is often at the nanoscopic level;
3. Asymmetric membranes often have nanoscale void structure;
4. Miniemulsion particles are below 100 nm;

**TABLE 7.1**

**List of Experiments to Form Materials with Predetermined Morphology**

| Entry | $\varphi_1$ | $\varphi_2$ | $\varphi_{1A}$ | $\varphi_{2A}$ | $\varphi_3$ | $\varphi_{2A}+\varphi_3$ | Morphology |
|---|---|---|---|---|---|---|---|
| 1 | 0.9 | 0.1 | 0.81 | 0.09 | 0.1 | 0.19 | Nanospheres |
| 2 | 0.9 | 0.1 | 0.63 | 0.07 | 0.3 | 0.37 | Nanolamellae |
| 3 | 0.9 | 0.1 | 0.54 | 0.06 | 0.4 | 0.46 | Nanolamellae |
| 4 | 0.9 | 0.1 | 0.27 | 0.03 | 0.7 | 0.73 | Cylindrical nanopores |
| 5 | 0.9 | 0.1 | 0.18 | 0.02 | 0.8 | 0.82 | Spherical nanopores |
| 6 | 0.9 | 0.1 | 0.09 | 0.01 | 0.9 | 0.91 | Spherical nanopores |
| 7 | 0.7 | 0.3 | 0.63 | 0.27 | 0.1 | 0.37 | Nanolamellae |
| 8 | 0.7 | 0.3 | 0.56 | 0.24 | 0.2 | 0.44 | Nanolamellae |
| 9 | 0.7 | 0.3 | 0.49 | 0.21 | 0.3 | 0.51 | Nanolamellae |
| 10 | 0.7 | 0.3 | 0.42 | 0.18 | 0.4 | 0.58 | Nanolamellae |
| 11 | 0.7 | 0.3 | 0.35 | 0.15 | 0.5 | 0.65 | Nanolamellae |
| 12 | 0.7 | 0.3 | 0.28 | 0.12 | 0.6 | 0.72 | Cylindrical nanopores |
| 13 | 0.7 | 0.3 | 0.21 | 0.09 | 0.7 | 0.79 | Cylindrical nanopores |
| 14 | 0.7 | 0.3 | 0.14 | 0.06 | 0.8 | 0.86 | Spherical nanopores |
| 15 | 0.7 | 0.3 | 0.07 | 0.03 | 0.9 | 0.93 | Spherical nanopores |
| 16 | 0.5 | 0.5 | 0.45 | 0.45 | 0.1 | 0.55 | Nanolamellae |
| 17 | 0.5 | 0.5 | 0.4 | 0.40 | 0.2 | 0.6 | Nanolamellae |
| 18 | 0.5 | 0.5 | 0.35 | 0.35 | 0.3 | 0.65 | Nanolamellae |
| 19 | 0.5 | 0.5 | 0.3 | 0.3 | 0.4 | 0.70 | Cylindrical nanopores |
| 20 | 0.5 | 0.5 | 0.25 | 0.25 | 0.5 | 0.75 | Cylindrical nanopores |
| 21 | 0.5 | 0.5 | 0.20 | 0.2 | 0.6 | 0.8 | Cylindrical nanopores |
| 22 | 0.5 | 0.5 | 0.15 | 0.15 | 0.7 | 0.85 | Spherical nanopores |
| 23 | 0.5 | 0.5 | 0.10 | 0.1 | 0.8 | 0.9 | Spherical nanopores |
| 24 | 0.5 | 0.5 | 0.05 | 0.05 | 0.9 | 0.95 | Spherical nanopores |
| 25 | 0.3 | 0.7 | 0.27 | 0.63 | 0.1 | 0.73 | Cylindrical nanopores |
| 26 | 0.3 | 0.7 | 0.24 | 0.56 | 0.2 | 0.76 | Cylindrical nanopores |

**TABLE 7.1 (CONTINUED)**
**List of Experiments to Form Materials with Predetermined**
**Morphology**

| Entry | $\varphi_1$ | $\varphi_2$ | $\varphi_{1A}$ | $\varphi_{2A}$ | $\varphi_3$ | $\varphi_{2A}+\varphi_3$ | Morphology |
|---|---|---|---|---|---|---|---|
| 27 | 0.3 | 0.7 | 0.21 | 0.49 | 0.3 | 0.79 | Cylindrical nanopores |
| 28 | 0.3 | 0.7 | 0.18 | 0.42 | 0.4 | 0.82 | Spherical nanopores |
| 29 | 0.3 | 0.7 | 0.15 | 0.35 | 0.5 | 0.85 | Spherical nanopores |
| 30 | 0.3 | 0.7 | 0.12 | 0.28 | 0.6 | 0.88 | Spherical nanopores |
| 31 | 0.3 | 0.7 | 0.09 | 0.21 | 0.7 | 0.91 | Spherical nanopores |
| 32 | 0.3 | 0.7 | 0.06 | 0.14 | 0.8 | 0.94 | Spherical nanopores |
| 33 | 0.3 | 0.7 | 0.03 | 0.07 | 0.9 | 0.97 | Spherical nanopores |
| 34 | 0.1 | 0.9 | 0.09 | 0.81 | 0.1 | 0.91 | Spherical nanopores |
| 35 | 0.1 | 0.9 | 0.08 | 0.72 | 0.2 | 0.92 | Spherical nanopores |
| 36 | 0.1 | 0.9 | 0.07 | 0.63 | 0.3 | 0.93 | Spherical nanopores |
| 37 | 0.1 | 0.9 | 0.06 | 0.54 | 0.4 | 0.94 | Spherical nanopores |
| 38 | 0.1 | 0.9 | 0.05 | 0.45 | 0.5 | 0.95 | Spherical nanopores |
| 39 | 0.1 | 0.9 | 0.04 | 0.36 | 0.6 | 0.96 | Spherical nanopores |
| 40 | 0.1 | 0.9 | 0.03 | 0.27 | 0.7 | 0.97 | Spherical nanopores |
| 41 | 0.1 | 0.9 | 0.02 | 0.18 | 0.8 | 0.98 | Spherical nanopores |

5. Interfacial phenomena in blends and composites involve nanoscale dimensions;
6. Carbon black reinforcement of elastomers, colloidal silica modification, and asbestos–nanoscale fiber diameter reinforcement are subjects that have been investigated for decades.

Recent developments in polymer nanotechnology include the *exfoliated clay nanocomposites*, CNTs, carbon nanofibers, exfoliated graphite, nanocrystalline metals, and a host of other filler modified composite materials. Polymer matrix-based nanocomposites with exfoliated clay are discussed in this section. Performance enhancement using such materials include increased barrier properties, flammability resistance, electrical/electronic properties, membrane properties, and polymer blend compatabilization. There is a synergistic advantage of nanostructuring that can be called the "nano effect." It was found that exfoliated clays could yield significant mechanical property advantages as a modification of polymeric systems. Crystallization rate and degree of crystallinity can be influenced by crystallization in confined spaces. The dimensions available for spherulitic growth are confined such that primary nuclei are not present for heterogeneous crystallization and thus homogeneous crystallization results. This occurs in the limit of $n$ in the Avarami equation approaching 1, and often leads to reduced crystallization rate, the degree of crystallinity, and the melting point. This was found by Loo et al. [10] in phase-separated block copolymers. Confined crystallization of linear polyethylene in nanoporous alumina showed homogeneous nucleation with pore diameters of 62–110 nm and heterogeneous nucleation for 15–48 nm diameter pores. When a nanoparticle gets attached to the polymer matrix, similarities to confined crystallization exist, as well as nucleation effects and disruption of attainable spherulite size. Nucleation of crystallization can be seen by the onset of temperatures of crystallization, and upon elapse of enough time past the half-life of crystallization, as seen in a number of nanocomposites [11] such as nanoclay-poly-ε-caprolactone, polyamide-66-nanoclay, polylactide-nanoclay, polyamide-6-nanoclay, polyamide 66-multiwalled carbon nanotube, polyester-nanoclay, and polybutylene terepthalate–nanoclay.

The clay called montmorillonite consists of platelets with an inner octahedral layer sandwiched between two silicate tetrahedral layers. The octahedral layer can be viewed as an aluminum oxide sheet with some Al atoms replaced with Mg atoms. The difference in valencies of Al and Mg creates negative charges distributed within the plane of the platelets that are balanced by positive counterions, typically sodium ions are located between the platelets or in the galleries. In its natural state, the clay exists as stacks of many platelets. Hydration of the sodium ions causes the galleries to expand and the clay to swell. The sodium ions can be exchanged with organic cations, with the platelets completely dispersed in water. The organic cations can be those from an ammonium salt leading to the formation of an organoclay. The ammonium cation may have hydrocarbon tails and other groups attached and can be called a surfactant due to its amphiphilic nature. The extent of the negative charge of the clay is characterized by the cation exchange capacity. The x-ray diffraction (XRD) lattice parameter of dry sodium montmorillonite is 0.96 nm, and the platelet is about 0.94 nm thick. The gallery expands when the sodium is replaced with larger organic surfactants, and the lattice parameter may increase by two- to threefold. The lateral dimensions of the platelet are not well defined. The thickness of montmorillonite platelets is a well-defined crystallographic dimension. It depends on the mechanism of the platelets' growth from solution in the geological formation process.

Nanocomposites can be formed from organoclays, in situ solution, and emulsion polymerization. The melt processing is of increased interest. Clays that are exfoliated are completely separated by polymers between them. This desired morphology is achieved to varying degrees in practice.

Three types of morphologies are possible during the preparation of clay nanocomposites: (1) immiscible nanocomposites, (2) intercalated nanocomposites, and

(3) exfoliated nanocomposites. Representative schematics describing the three different morphologies are shown in Figures 7.3–7.5. As shown in Figure 7.3, in the case of immiscible morphology, organoclay platelets exist in particles. The particles comprise tactoids or aggregates of tactoids, more or less as they were in the organoclay powder, that is, with no separation of platelets. The wide-angle x-ray scattering scan (WAXS) of the nanocomposite is expected to appear essentially the same as that obtained for the organoclay powder. There is no shift in the x-ray lattice parameter reading. Such scans are made over a low range of angles, $2\theta$, such that any peaks forming a crystalline polymer matrix are not seen because they occur at higher angles. For completely exfoliated organoclay, no wide-angle x-ray peak is expected for the material because there is no regular spacing of the platelets, and the distances between platelets would be larger than the WAXS could detect.

Often WAXS patterns of polymer nanocomposites reveal a peak similar to the peak of the organoclay but shifted to a lower $2\theta$ or larger lattice parameter spacing. The peak height indicates that the platelets are not exfoliated, and its location indicates that the gallery has expanded; it is often implied that polymer chains have entered or have been intercalated in the gallery. Placing polymer chains in confined spaces involves a significant entropy penalty that must be driven by an energetic attraction between

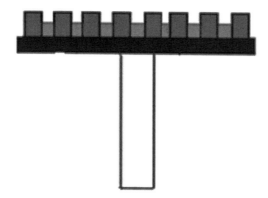

**FIGURE 7.2** Polymer film structure formation from spin-coating solvent.

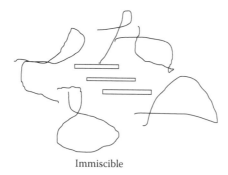

Immiscible

**FIGURE 7.3** Morphology of immiscible nanocomposite.

**FIGURE 7.4**  Morphology of intercalated nanocomposite.

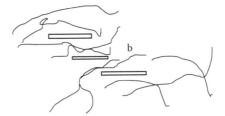

**FIGURE 7.5**  Morphology of exfoliated nanocomposites.

the polymer and organoclay. The gallery expansion in some cases may be caused by intercalation of oligomers. Intercalation (Figure 7.4) may be a prelude to exfoliation. Exfoliation morphology (Figure 7.5) may be achieved by the melt processing.

There is increased interest in relating morphological structure to the performance properties of nanocomposites. WAXS is frequently used. The interpretation of WAXS spectra needs to be fine tuned. The exfoliated and immiscible morphologies have to be picked up by the WAXS spectra. Shifts in the peaks of the organoclay should be interpreted with respect to the morphology of the nanocomposite. Small angle x-ray scattering (SAXS) may be more quantitative and informative. Solid state nuclear magnetic resonance (NMR), and neutron scattering are used by some investigators. Transmission electron microcopy (TEM) is used to visualize the morphology of nanocomposites. The downside of the use of TEM is that it is confined to a small region of the material. More scans may be needed at different magnifications so that it becomes representative of the material considered. No staining is needed, as the organoclay is sufficiently different in characteristics compared with that of the polymer matrix. In nylon-6 nanocomposites, ~ 1 nm thick clay platelets can be seen as dark lines when the microtone cut is perpendicular to the platelets. Image analysis can be used to quantitate the distribution of platelet lengths. In order to draw conclusions from the analyses, hundreds of particles need to be studied. Aspect ratio distributions of the platelets can be obtained. Often, in practice, the exfoliation is never complete.

The addition of fillers to polymers is to obtain an increase in the modulus or stiffness by reinforcement mechanisms. Stiffness of the material can be increased by proper dispersion of aligned clay platelets. This can be seen by the increase in the tensile modulus of the material with the addition of the filler to the product. Increase of modulus by a factor of two is accomplished using montmorillonite, with one third of the material needed when glass fibers are used, and a quarter of the material when talc is used with thermoplastic polyolefins (TPO). Addition of montmorillonite to a nylon-6 matrix can result in increases of 100°C in heat distortion temperature (HDT). Depending on the level of adhesion between the dispersed phase and the continuous phase, the fillers can also increase strength as well as modulus. For glass-fiber composites, chemical bonding is used to increase the interfacial adhesion. Addition of fillers results in a decrease in ductility of the material. Melt rheological properties of polymers can be drastically modified upon addition of fillers in the low shear rate region. Clay particles serve as effective nucleating agents that greatly change the crystalline morphology and crystal type for polymers like nylon-6 or polypropylene (PP).

Nanocomposites, like plastics, have a high coefficient of thermal expansion, unlike metals which have a low coefficient of thermal expansion. Addition of fillers can result in lowering the coefficient of thermal expansion (CTE). This can become an increased concern in automotive applications. The barrier properties of polymers can be significantly altered by inclusion of inorganic platelets with sufficient aspect ratio in order to alter the diffusion path of penetrant molecules. Exfoliated clay modified polyethylene terepthalate (PET) has been investigated for offering improved barrier properties. In situ polymerized PET–exfoliated clay composites were noted to show a factor of two reduction in barrier property at 1% clay. A 10- to 15-fold reduction in oxygen permeability with 1–5 wt % clay has been reported for PET-clay nanocomposites. The permselectivity improvement needed for better membrane separation efficiency is achieved by molecular sieve inclusions in a polymer film. Increased flammability resistance can be achieved by nanoplatelet/nanofiber modification of polymeric matrices. With nanofiller incorporation the maximum heat release rate is reduced. Flame propagation to adjacent areas is minimized.

Addition of nanoparticles results in less coalescence of particles during the melt processing, causing improved compatibilization. For example, exfoliated clay compatibilization, such as in a polycarbonate/polymethylmethacrylate (PC/PMMA) system, polyphenylene oxide/polyamide (PPO/PA), polyamide/ethylene propylene diene elastomer (PA/EPDM) rubber, polystyrene/polymethylmethacrylate (PS/PMMA), and polyvinyl fluoride/polyamide-6 (PVF/PA) blends, is affected by lowering the interfacial tension between the two phases that are phase separated.

There are a number of biomedical applications for polymer nanocomposites. Electrospinning for producing bioresorbable nanofiber scaffolds is an active pursuit of scientists for tissue engineering applications. This is an example of a nanocomposite, as the resultant scaffold allows for cell growth, yielding a unique composite system. Poly(lactic acid)/exfoliated montmorillonite clay/salt solutions were electrospun, followed by salt leaching/gas forming. The resultant scaffold structure contained both nanopores and micropores, offering a combination of cell growth and blood vessel invasion microdimensions, along with nanodimensions for nutrient and metabolic waste transport. Electrically conducting nanofibers based on conjugated polymers for regeneration of nerve growth in a biological living system are another example. Antimicrobial/biocidal activity can be achieved by introducing nanoparticle Ag, salts, and oxide of silver into polymer matrices. Hydroxyapetite-based polymer nanocomposites are used in bone repair, and implantation. Some drug delivery applications use polymer nanocomposites. Nanoparticles can result in more controlled release and reduced swelling, and will improve mechanical integrity.

Some commercial applications for polymer nanocomposites include Toyota's nylon/clay nanocomposite timing belt cover, General Motors' TPO/organoclay nanocomposite for exterior trim of vans, Babolat's epoxy/CNT nanocomposite for tennis rackets, Hybtonite's epoxy/CNT nanocomposite for hockey sticks, Inmat's polyisobutylene/exfoliated clay nanocomposite for tennis balls, styrene butadiene rubber (SBR)/carbon black nanocomposite for tires, Hyperions's polymer/MWNT nanocomposite for electrostatic dissipation, Curad's polymer/silver nanocomposite for bandages; Nanocor's nylon/clay nanocomposite for beverage containers, Pirelli's SBR/dispersant nanocomposite for winter tires, and Ube's polyamide/clay nanocomposite for auto fuel systems.

## 7.12  SUMMARY

The topic of polymer nanocomposites pertains to the synthesis, characterization, and applications of polymer materials with at least one or more dimensions less than 100 nm. According to the Rayleigh criterion, the maximum resolution or the minimum detectable size achievable was believed to be half the wave length of light, that is, 200 nm. The first clay nanocomposite was commercialized by Toyota with a belt cover made of a nylon-6 matrix filled with 5% clay. General Motors commercialized the first exterior trim application of nanocomposite in their 2002 mid-sized vans.

Thermodynamic equilibrium and stability are important considerations in the formation of nanocomposites. LCST and UCST behavior can be seen in polymeric systems. The four thermodynamically stable forms of carbon are diamond, graphite, $C_{60}$ Buckminister fullerenes, and carbon nanotubes. At some point during miniaturization, atom-by-atom quantum analysis would be more applicable compared with macroscale Newtonian mechanical treatment. Layer rearrangement can happen due to Marongoni instability.

Feynman's vision of miniaturization and the Drexler-versus-Smalley debate on feasibility of mechanosynthetic reactions using molecular assemblers were discussed. Fullerenes are the third allotropic form of carbon. Soccer-ball-structured $C_{60}$ with a surface filled with hexagons and pentagons satisfies Euler's law. Howard patented the first generation combustion synthesis method for fullerene production. The projected price of the fullerenes has decreased from \$165,000 per kg to \$200 per kg in the second-generation process. Fullerenes can also be synthesized using chemical methods, a supercritical extraction method, and the electric arc process. Applications of fullerenes include high temperature superconductors, bucky onion catalysts, advanced composites and electromechanical systems, synthetic diamonds.

Five methods of synthesis of CNTs are (1) arc discharge, (2) laser ablation, (3) CVD, (4) the HIPCO process, and (5) surface-mediated growth of vertically aligned tubes. CNTs can have different morphologies such as SWNT, DWNT, MWNT, nanoribbon, nanosheet, nanopeapods, linear and branched CNTs, conically overlapping bamboo-like Y-shaped tubules, nanopores, nanovoids, nanowire, and nanofiber.

Phase-separated copolymers/stabilized blends of polymers can be pyrolyzed to form the carbon tubular morphology. There are 20 different precursor types that can lead to 20 different morphologies of product CNTs. Typical ones are spherical, cylindrical, and lamellar.

Nanostructuring operations include vacuum synthesis, triangular prism by exposure to light, lithography, etching, galvanic fabrication, dip-pen lithography, SAM, hot embossing, nanoimprint lithography, electron beam lithography, quantum dot structuring, the sol–gel method, cryogenic milling, atomic lithography, plasma compaction, and direct-write lithography.

When polymers undergo phase separation in thin films, the kinetic and thermodynamic effects are expected to be pronounced. Phase morphology with a single characteristic length scale can be synthesized by quenching a partially miscible polymer blend below the critical temperature of demixing.

Immiscible nanocomposites, intercalated nanocomposites, and exfoliated nanocomposites offer different morphologies. There are some interesting biomedical applications of nanocomposites. Electrospinning for producing a bioresorbable nanofiber scaffold is an active pursuit of scientists for tissue engineering applications. Hydroxyapetite-based polymer nanocomposites are used in bone repair and implants. Some commercial applications for polymer nanocomposites include Toyota's nylon/clay nanocomposite timing belt cover, General Motors' TPO/organoclay nanocomposite for the exterior trim of vans, Babolat's epoxy/CNT nanocomposite for tennis rackets and nitro Hybtonite's epoxy/CNT nanocomposite for hockey sticks.

## EXERCISES

1. What are the conditions favoring spherulitic growth?
2. What are the characteristic features of exfoliated clay nanocomposites?
3. What is meant by confined crystallization?
4. Show by using a neat schematic the different layers with dimensions that constitute a montmorillonite clay intercalated by polyolefins.
5. What are the differences between exfoliated and intercalated nanocomposites?
6. What are the differences in morphology of immiscible, intercalated and exfoliated nanocomposites?
7. How is the WAXS spectra used to distinguish between exfoliated and intercalated nanocomposites?
8. Addition of fillers to polymers results in _____ in the modulus or stiffness by reinforcement mechanisms.
9. What happens to heat distortion temperature when montmorillonite is added to the nylon-6 matrix?
10. Comment on the coefficient of thermal expansion property of nanocomposites.
11. What happens to the barrier properties of exfoliated clay, modified PET, and polyethylene terepthalate?
12. How is the improvement in permselectivity of membranes effected by changing the parameters of nanocomposites?
13. How are nanopores generated in scaffold structures that are bioresorbable?
14. What nanocomposites were introduced by Toyota as belt covers?

15. What nanocomposites were developed by General Motors for use in exterior trim of vans?
16. What nanocomposites are used as
    a. Tennis rackets
    b. Hockey sticks
    c. Tennis balls
    d. Electrostatic dissipation
17. What nanocomposites were introduced as
    a. Bandages
    b. Beverage containers
    c. Winter tires
    d. Auto fuel systems
18. Find the side of a cube that contains 1 attomole of silver.
19. How many grams of silica are present in a sphere with a diameter of 5 nm?
20. How many molecules of copper are present in a cube of side 50 nm?
21. Explain the Rayleigh criterion for minimum resolution size?
22. What did Feynman mean by miniaturization of computers?
23. Does a circuit of seven atoms satisfy the laws of quantum mechanics?
24. What is a nanostructure?
25. When is something called a nanodevice?
26. What is the ion microscope?
27. Why is a scanning probe microscope needed?
28. What is the diameter of an electron?
29. Can atoms be rearranged at will?
30. How many lathes can be made with 2% of the material from a regular size lathe?
31. What is pantograph?
32. What did Feynman illustrate by taking lubrication as an example?
33. What is miniaturization by evaporation?
34. What is the $f$ value in the lens of a electron microscope?
35. What is the resolving power of the eye?
36. What is the "fat fingers" or "sticky fingers" objection of Smalley?
37. What are nanobots and how is it an impossibility to control chemistry with them?
38. Discuss the self-replicating nanoassembler.
39. What are the implications of a self-replicating nanoassembler?
40. Why did Drexler discuss enzymes and ribosome in the debate?
41. Why is it infeasible for complex chemistry to be performed by molecular assemblers?
42. Discuss the use of CNTs in solar cells.
43. Discuss the use of nanostructures in disease diagnosis.
44. What is the role of nanostructures in drug discovery?
45. What is "pixie dust"?
46. What is a quantum dot?
47. What is the fastest computer possible?
48. What is the smallest gate size achievable in silicon chips?
49. What is the difference between mechanosynthetic reactions and traditional chemistry?
50. Can the use of conveyors and positioners of atoms lead to the void of undesired side reactions?
51. What is the role of water in reactions?

52. How did Feynman, in his visionary talk, explain the demagnification concept?
53. Writing using a light beam or ion beam in small letters was suggested by Feynman to whom?
54. How is nanotechnology used in sunscreens?
55. What is the diameter of a DNA molecule and that of a carbon nanotube?
56. Compare the length of the bacterium with the width of a house?
57. What does the movie *Fantastic Voyage* have to do with nanotechnology?
58. Discuss some features of the first commercial nanocomposite products put forward in the market by Toyota and General Motors?
59. How are thermal batteries improved by the development of nanotechnology?
60. How does nanotechnology better the lives of painters?
61. When was the first book published on nanotechnology?
62. When was the first journal introduced covering nanotechnology?
63. Which U.S. president first allocated research funds for nanotechnology?
64. What is NNI?
65. Are molecular assemblers feasible?
66. Are self-replicating nanobots feasible?
67. What are multilayered ceramic capacitors and how are they used in enhancement of thermal battery performance?
68. What are polymer lithium ion batteries and how is nanotechnology research improving their performance?
69. Why did Feynman call for improvement in SEM?
70. Can biological systems produce crystalline silicon, copper, or steel?
71. What are dendrimers and where are they used?
72. How is nanotechnology used in GMRs?
73. What is Kaposi's sarcoma?
74. What is metastability?
75. Can nanotubes be synthesized using material other than carbon?
76. Can particles with spherical morphology be made with less than 5 nm in diameter?
77. What is the difference between metastability and instability?
78. Name two challenges of nanotechnology in the future.
79. What are the issues involved in characterization of nanostructures?
80. How are the 20 different morphologies formed by self-assembly from block copolymers?

# REFERENCES

1. K. R. Sharma, *Nanostructuring Operations in Nanoscale Science and Engineering*, McGraw Hill Professional, New York, 2010.
2. K. Matyjaszewski, T. Kowalewski, D. N. Lambeth, J. T. Spanswick, and N. V. Tsarevsky, Process for the Preparation of Nanostructured Materials, US Patent 7,056,455, Carnegie Mellon University, Pittsburgh, PA, 2006.
3. M. Yudasaka, S. Iijima, Process for Producing Single Wall Carbon Nanotubes Uniform in Diameter and Laser Ablation Apparatus Used Therein, 2001, NEC Corp. Tokyo, Japan.
4. K. R. Sharma, Thermodynamic Stability of Nanocomposites, Interpack 99, 1999, Mauii, HI.
5. Sir Harold W. Kroto, Symmetry, Space, Stars and $C_{60}$, *Nobel Lecture*, http://nobel.se, Dec' 1996.

6. U. Steiner, Structure Formation in Polymer Films: From μm to sub 100 nm Length Scales, in *Nanoscale Assembly—Chemical Techniques (Nanostructure Science and Technology)* Ed. W. T. S. Huck, Springer, New York, 2005.

8. K. R. Sharma, Change in Entropy of Mixing and Two Glass Transitions for Partially Miscible Blends, 228th ACS National Meeting, Polymeric Materials: Science and Engineering, 91, 755, Philadelphia, PA, August 2004.

9. J. N. Cha, J. L. Hedrick, H. C. Kim, R. D. Miller, and W. Volksen, Materials Having Predefined Morphologies and Methods of Formation Thereof, US Patent 7,341,788, International Business Machines, Armonk, New York, 2008.

10. Y. L. Loo, R. A. Register, and A. J. Ryan, Polymer crystallization in 25 nm spheres, *Phys. Rev. Letters.*, 84, 4120–4123, 2000.

11. D. R. Paul and L. M. Robeson, Polymer nanotechnology: Nanocomposites, *Polymer*, 49, 3187–3204, 2008.

# 8 Polymer Alloys

## LEARNING OBJECTIVES

- Use of compatibilizer in polymer alloys
- PC/ABS alloys
- Nylon/ABS alloys
- PVC/ABS alloys
- PE-based alloys
- Natural polymer alloys

## 8.1 INTRODUCTION

Polymer alloys are a commercial polymer blend with improvement in property balance with the use of *compatibilizers*. They exhibit an interface and show varied physical characteristics. Sometimes they have excellent physical properties in one area but possess poor physical properties in others. For example, silicone rubber has poor oil and abrasion resistance but possess excellent heat resistance. A product solution in this regard would be to obtain a polymer blend with constituents possessing physical properties that complement each other such that the resultant polymer blend would exhibit superior physical properties compared with the components of the blend.

Modern structural applications of polymeric systems require both *impact-resistant* as well as *scratch-resistant* materials. The automotive industry needs such materials for bumpers and fenders. In practice, improving the scratch resistance generally reduces impact resistance. Rather than using additives and fillers to achieve these objectives, a compatibilized blend or polymer alloy design may be the solution to this problem. Texas A&M University [1] patented a compatibilizer that can be used to form a polypropylene. The resulting product has both high impact resistance as well as scratch resistance. The addition of a modifier increases the Young's modulus of the polypropylene and creates a blend with a stable morphology contributing to scratch resistance. In the resultant polymer system, the modifier dissipates fracture energy in part by promoting crazing around crack tip damage zones to reduce the development of large cracks.

The modifier forms a particulate dispersion in the polymer phase, and under stress, crazing develops around modifier particles. Moreover, the addition of a compatibilizer improves particle dispersion and interfacial adhesion, resulting in engaged toughness. Rubber modified polystyrene, high-impact polystyrene (HIPS), or polyphenylene oxide can be used as a modifier. The compatibilizer can be a triblock copolymer of styrene–ethylene–propylene. The copolymer can even be random.

## 8.2  PC/ABS ALLOYS

Deficiencies in polymer blends in terms of some properties of interest can be overcome by making a polymer alloy. For example, a polymer blend extensively practiced in the industry is the PC/ABS blend—polycarbonate, acrylonitrile butadiene, and styrene engineered thermoplastic. PC/ABS blends possess higher heat resistance, reduced notched sensitivity, and undesirable flow properties in many injection-molding applications.

Udipi [2] has discovered that polymer blends comprising a PC, amorphous polyester such as PETG, poly(ethylene terepthalate), and a nitrile rubber can form an alloy with superior balance of properties. Improved melt flow and spiral flow was demonstrated in injection molding applications. Udipi prepared a polymer alloy with 30% aromatic polycarbonate, 30% of PETG made of a condensation copolymer of terepthalic acid and a mixture of ethylene glycol and 1,4-cyclohexanedimethanol, and 3% nitrile rubber. This blend has a spiral flow of 27 cm, heat distortion temperature of 75°C, and Izod impact resistance of 130 J/m.

Blends of PC and ABS engineering thermoplastic are commercially important. One of the reasons why the PC/ABS blends are successful is due to the complementary properties of the components. Good thermal and mechanical properties are found in PC, and ABS has ease of processability and reliable notched impact resistance.

The control of morphology is a salient consideration in design of these blends, especially in order to obtain optimal performance. The morphology of the blend depends on a number of parameters including surface tension, viscosity of the components, and processing temperature and pressure. Addition of appropriate graft and block copolymers is an efficient method to control the interface of the morphology. The addition of compatibilizer improves the degree of dispersion possible. It offers stability of the morphology and strength of the interface. The compatibilizer can be presynthesized or prepared in situ during blending. Wildes et al. [3] studied the factors that affected the morphology of PC/S blends. These factors include (i) molecular weight of PC, (ii) composition of AN in SAN copolymer, (iii) dispersed phase volume fraction, and (iv) processing apparatus. These can serve as reasonable models for PC/ABS commercial blends. Reactive compatibilization using an SAN–amine compatibilizer on the morphology was also examined. The competing processes of drop break-up and coalescence during melt processing can determine the final morphology of the blend.

Taylor's theory [4] for the break-up of individual droplets for Newtonian fluids has been found applicable to better understand the morphology formed in polymer blends. The stable size of a drop ($d_{crit}$) against break-up in a simple shear field is given by

$$d_{crit} = \frac{4\sigma(\mu_r + 1)}{\gamma\mu_m(4.75\mu_r + 4)},$$
(8.1)

where $\sigma$ is the surface tension and $\mu_r$ is the viscosity ratio of the dispersed phase and matrix phase. When $\mu_r > 2.5$, drop split is obviated. Equation (8.1) has been verified with experimental data. The flow field in extruders and compounders is a complex hybrid between shear and extensional flow. Different relations for drop coalescence have been developed in the literature. Often, for a given polymer blend application the drop size calculated from drop split or drop coalescence is chosen to give better fit with experimental observations.

Attempts have been made to predict the rate of coalescence. The particle size observed in practice can be taken to be the result of the dynamic equilibrium between drop coalescence and drop split. With an increase in applied stress and a decrease in surface tension, the drop size is expected to decrease. As the volume fraction of the dispersed phase increases, the dispersed phase particle size is expected to increase. As the shear rate increases, investigators have found that the dispersed phase particle size increases, decreases, or shows complex nonmonotonic behavior, depending on the system studied and experimental operating conditions used. This observation may be attributed to increased particle–particle contacts, shorter contact times, or changes in elasticity of the blend components at high shear rates.

Compatibilization improves the morphological stability of polymer blends when block or graft copolymers are used by the introduction of steric hindrance to drop coalescence. A chemical approach is possible for formation of grafted PC with SAN chains at the PC/SAN interface. SAN–amine copolymer is prepared by reaction of a terpolymer of styrene, maleic anhydride, and acrylonitrile with 1-(2aminoethyl) piperazine. The SAN–amine polymer is miscible with the SAN matrix of ABS and reacts with PC. The formation of graft polymer has resulted in the reduction of SAN dispersed-phase particle size. The blend morphology has been found to stabilize in the mixing process.

Transmission electron microscopy (TEM) analysis was performed to study the blends of PC/SAN and PC/SAN/SAN–amine. Specimens 20-nm thick were microtoned at a temperature of –10°C using a diamond knife. The PC phase of the blends was stained by exposing ultrathin sections to the vapor of a 0.5 % aqueous solution of Ruthenium tetraoxide at room temperature. TEM imaging was performed on a JEOL 200 CX microscope operating at 120 keV. The apparent particle size was determined by digitizing TEM photomicrographs. Addition of a SAN–amine compatibilizer resulted in a reduction in particle size of SAN phase from 1 μm to 500 nm. The AN content in the SAN copolymer was found to affect its interactions with both PC and SAN–amine compatibilizer. Morphology of the blend was found to change with the choice of processing apparatus used to obtain the blend, such as Brabender versus a twin screw extruder.

PC/SAN blends have been found to exhibit maximal properties at 25–27 wt% AN. This was evidenced by the improvement in lap shear adhesion of compression-molded laminated sheets of PC and SAN copolymers, and in tensile modulus and elongation at break, improved notched impact Izod strength, and inward shifts of the glass transition temperatures of PC and SAN blend components with respect to their homopolymer values. PC and SAN were found to be partially miscible. The 25–27 wt% AN was found close to the azeotropic composition of SAN.

Mendelson [5] studied the miscibility of SAN with PC and several other polymers. PC and SAN were found to be miscible over a broad range of AN composition (23–70%). From mechanical tests Mendelson [5] have found that the amount of material deformed by crazing was found to decrease with increasing the PC content in the blend. The phase boundary observed by TEM was found to be rather diffuse, indicating a certain degree of interpretation between the PC and SAN domains.

In PC/SAN blends, other investigators have found linear dependence with blend composition of several parameters such as strength, modulus, and the heat distortion temperature under load. The elongation at break and the dart impact energy fell from the PC level down to the SAN level at a given SAN amount. The Izod impact toughness followed the same trend but the drop occurred at a lower SAN content. The blend behavior *depends* on the processing conditions. The blends tend to be viscoelastic. The micromechanisms acting in slow or high speed deformation processes are not clear.

Tensile tests that include yield stress and elongation at break usually decrease with the ABS addition of the PC value to that of the ABS showing a minima at intermediate composition range. PC and ABS were found to be compatible. At high PC contents, synergistic effects were found in the impact strengths with pronounced extrema at maximum stress. Fractography was used to elucidate the deformation mechanisms.

## 8.3 NYLON/ABS ALLOYS

Thermoplastic polyamides generally possess good elongation and energy to break as seen in tensile tests. They have high tensile impact strength and high energy absorption as seen in falling dart tests such as the Gardner impact test. The polyamides are found to be deficient in crack propagation resistance. Notched sensitivity, brittle breaks, and catastrophic failure of molded extruded parts is evidence of this. Breakage of nylon in a ductile manner limits its end use applications.

A polymer alloy was prepared by Lavengood, Patel, and Padwa [6]. This alloy is comprised of ABS with polybutadiene rubber, a polyamide such as nylon-6 or nylon-11, and a compatibilizer. The polyamide and ABS are immiscible. The compatibilizer is selected in a fashion that it is either partially or completely miscible with the graft copolymer, and has acid functional groups that can be made to react with the end groups of polyamides. This can be a terpolymer of styrene, acrylonitrile, and maleic acid. The resulting polymer alloy has been a successful product in the commercial arena under the name of Triax 1000. The performance properties were a step change improvement.

Various approaches have been undertaken for reactive compatibilization of polyamide/ABS alloys. Maleic anhydride can be grafted to the ABS. Styrene maleic anhydride (SMA) copolymers have been employed as compatibilizers for polyamide/ABS blends. SMA and SAN copolymers are miscible when the AN and maleic anhydride (MA) contents are equal. The impact strength of these blends has been found to be sensitive to the amount and composition of the SMA copolymer. Addition of SMA to SAN/polyamide blends was found to enhance the tensile and impact properties of these blends. Imidized acrylic polymers have been used as compatibilizers for nylon-6/ABS blends. Glycidyl methacrylate and methyl methacrylate (GMA/MMA) copolymers are used as compatibilizing agents. The epoxide functionality in GMA is capable of reaction with polyamide end groups. GMA/MMA copolymers can be shown to be miscible with SAN over the range of AN content of ABS. Styrene/GMA copolymers have been reported to be used as compatibilizers for polymer pairs such as

1. Polystyrene/polyamide
2. Polystyrene/polybutylene terepthalate (PBT)

3. Polystryene/polyethylene terepthalate (PET)
4. Polyphenylene oxide (PPO)/polybutylene terepthalate (PBT)

Terpolymers of styrene, acrylonitrile, and glyciedal methacrylate have been suggested as a compatibilizing agent for PPO/PBT blends. PBT/polyolefin blends have been compatibilized using random or graft copolymers of GMA.

## 8.4  PVC ALLOYS

Polymer electrolytes are used in lithium ion rechargeable batteries. Pure polymer electrolyte systems include polyethylene oxide (PEO), polymethylene–polyethylene oxide (MPEO), or polyphosphazenes. Chlorinated PVC blended with a terpolymer comprising vinylidene chloride/acrylonitrile/methyl methacrylate can make a good polymer electrolyte. Rechargeable lithium ion cells use solid polymer electrolytes. Plasticized polymer electrolytes are safer than liquid electrolytes because of a reduced amount of volatiles and flammables. The polymer membrane can conduct lithium ions. The polymer membrane acts as both the separator and electrolyte [7].

PVC can be blended with polyketones (PK). Polyketones are linear polymers having an alternating structure of CO (carbon monoxide), and olefenic repeat units. Thus, for example, the repeat unit of one polyketone is as follows:

$$CH_2 - CH_2 - \overset{\overset{\displaystyle O}{\|}}{C}$$

Polyketones prepared using ethylene and CO have high melting points close to the temperature at which they undergo chemical degradation. PVC, especially uPVC (unplasticized PVC), has a relatively low melting and softening point. Certain applications using PVC require higher service temperature. Stiffness, strength, and toughness of PVC are acceptable for most applications; PVC is not used in some applications because its softening point is too low. In order to increase the softening point of the product while maintaining good physical properties, blends of PVC and PK are prepared. PVC and PK were found to be compatible. PKs were used as plasticizers. They can also be used as a blend constituent. BP chemicals [8] claimed the patent for a polymer blend composition with PK and PVC. The PK has a number averaged molecular weight in the range of 40,000 to 1 million. The blend had a melt flow rate with 5 kg load at 240°C in the range of 5–150 gram/min. The PK can be a terpolymer made of ethylene, propylene, and carbon monoxide. The PK forms 10% of the binary blend. The blends can be prepared in a twin screw extruder.

Chlorinated polyvinyl chloride (CPVC) possess excellent high-temperature performance and other desirable properties. Typical CPVC has about 57% chlorine and is prepared by chlorination of PVC. It has outstanding flame and smoke properties, chemical inertness, and low sensitivity to hydrocarbon feedstocks. The glass transition temperature of the polymer, $T_g$, increases with increase in chlorine content in the polymer.

CPVC has poor melt processability. This is evidenced in the milling of CPVC on a roll mill that results in high torque and high temperatures, as well as decomposition of the CPVC. Softening additives or plasticizers have been added to CPVC in order to improve its processability but this has resulted in deterioration of other properties.

Blends of CPVC and PC can result in a product with improved physical properties such as heat distortion temperatures, impact strength, and processability. An impact modifier can be added as a compatibilizer to the blend. The product can be designed to eliminate degradation of CPVC with high heat distortion temperatures, impact resistance, flame retardancy, etc. B. F. Goodrich claims a patent [9] with 58–70 wt% CPVC. A 10–30 wt% aromatic PC, with a molecular weight of 10,000–30,000, core-shell impact modifier with the core polymer cross-linked and alkyl methacrylate shell, and other additives such as vinyl stabilizers, antioxidants, lubricants, heat stabilizers, and pigments such as titanium oxide and carbon black. CPVC can be synthesized from a solution process or from a fluidized bed process, water slurry process, thermal process, or by a liquid chlorine process. The $T_g$ of CPVC can range from 95°C–200°C, depending on the degree of chlorination, and the density is between 1.55–1.60 gm/cc.

Soft PVC was readily processable with its content of plasticizers. High molecular weight PVC was difficult to process due to its thermal instability and rheological properties. Lubricants can be added to alleviate this problem. Lubricants decrease the friction of the PVC particles and prevent generation of heat from friction. Processing aids based on PMMA are used to shorten the plasticizing process. Structural materials for manufacturing cases are made from PVC/ABS blends. Large macroscopic regions of heterogeneity were found during processing of PVS/ABS blends, especially when formed into parts using injection molding. PVC/SAN blends were found to be miscible in some AN compositional window in the SAN copolymer. The phase behavior of PVC/SAN and PVC/AMS–AN blends were discussed in Chapter 3. SAN containing about 12–26 wt% AN is miscible with PVC and is immiscible outside this range. From observations of the impact strength behavior of the ABS/PVC blends, a large negative deviation from the additive value of component polymers was found for systems when the AN content of matrix ABS was 35 wt%. In order to enhance the impact strength of this blend, SAN with 25 wt% AN was selected as candidate material [10]. When some of the matrix SAN35 in an ABS/PVC blend was replaced with SAN25, the compatibility was enhanced. About a twofold increase of impact strength was attained. The impact strength of PVC/ABS blend showed a negative deviation from simple addition of properties from polymer components. This can be attributed to incompatibility between SAN35 and PVC. Further, from the thermodynamics of the PVC/SAN blend in that composition range, PVC and SAN are found to be immiscible. PVC and SAN with a 25% AN composition are thermodynamically miscible as well as compatible. When some of SAN35 was replaced by SAN25, a twofold increase of impact strength and positive deviation from the simple additive values was observed. The stress whitening of all fractured surfaces after the impact test was observed. Phase equilibrium information in partially miscible multiphase polymer systems can be obtained from thermal properties such as $T_g$ or $\Delta C_p$. The blends exhibit two glass transition temperatures for the PVC-rich phase and SAN-rich phase. SAN25 was found to have a compatibilizing effect.

## 8.5 POLYOLEFIN ALLOYS

Polycarbonate can be blended with polyethylene (PE). Blending PC with PE has found many uses with a combination of a high level of heat resistance and dimensional stability as well as noncorrosive properties and ease of moldability. It is deficient in its tendency to craze and crack under the effects of contact with organic solvents such as gasoline. The failure mechanism of a crazed PC would be *brittle fracture* rather than *ductile fracture*. Blending PC with low density polyethylene (LDPE) and with linear low density polyethylene (LLDPE) or with ethylene propylene diene monomer (EPDM) rubber improves the chemical resistance of the product to solvents.

Dow Chemical claims a patent on a blend [11] with a decreased tendency toward delamination, which exhibits a desirable balance of properties, a high level of impact resistance, solvent resistance, and processability. The product can be blown into films, spun into fibers, extruded into sheets, formed into multilayer laminates, and molded into any desired shape and can be used in data storage apparatus, appliance and instrument housings, motor vehicle body panels, and other parts and components for use in electrical, automotive, and electronic industries.

The composition of the blend consists of 60–70% PC, 1–20% LDPE, 5–10% styrenic copolymer, 0.1–15% impact modifier, and 5–10% of another molding polymer. The low density polypropylene (LDPP) is homogeneously branched. The PC can be prepared from bisphenol A monomer, aromatic PC, or even a copolycarbonate. The impact modifier can be a core-shell grafted copolymer elastomer such as ABS/MBS engineered thermoplastic. The ABS can be made of polybutadiene for the rubber phase. The styrenic copolymer can be SAN copolymer. The $\alpha$-olefin can have 3–20 carbon atoms. A measurable high density, crystalline polymer fraction is missing in the olefenic constituent of the blend. The branching can be characterized using the short-chain branch distribution index (CBDI). The CBDI of a polymer can be estimated by employing temperature rising elution fractionation. The CBDI of LDPP used in the Dow patent is greater than 30%. Long chain branching of polyolefins can be measured using $^{13}C_{NMR}$, or isotopic carbon nuclear magnetic resonance spectroscopy. Long-chain branching is defined as more than six carbon atoms. The homogenously branched polyolefin has a sharp melting peak as measured by the differential scanning calorimetry (DSC), and is about $-30°C–150°C$. Heterogeneously branched polyolefin would exhibit two or more melting peaks. The density of linearly branched polyolefin is less than 0.88 gm/cc. Melt index was used to characterize the molecular weight of the polyolefin. Gel permeation chromatography (GPC) can also be used to measure the molecular weight.

Although PC possesses good toughness and weldline properties, it has poor impact resistance at low temperatures. Use of methacrylate butadiene and styrene (MBS), engineered thermoplastic, as an impact modifier can lead to a product with improved *low-temperature impact resistance*. This comes with some sacrifice to toughness as evidenced by a significant reduction in tensile strength. Ignition resistance additives such as halogenated hydrocarbons can be used in the product. Fillers are also sometimes added.

The impact toughness and stiffness of polypropylene has important practical significance in extending its applications. For particulate-filled polypropylene (PP) composites or blends, the ability of inclusions to play a role in agent-induced crazing, shear yielding of the matrix, and ended crack propagation is key to toughening brittle or quasi-brittle polymeric materials. The impact toughness of PP may be enhanced by rubber modification. Stiffness of blend decreases with increasing filler content. Because of the addition of rubber into PP, a variation of molecular structure can be induced, such as molecular weight, as well as crystallinity and miscibility between the matrix and filler. The factors affecting the toughening of the rubber modification of PP are more complex than those compared with inorganic particle-filled PP composites.

The major theories used to interpret the toughening mechanisms of rubber-modified PP are multiple-crazing, damage competition, shear yielding theories, and microvoid and cavitation mechanisms. The rubber particles in a PP matrix have to be uniform in size, and the interfacial adhesion and morphological structure between the matrix and filler have to be excellent.

PP homopolymer is used in packaging, textiles, and the automobile industry. It possesses excellent processability. Due to its poor impact resistance, it cannot be used as engineering thermoplastic. Toughened PP can be prepared by a process of filling, compounding, or blending. Four types of blending and reinforcing PP are as follows;

1. Rigid organic particle-filled (ROP) PP
2. Blending PP with rubber
3. Rigid inorganic particle-filled (RIP) PP
4. ROP- or RIP-filled PP–rubber blends

RIP- and ROP-filled brittle polymers toughen by a mechanism of nonelastic toughening. Studies on toughening mechanisms for PP–rubber blends are divided into four categories:

1. Structure/property relations
2. Content, shape, size, and distribution of rubber fillers
3. Influence of interfacial adhesion
4. Effects of processing and testing

Material properties depend on their micromorphological structure under a given set of experimental conditions. The morphology of PP–rubber-based blends has been found to be related to the compatibility between the continued phase and the dispersed phase. The effect of processing and shaping conditions on morphology is significant for crystalline polymers. The shape, content, size, and size distribution of the dispersed-phase particles are important factors affecting the toughening effect of PP–rubber blends. TEM and computer image analysis can be used to provide rubber particle size information and its effect on crazing.

## 8.6  NATURAL POLYMER ALLOY

Due to increased awareness of biodegradable plastic, polymer blends with lignin from cellulose as one of the constituents has been developed. When synthetic polymers are used as constituents of polymer blend, it results in an increase in recycling costs. Recovery from recycling of plastics often involves separation and purification processed that are difficult to perform and are expensive. The mineral oil resources from which plastics are synthesized are limited. Blend constituents can be replaced with natural polymers.

Blends based on natural polymers are modified by oxidation process, enzymatic degradation, etc.—for example, duroplasts made from casein or thermoplasts made from cellulose nitrates, acetates, esters, and ethers. Lignin is a natural polymer with improved material properties such as strength, rigidity, impact strength, and ultraviolet stability. It can be used for sound insulation materials and for thermal insulation as well. It is a high molecular polyphenolic macromolecule that fills the spaces between the cell membranes of ligneous plants. The phenyl groups of the lignin may be substituted by methoxy groups, depending on the wood type. Bulk lignin is obtained as a by-product of cellulose production. Degradation of wood results in lignosulfonic acids as part of the sulfite waste liquor. Treatment with sulfuric acid and $CO_2$ (carbon dioxide) results in the precipitation of lignin.

Alkali lignin is used as a binding agent in the cellulose industry to prepare hardboard made from wood and cellulose. It is used as a stabilizer in asphalt emulsion.

A patent from Germany [12] describes a method for producing construction materials made from a polymer blend. The blend consists of alkali lignin and protein. The product is used to house electrical and electronic devices. Polylactide with high impact strength is used. Chitin or natural starch may also be used.

Hemasure Inc. (Marlborough, Massachusetts) [13] developed a process for modification of hydrophobic polymer blend surfaces. An isotropic microporous membrane is produced using a process that taps into the phase inversion properties of a quaternary blend and by use of an spinnerette assembly. The process makes use of functionalizable chain ends of the polymer. The thermal phase inversion boundaries at LCST and UCST for the quaternary blend is shown in Figure 8.1. The manufacturing process uses a temperature-regulated antisolvent phase inversion phenomenon, as well as freezing or precipitating out and preservation of the ensuing microporous structure.

The dope composition consists of two polymer components and two solvent components for each polymer, respectively. Polyether sulfone (PES) and PEO are selected as the two polymer components. Glycerin can serve as a solvent for one of

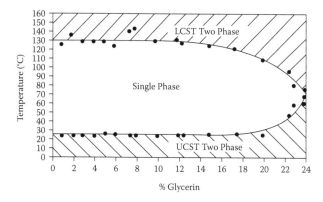

**FIGURE 8.1**  Phase diagram of quaternay blend of PES/PEO/glycerin/nonsolvent blend.

the polymers and a nonsolvent for another. The final membrane desired is formed from phase separation of the solution. PES forms the primary component of the membrane. The polymer that is soluble in both solvents possesses a hydrophilic character. Water soluble polymers at high molecular weight are more likely to be retained in the final membrane by chain entanglements compared with polymers at low molecular weight when an aqueous quench is employed during preparation. A certain degree of hydrophilicity and wettability is imparted to the otherwise hydrophobic membrane surface. The preferred molecular weight of PEO is at least 100,000 gm/mole. As PESs possess a high $T_g$ and high melting point, the membranes can be autoclaved frequently without deteriorating the membrane properties. Autoclaving can result in higher tensile strength of PES/PE fibers. The second solvent can be $N$-methyl-2-pyrrolidone (NMP).

Flat sheet membranes with isotropic pore structure in the micron range can be prepared by subjecting the blend to an abrupt change in temperature above the LCST, causing precipitation by addition of antisolvent for the primary polymer, PES. When the quench temperature was below LCST, the pore structure formed was found to be anisotropic. This technique taps into thermal-phase inversion phenomena.

## 8.7  SUMMARY

Polymer alloys are commercial polymer blends with improvement in property balance with the use of *compatibilizers*. Texas A&M University [1] has patented a compatibilizer that can result in a product with high impact resistance as well as scratch resistance. The blend is composed of HIPS or polypropylene (PP) and a compatibilizer made of a triblock copolymer of styrene–ethylene–propylene. Udipi [2] discovered that polymer blends composed of PC, a copolyester of PETG, and nitrile rubber exhibit a superior balance of properties. Reactive compatibilization of PC/SAN blends at various AN compositions were conducted by Wildes et al. [3] using a SAN–amine compatibilizer. PC and SAN were found to be miscible over a range of AN composition by Mendelson [5]. Nylon/ABS blends can be compatibilized by use of SAN–maleic acid (Lavengood et al. [6]). Styrene–GMA copolymers can be used as compatibilizers for PS/PA, PS/PBT, PS/PET, and PPO/PBT blends.

Chlorinated PVC blended with a terpolymer comprising vinylidene chloride/acrylonitrile/methyl methacrylate can make a good polymer electrolyte. PVC/PK blends are found to be compatible. CPVC and PVC blends can result in improved properties.

PVC/SAN blends were found to be miscible and not compatible at certain AN compositions. Miscible and compatible PVC/SAN blends with better properties can be prepared at a different AN composition. PC and LDPE can be blended with each other with EPDM as the compatibilizer. LDPP and PC can be blended together, with ABS used as the modifier for the alloy. The impact toughness of PP may be enhanced by rubber modification of PP. Stiffness of blend decreases with increasing filler content.

Polymer alloys using natural polymers can be prepared. Lignin and protein can form interesting blends. An isotropic porous membrane can be prepared using a quaternary blend taking advantage of the LCST/UCST behavior.

## EXERCISES

1. What is the use of a compatibilizer in polymer alloys?
2. What is the role of a modifier in polymer alloy?
3. Do the properties of polymer alloys change with temperature?
4. Where are compatibilizers more useful: (a) with miscible blends or (b) immiscible blends?
5. Do polyamide and SAN form miscible blends over certain AN composition?
6. What happens when PVC/SAN blends at certain AN compositions are found to be miscible and incompatible?
7. What happens when PVC/SAN blends at certain AN compositions are found to be miscible and compatible?
8. When polymer A and polymer B form a compatible blend, and polymer A and polymer C forms a compatible blend, will polymer B and polymer C necessarily form a compatible blend?
9. Do polymer alloy properties change with pressure?
10. What are the differences between reactive compatibilization and a pre-synthesized compatibilizer?
11. What are the differences in the properties of polymer alloys made from synthetic polymers and from polymer alloys made from natural polymers?
12. Do lignin and protein form a miscible blend?
13. What are the advantages of blending of PC and SAN?
14. Can PVC be replaced with PTFE in PVC-based alloys?
15. Can styrene be replaced with alpha-methylstyrene in polymer alloys where one constituent has styrene as a comonomer?
16. What is the role, if any, of a plasticizer in a polymer alloy?
17. Can constituents of a polymer blend be made to react with each other instead of using a compatibilizer?
18. What is the effect of chain branching in polyolefin-based alloys?
19. What is the purpose of rubber modification of polypropylene?
20. What is the role of fillers in PP-based alloys?

## REFERENCES

1. H. J. Sue and G. X. Wei, Method for Improving the Impact Resistance and Scratch Resistance of Polymeric Systems, US Patent 6,423,779, Texas A & M University System, College Station, TX, 2002.
2. K. Udipi, Rubber-Modified Polymer Blends of Polycarbonate and PETG, US Patent 5,187, 230, Monsanto Co., St. Louis, MO, 1993.
3. G. Wildes, H. Keskulla, and D. R. Paul, Morphology of PC/SAN blends: Effect of reactive compatibilization, SAN concentration, processing and viscosity ratio, *J. Polym. Sci.: Part B: Polym. Phys.*, 37, 71–82, 1999.
4. G. I. Taylor, The spectrum of turbulence, *Proc. Royal Society, London*, A164, 476, 1938.
5. R. A. Mendelson, Miscibility and deformation behavior in some thermoplastic polymer blends containing poly(styrene-co-acrylonitrile), *J. Polym. Sci. Polym. Phys. Ed.*, 23, 1975, 1985.
6. R. E. Lavengood, R. Patel, and A. R. Padwa, Rubber-Modified Nylon Composition, US Patent 4,777,211, Monsanto Co., St. Louis, MO, 1988.

7. Y. H. Chia, J. J. Coleman, M. Parsian, and K. A. Snyder, Lithium Ion Rechargeable Batteries Utilizing Chlorinated Polymer Blends, US Patent 6,617, 078, Delphi Technolgies Inc., Troy, MI, 2003.

8. J. G. Bonner, Polymer Blends of PVC and Polyketones, US Patent 5,610,236, BP Chemicals Lmt., London, U.K., 1997.

9. D. I. Lawson, Chlorinated Polyvinyl Chloride/Polycarbonate Blend, US Patent 5,268,424, B. F. Goodrich, Akron, OH, 1993.

10. D. W. Jin, K. H. Shon, B. K. Kim, and H. M. Jeong, Compatibility enhancement of ABC/PVC blends, *J. Appl. Polym. Sci.*, Vol. 70, 4, 705–709, 1998.

11. F. M. Hofmeister, M. M. Hughes, H. Farah, and S. R. Ellebracht, Blends of Polycarbonate and Linear Ethylene Polymers, US Patent 5,786,424, Dow Chemical Co., Midland, MI, 1998.

12. H. Nagele, J. Pfitzer, B. Eisenreich, P. Everer, P. Elsner, and W. Eckl, Plastic Material Made from a Polymer Blend, US 6,509,397, Fraunhofer-Gesselschaft zur Forderung der angewandten Forscung, e.V., Munchen, DE, 2003.

13. A. R. M. Azad and R. A. Goffe, Covalent Attachment of Macromolecules to Polysulfones or Polyethersulfones Modified to Contain Functionalizable Chain Ends, US Patent 5,462,867, Hemasure, Marlborough, MA, 1995.

# 9 Binary Diffusion in Polymer Blends

## LEARNING OBJECTIVES

- Importance of mutual diffusion
- Phase separation, polymer blend processing
- Relation to interaction parameter
- Transient diffusion
- Fick's law of diffusion
- Damped wave diffusion and relaxation
- Semi-infinite medium, finite slab
- Periodic boundary condition

## 9.1 INTRODUCTION

Self-diffusion in polymer melts has been modeled using the reputation model. Polymer–polymer binary diffusion in miscible blends of polymers with different repeat units in their chains during miscibilization and phase separation is very important. In such blends, attractive interactions between the monomers, when summed over a polymer chain, may lead to large enthalpic driving forces favoring mixing. This in turn results in a mutual diffusion rate that is rapid compared with entropically driven self-diffusion. This phenomenon is dependent on composition. Binary diffusion coefficients between miscible blends of PVC (polyvinyl chloride) and PCL (polycaprolactone) were measured by Jones et al. [1]. Diffusion coefficients were obtained from measurements of transient concentration profile from an initial step change in concentration. A sharp interface was set up between blends of PVC and PCL, differing in concentration by 10%. The sample was then held at a temperature of 91.5 ± 0.5°C in a vacuum oven for known times of 8 h to 7 days. The final distribution of the two polymer species after diffusion broadening across the interface was then determined using x-ray microanalysis in a scanning electron microscope in which x-rays from inner-shell transitions excited by the electron beam and characteristic of the element producing them were counted by a detector. The concentration of chlorine atoms in PVC was measured across the sample. The spatial resolution of this technique was 3 μm. With care during sample preparation, a concentration resolution of 1% is possible.

The diffusion broadening was analyzed using solutions from the Fick model for diffusion. The binary diffusion coefficient was obtained at the average concentration of C of the diffusion couple. The binary diffusion coefficient may be expressed as a product of a mobility term and a thermodynamic term expressing the departure from ideality of mixing. The thermodynamic term may be evaluated using the Flory–Huggins theory for free energy of mixing of a polymer blend. For polymers of identical chain lengths:

$$D = D_0 (1 - 2\chi n\phi_1 (1 - \phi_1)), \tag{9.1}$$

where $\phi_1$ is the volume fraction of the first polymer in a blend of two polymers, $D_0$ is the diffusion coefficient expressing the mobility of the components in each other, and $\chi$ is the interaction parameter of the blend. For miscible blends, $\chi < n/2$. $\chi$ was measured by melting-point depression and can be expressed per monomer unit of PCL as $\chi = -0.38$. The presence of the polymerization index $n$ ensures that the second term in Equation (9.1) dominates for all but the smallest concentrations. The diffusion is driven by *enthalpy* and not *entropy*. This can be used to explain the concentration dependence of the diffusion coefficient.

As diffusion phenomena are extremely important in miscible blend formation and phase separation of polymer blends, these phenomena are discussed in detail in this chapter.

## 9.2   DIFFUSION PHENOMENA

Diffusion is a phenomenon of migration of a species from a region of a higher concentration to a region of a lower concentration under the driving force of a concentration gradient [1a]. There can be other driving forces such as temperature difference, the large concentration gradient of a second species, osmotic potential, steam sweep, centripetal forces, pressure drop, electromotive forces, surface tension gradient, and surface forces, that can cause the transfer of species from one point to another, often in a secondary manner.

The term *molecular diffusion* refers to the Brownian motion of molecules as observed by Einstein [2], and movement from a region of higher concentration to one of lower concentration. This is in accordance with the second law of thermodynamics, the Clausius inequality. The movement of species from a region of lower concentration to a region of higher concentration in a spontaneous manner is infeasible. This is because not all heat can be converted to work without some heat being dissipated into the atmosphere. Heat always flows from a higher temperature to a lower temperature by molecular conduction. By analogy, molar species also move from a region of higher chemical potential to one of lower chemical potential. The direction of transfer is to equalize the concentration.

There can be no negative concentration, as the third law of thermodynamics states that the lowest attainable temperature is 0 K. By analogy, the lowest concentration attainable is 0 mol/m³. This will be used in later discussions to obtain a point in the space coordinate system beyond which zero concentration of the migrating species can be expected. This is obtained by realization that there can be no negative concentration from the third law of thermodynamics. Sometimes during model development mathematical expressions arise that indicate negative concentration values. These are mathematical artifacts and will not occur in practice. In practice there will be a certain penetration length beyond which there is no effect of the surface disturbance in concentration of the migrating species.

Diffusion is central to separation operations widely used in the chemical and bio-technology industries. It is used to better understand the transport of solutes in living cells and to design artificial organs. It plays a pivotal role in sequence distribution analysis in genome projects. The efficiency of distillation and dispersal of pollutants can be derived from principles of diffusion. As cities around the world face a drought crisis, the desalination of sea water for potable water needs is going to be increas-ingly relied upon. This is the method of choice in the deserts of the Middle East, where energy is cheap and abundant, and drinking water scarce. In order to desali-nate sea water, several transfer operations such as reverse osmosis, electrodialysis, ion exchange, extraction, flash vaporization, molecular sieve filtration, and pervapo-ration can be used. The principles of diffusion and mass transfer can help evaluate the technical feasibility of each operation at large scale at the lowest possible cost without inflicting much harm to the environment. The chemical reactions performed on a large scale during commercial manufacture of products are often conducted in the presence of a catalyst. During the reaction, the critical reactant has to diffuse through the catalyst, approach the active site for reaction, and encounter the other reactant prior to reacting and forming the product. Diffusion in the catalyst needs to be understood for better design. The useful product has to be separated from the unreacted reactants and other by-products using mass transfer separation operations, where diffusion is a critical governing phenomena.

Albert Einstein observed that a cube of sugar placed at the bottom of a cup full of hot tea diffuses, and a uniform concentration of sugar throughout the entire cup results at the final state. The term *thermophoresis* refers to a process in which the primary driving force of diffusion is the temperature difference. *Diffusophoresis* is when a large drop in concentration of a second species drives the transfer of the first species. *Osmosis* is where osmotic potential drives the flow of solvent from a region of low solute concentration to a region of higher solute concentration. In *reverse osmosis*, the solvent is pumped from a region of higher solute concentration to one of lower solute concentration. The wilting of lettuce when salted is a good example of osmosis; as water oozes out of the leafy vegetable and the turgor pressure reduces, a shrunken mass results. In *sweep diffusion*, steam sweeps away the solute with it. The centripetal forces during *centrifugation* result in different forces upon different masses, resulting in separation. *Pressure diffusion* is characterized by a pressure drop, $\Delta P$, in the direc-tion of transfer. *Electrolysis* refers to the movement of charged particles subject to an electromotive force. *Surface diffusion* is the movement of species of interest on the sur-face of the solid. Surface tension gradient can be utilized in separation by *foaming*.

## 9.3   FICK'S FIRST AND SECOND LAWS OF DIFFUSION

In the mid-1800s, Fick [3,4] introduced two differential equations that provide a math-ematical framework to describe the otherwise random phenomenon of molecular diffusion. The flow of mass by diffusion across a plane was proportional to the concen-tration gradient of the diffusant across it. The components in a mixture are transported by a driving force during diffusion. The molecular motion is Brownian. The ability of the diffusant to pass through a body is dependent on both the diffusion coefficient, $D$, and the solubility coefficient, $S$. The permeability coefficient, $P$, is given by

$$P = D * S \tag{9.2}$$

Fick's laws of diffusion were proposed in the year 1855. Adolf E. Fick, the youngest of five children, was born on September 3, 1829. His father was a civil engineer. During his secondary schooling, Fick was interested in mathematics and was enamored of the work of Poisson. His brother, a professor of anatomy at the University of Marlburg, persuaded him to switch to medicine from mathematics. Carl Ludwig was Fick's tutor at Marlburg. Ludwig strongly believed that medicine and life itself have a basis in mathematics, physics, and chemistry. His thesis dealt with the visual errors caused by astigmatism. Ironically, most of Fick's accomplishments do not depend on diffusion studies at all, but on his more general investigations of physiology. He did outstanding work in mechanics of hydrodynamics and hemorheology, and in the visual and thermal functioning of the human body.

In his first diffusion paper [3], Fick interpreted the experiments from Graham with interesting theories, analogies, and quantitative experiments. He showed that diffusion could be described on the same mathematical basis as Fourier's law of heat conduction [5] and Ohm's law of electricity. Fick's first law of diffusion can be written as

$$J = -AD \frac{\partial C}{\partial x}, \tag{9.3}$$

where $J$ is defined as the one-dimensional molar flux. The diffusivity is the proportionality constant that depends on the material under consideration.

Fick's second law of diffusion can be derived by considering a thin shell of thickness $\Delta x$ with constant cross-sectional area $A$ across which the diffusion is considered to occur. A mass balance in the incremental volume considered, $A\Delta x$, for a incremental time, $\Delta t$, neglecting any reaction or accumulation of the species, can be written as

$$(\text{mass in}) - (\text{mass out}) \pm (\text{mass reacted/generated}) = \text{mass accumulated} \tag{9.4}$$

$$\Delta t(J_x - J_{x+\Delta x}) = A\Delta x \, \Delta C. \tag{9.5}$$

Dividing Equation (9.5) throughout by $A\Delta x\Delta t$ and obtaining the limits as $\Delta x$ and $\Delta t$ goes to zero:

$$-\frac{\partial J}{\partial x} = A\frac{\partial C}{\partial t}. \tag{9.6}$$

Combining Equations (9.6) and (9.3), the governing equation for the diffusing species is obtained when the area across which the diffusion occurs is a constant:

$$D\frac{\partial^2 C}{\partial x^2} = \frac{\partial C}{\partial t}. \tag{9.7}$$

Equation (9.7) is sometimes referred to as Fick's second law of diffusion [4]. This is a fundamental equation that describes the transient, one-dimensional diffusion of diffusing species. Fick attempted to integrate Equation (9.7) and was discouraged by the numerical effort needed. He found the second derivative difficult to measure experimentally, and he found that the second difference exaggerates the effect of experimental errors. Finally, he demonstrated in a cylindrical cell the steady-state linear concentration gradient of sodium chloride. He used a glass cylinder containing crystalline sodium chloride at the bottom and a large volume of water at the top. By periodically changing the water in the top volume, he was able to establish a steady-state concentration gradient in the cylindrical cell. He confirmed his equation from this steady-state gradient.

In three dimensions, Fick's first law of diffusion can be written as

$$J'' = -D \, \nabla C. \tag{9.8}$$

where the differential operator $\nabla$ is given by

$$\nabla = i \frac{\partial}{\partial x} + j \frac{\partial}{\partial y} + k \frac{\partial}{\partial z} \tag{9.9}$$

such that

$$\nabla C = i \frac{\partial C}{\partial x} + j \frac{\partial C}{\partial y} + k \frac{\partial C}{\partial z}. \tag{9.10}$$

In the special case of Equation (9.8), the one-dimensional case of Equation (9.3) results. Diffusion may be viewed as a process by which molecules intermingle as a result of their kinetic energy of random motion.

In liquid systems on earth, density differences within a liquid system often result in convective mixing of the fluid. This gravity-induced convection, coupled with gravity-independent diffusion, contributes to the overall mass transfer within the system. In space, the convective contribution is greatly reduced, and a closer examination of the diffusion contribution can be observed.

## 9.4  SKYLAB DIFFUSION DEMONSTRATION EXPERIMENTS

The Skylab science demonstration was the first in a series of investigations designed by Fascimire [6] to study low-gravity diffusive mass transfer. Skylab pilot Jack Lousma filled a ½ in. diameter, 6 in. long transparent tube three-quarters full of water. A highly concentrated tea solution was then delivered to the water surface (via a 5 cc syringe) through a synthetic fiber pad. The tube was then capped. The fiber pad was employed to try to bring the tea and water in contact without entrapped air. Three attempts to produce the pad were unsuccessful. During the fourth pad/attempt, a "a good bubble-free interface" was realized. The next day, Lousma reported that no diffusion of the tea in the liquid had occurred. Thus, the experiment was initiated again.

During this new experimental run, the wad was removed, and the tea was delivered on top of the water. After an air bubble between the tea and water was removed via the syringe, a "smooth continuous interface" was achieved. The tea was allowed to diffuse during the next 3 days. Post flight, 16 mm photographs of the diffusion were analyzed. In 51.15 h, the visible diffusion front advanced 1.96 cm. It was noted that the diffusion front became increasingly parabolic during the demonstration. It was also noted that very little diffusion occurred near the container wall. A similar ground-based experiment was performed for comparison with the space investigation. After 45.5 h, three different zones were visible: (a) a dark area, (b) an area of medium darkness, and (c) a very light area. The medium-colored area had advanced 1.6 cm in 45.5 h.

If a few crystals of $K_2CrO_4$ (potassium chromate) are placed at the bottom of a tall bottle filled with triple-distilled water, the yellow color will slowly spread through the bottle. At first the color will be concentrated in the bottom of the bottle. After a day, it will penetrate upward a few centimeters. After several years, the solution will appear homogeneous. The process responsible for the movement of the colored material is diffusion. Diffusion was studied by Albert Einstein. He noted that as sugar dissolves in water, the viscosity of the solution increases. The Stokes–Einstein equation is used for estimating diffusion coefficients for molecules in the liquid phase. Diffusion is a molecular phenomenon. At a microscopic level, molecules do undergo Brownian random motion. However, the driving force imparts some direction to the transfer of the species under consideration. In gases, diffusion progresses at a rate of about 10 cm/min; in liquids, its rate is about 0.05 cm/min; and in solids, its rate is about 100 nm/min [7–13].

Diffusion varies less with temperature, although for polymers, Arrhenius relationships have been reported for change of diffusion coefficient with temperature. The slow rate of diffusion makes it a rate-limiting step in cases where it occurs sequentially with other phenomena. The rates of distillation are limited by diffusion and that of industrial reactions on porous catalysts. Diffusion limits the speed of absorption of nutrients in the human intestine; the control of microorganisms in the production of penicillin is a diffusion limited process. The rate of corrosion of steel, splat cooling of metallic glasses, dopant diffusion in silicon chip manufacture, the release of flavor from food, and delivery of drugs to tumor cells are limited by diffusion. The equalizing effect of diffusion needs to be distinguished from other methods of producing a uniform mixture, such as bulk convective mixing. Agitation also is used in homogenization. The energy for movement from diffusion comes from the thermal energy of the molecules. The rate of evaporation of water at 25°C in a complete vacuum was calculated as 3.3 $kg/m^{-2}/s^{-1}$. Placing a layer of stagnant air at 1 standard temperature and pressure (STP) and 100 micron thickness above the water surface reduces the rate of evaporation by a factor of about 600.

## 9.5  BULK MOTION, MOLECULAR MOTION, AND TOTAL MOLAR FLUX

Consider two containers of $CO_2$ gas and He gas separated by a partition as shown in Figure 9.1. The molecules of both gases are in constant motion and make numerous collisions with the partition. If the partition is removed, the gases will mix due to the random velocities of their molecules. Given sufficient time, a uniform mixture of $CO_2$ and He molecules will result in the container.

The helium molecules will move to the bottom, and the carbon dioxide molecules will move to the top. If the two species are denoted by A and B, and the total fluxes as $N_A$ and $N_B$, then the net flux $N$ can be written as

$$N = N_A + N_B. \tag{9.11}$$

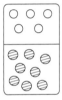

**FIGURE 9.1**   Container with a partition separating helium and carbon dioxide.

The movement of A comprises two components, one due to the bulk motion $N$ and the fraction $x_A$ of $N$ which is A, and the second component resulting from the diffusion of A, $J''_A$.

$$N_A = Nx_A + J''_A$$

$$N_B = Nx_B + J''_B.$$

(9.12)

Adding $N_A$ and $N_B$ in Equation (9.12) leads to

$$N = N + J''_A + J''_B.$$

(9.13)

Or

$$-D_{AB}\frac{\partial C_A}{\partial z} = D_{BA}\frac{\partial C_B}{\partial z}.$$

(9.14)

If $C_A + C_B$ is constant, then $D_{AB} = D_{BA}$.

## Example 9.1:   Unimolar Diffusion

Consider the diffusion of a liquid A evaporating into a gas B in a partially filled tall tube. Assume that the liquid level is maintained at $z = z_1$. At the top of the tube at $z = z_2$, a stream of gas, a mixture of A and B, flows steadily past, thereby maintaining the mole fraction of A at $X_{A2}$. At the liquid–gas interface, the gas-phase concentration of A expressed as mole fraction is $X_{A1}$. This is taken to be the gas-phase concentration of A corresponding to the equilibrium with the liquid at the interface. That is, $X_{A1}$ is the vapor pressure of A divided by the total pressure, $p_A{}^{vap}/P_{tot}$, provided that A and B form an ideal gas mixture. It is further assumed that the solubility of B in liquid A is negligible. The entire system is presumed to be held at constant temperature and pressure. Gases A and B are assumed to be ideal. When this evaporating surface attains steady state, there is a net motion of A away from the evaporating surface, and vapor B is stationary. Obtain the concentration profile at steady state for A in the head space.

$$N_B = 0.$$

(9.15)

$$N_A = N_A x_A - D_{AB}\frac{\partial C_A}{\partial z}.$$

(9.16)

In terms of mole fractions:

$$N_A(1 - x_A) = -C_{tot}D_{AB}\frac{\partial x_A}{\partial z}.$$

(9.17)

A mass balance over an incremental volume of height $\Delta z$ at steady state across a constant cross-sectional area, A is shown by:

$$-\frac{\partial N_A}{\partial z} = 0. \tag{9.18}$$

Combining the previous two equations and integrating the resulting second-order differential equation with respect to $z$ gives

$$\frac{1}{(1-x_A)}\frac{\partial x_A}{\partial z} = c_1. \tag{9.19}$$

A second integration then gives

$$-\ln(1-x_A) = c_1 z + c_2. \tag{9.20}$$

The two integration constants can be solved for by use of the information given as boundary conditions at locations 1 and 2:

$$\frac{\ln(1-X_{A1})}{(1-x_A)} = \frac{(z-z_1)}{(z_2-z_1)}\frac{\ln(1-X_{A1})}{(1-X_{A2})}. \tag{9.21}$$

$$\frac{(1-X_{A1})}{(1-x_A)} = \left(\frac{1-X_{A1}}{1-X_{A2}}\right)^{\left(\frac{z-z_1}{z_2-z_1}\right)}. \tag{9.22}$$

$$N_{Az} = \left(\frac{cD_{AB}}{z_2-z_1}\right)\ln\left(\frac{X_{B2}}{X_{B1}}\right). \tag{9.23}$$

The foregoing expressions are used during the experimental measurement of gas diffusivities.

## 9.6 STOKES–EINSTEIN EQUATION FOR DILUTE SOLUTIONS

The Stokes–Einstein equation can be used to calculate diffusion coefficients in liquids:

$$D = \frac{k_B T}{f} = \frac{k_B T}{6\pi\mu R_0}, \tag{9.24}$$

where $k_B$ is the Boltzmann constant, $f$ the frictional drag coefficient, $T$ the temperature, $\mu$ is the viscosity of the surrounding medium and $R_0$ is the radius of the solute that is diffusing. Equation (9.24) can be derived as follows. A rigid solute sphere is assumed for the molecule diffusing in a common solvent. The frictional drag force acting on the molecule opposing its motion is proportional to the velocity of the sphere:

$$\text{Drag force} = f v_1. \tag{9.25}$$

where $v_1$ is the velocity of the molecule. From Stokes law [14], for a sphere moving in a fluid, $f = 6\pi\mu R_0$. The driving force was taken by Einstein [2] to be the negative of the chemical potential gradient $(-\nabla\mu_A)$ defined per molecule:

$$-\nabla\mu_A = (6\pi\mu R_0)v_A. \tag{9.26}$$

Equation (9.26) is valid when the molecule reaches a steady-state velocity. This is when the net force acting on the molecule is zero. The solution is assumed to be ideal and dilute.

$$\mu_A = \mu_A{}^0 + k_B T \ln(x_A) = \mu_1{}^0 + k_B T \ln C_A - k_B T \ln C_B. \tag{9.27}$$

For dilute solutions, the concentration of the second species, $C_B$, far exceeds the solute concentration and can be taken as constant. The gradient at constant temperature, then is

$$\nabla\mu_1 = k_B T \frac{\nabla C_A}{C_A} = -(6\pi\mu R_0)v_A. \tag{9.28}$$

$$\frac{-k_B T}{(6\pi\mu R_0)}\nabla C_A = C_A v_A = J/A. \tag{9.29}$$

Comparing Equation (9.29) with Fick's law of molecular diffusion, the Stokes–Einstein relationship of Equation (9.24) results.

Equation (9.24) is valid only at steady state. Often, in transient applications, a sudden step change in concentration occurs; that is, a driving force is imposed on the system. The molecule will experience an accelerating regime prior to reaching steady state. During the accelerating regime:

$$-\nabla\mu_A - (6\pi\mu R_0)v_A = m\frac{dv_A}{dt}. \tag{9.30}$$

where $m$ is the mass of the molecule.

$$mC_A\frac{dv_A}{dt} = -(6\pi\mu R_0)C_A v_A. \tag{9.31}$$

or

$$-\frac{k_B T A}{6\pi\mu R_0}\nabla C_A = \frac{m}{(6\pi\mu R_0)}\frac{\partial J}{\partial t} + J. \tag{9.32}$$

Equation (9.32) is the generalized Fick's law of diffusion that accounts for the acceleration regime of the molecule as well as the steady-state regime. An expression for the relaxation time for molecular diffusion falls out of the analysis; that is,

$$\tau_r = \frac{m}{(6\pi\mu R_0)} = \frac{mD}{k_B T}. \tag{9.33}$$

In terms of $P_{tot}$, the system pressure for an ideal gas, the relaxation time can be written as

$$\tau_r = \frac{MD\rho_m}{P}, \tag{9.34}$$

where $\rho_m$ is the molar density of the species migrating. The velocity of mass diffusion is given by

$$V_m = \sqrt{\frac{D}{\tau_r}} = \sqrt{\frac{k_B T}{m}}. \tag{9.35}$$

Equation (9.35) can be rewritten in terms of the molar gas constant and molecular weight as

$$V_m = \sqrt{\frac{D}{\tau_r}} = \sqrt{\frac{RT}{M}}. \tag{9.36}$$

The kinetic representation of pressure can be written after observing that a molecule moving in a cube of dimensions $l$ with a velocity of $v_x$ undergoes a momentum change of $2mv_x$ upon a collision with the wall. The number of collisions on the wall can be estimated by first calculating the time taken by the molecule to perform the roundtrip from the wall after a collision with the opposite wall and return, as $2l/v_x$. The number of collisions undergone by a molecule is $v_x/2l$. The rate of transfer of momentum to the surface from the molecular collisions is then $mv_x^2/l$. The total force exerted by all the molecules colliding can be obtained by summing the contribution from each molecule, and the pressure is obtained by dividing the resulting sum by the area of the wall and is given by (Resnick and Halliday [15])

$$P_{tot} = \frac{m}{l^3}(v_{x1}^2 + v_{x2}^2 + v_{x3}^2 + \ldots\ldots). \tag{9.37}$$

Let $N_m$ be the number of molecules in the system and $n$ the number of molecules per unit volume. Then Equation (9.35) can be rewritten, after multiplying the numerator and denominator by $N_m$:

$$P_{tot} = mn <v_x^2> = \rho <v_x^2> = \frac{1}{3}\rho <v^2>. \tag{9.38}$$

Because the molecules treated as particles move at random, there is no preferred direction in the box. Hence, $v^2 = v_x^2 + v_y^2 + v_z^2$. The square root of $v^2$ is called the root-mean-square speed of the molecule and is a widely accepted average molecular speed. From the ideal gas law, $P_{tot} = \rho RT/M$. Combining this with Equations (9.8–9.18):

$$\frac{1}{3<v^2>} = \frac{RT}{M} = \frac{A_N k_B T}{M}. \tag{9.39}$$

Comparing Equations (9.36) and (9.38), it can be seen that the velocity of the mass is 1/3 of the root mean square velocity of the molecules. This could be because only one-dimensional diffusion has been considered. When all three dimensions are considered, these two velocities would be identical, although derived from different first principles.

The governing equation for concentration in Cartesian, cylindrical, and spherical coordinates, taking into account the generalized Fick's law of mass diffusion and relaxation, is given by the following equations:

$$\tau_{mr}\left[\frac{\partial^2 C_A}{\partial t^2} + v_x\frac{\partial^2 C_A}{\partial x\partial t} + v_y\frac{\partial^2 C_A}{\partial y\partial t} + v_z\frac{\partial^2 C_A}{\partial z\partial t}\right] + \frac{\partial C_A}{\partial t} + \frac{\partial C_A}{\partial x}\left[\tau_{mr}\frac{\partial v_x}{\partial t} + v_x\right]$$

$$+\frac{\partial C_A}{\partial y}\left[\tau_{mr}\frac{\partial v_y}{\partial t} + v_y\right] + \frac{\partial C_A}{\partial z}\left[\tau_{mr}\frac{\partial v_z}{\partial t} + v_z\right] = D\left[\frac{\partial^2 C_A}{\partial x^2} + \frac{\partial^2 C_A}{\partial y^2} + \frac{\partial^2 C_A}{\partial z^2}\right] + R_A. \tag{9.40}$$

$$\tau_{mr}\left[\frac{\partial^2 C_A}{\partial t^2} + v_r\frac{\partial^2 C_A}{\partial r\partial t} + \frac{v_\theta}{r}\frac{\partial^2 C_A}{\partial\theta\partial t} + v_z\frac{\partial^2 C_A}{\partial z\partial t}\right] + \frac{\partial C_A}{\partial r}\left[\tau_{mr}\frac{\partial v_r}{\partial t} + v_r\right]$$

$$+\frac{1}{r}\frac{\partial C_A}{\partial\theta}\left[\tau_{mr}\frac{\partial v_\theta}{\partial t} + v_\theta\right] + \frac{\partial C_A}{\partial z}\left[\tau_{mr}\frac{\partial v_z}{\partial t} + v_z\right] \tag{9.41}$$

$$+\frac{\partial C_A}{\partial t} = D\left[\frac{1}{r}\frac{\partial}{\partial r}\left(r\frac{\partial C_A}{\partial r}\right) + \frac{1}{r^2}\frac{\partial^2 C_A}{\partial\theta^2} + \frac{\partial^2 C_A}{\partial z^2}\right] + R_A.$$

$$\tau_{mr}\left[\frac{\partial^2 C_A}{\partial t^2} + v_r\frac{\partial^2 C_A}{\partial r\partial t} + \frac{v_\theta}{r}\frac{\partial^2 C_A}{\partial\theta\partial t} + v_\varphi\frac{1}{r\sin\theta}\frac{\partial^2 C_A}{\partial\varphi\partial t}\right] + \frac{\partial C_A}{\partial r}\left[\tau_{mr}\frac{\partial v_r}{\partial t} + v_r\right]+$$

$$+\frac{1}{r}\frac{\partial C_A}{\partial\theta}\left[\tau_{mr}\frac{\partial v_\theta}{\partial t} + v_\theta\right] + \frac{1}{r\sin\theta}\frac{\partial C_A}{\partial\varphi}\left[\tau_{mr}\frac{\partial v_\varphi}{\partial t} + v_\varphi\right] \tag{9.42}$$

$$+\frac{\partial C_A}{\partial t} = D\left[\frac{1}{r^2}\frac{\partial}{\partial r}\left(r^2\frac{\partial C_A}{\partial r}\right)\right] + \frac{1}{r^2\sin\theta}\frac{\partial}{\partial\theta}\left[\sin\theta\frac{\partial C_A}{\partial\theta}\right] + \frac{\partial}{r^2\sin^2\theta}\frac{\partial^2 C_A}{\partial\varphi^2} + R_{,}$$

Six reasons were listed to seek a generalized Fourier's law of heat conduction (Sharma [27]). By analogy, the generalized Fick's law of diffusion needs to be considered;

$$J'' = -D_{AB}\frac{\partial C_A}{\partial x} - \tau_r\frac{\partial J''}{\partial t} \tag{9.43}$$

The Stokes–Einstein formula for diffusion coefficients is limited to cases in which the solute is larger than the solvent. Predictions for liquids are not as accurate as for gases. The Wilke and Chang [16] correlation for diffusion in liquids is an empirical correlation and is given by

$$D = 7.4\ E{-}8\ (\varphi M_2)^{1/2}T/\mu V_1^{0.6}. \tag{9.44}$$

## Example 9.2    Effect of Temperature on Relaxation Time

Write an expression for the relaxation time during diffusion of the considered species in liquids. Combine this expression with that for the effect of temperature on diffusion coefficient and obtain the dependence of relaxation time on temperature.

Equation (9.33) can be multiplied by the Avogadro number:

$$\tau_r = DM/RT. \tag{9.45}$$

Combining Equations (9.44) and (9.45):

$$\tau_r = 7.4\ E{-}8\ (\varphi M_2)^{3/2}/R\mu V_1^{0.6}. \tag{9.46}$$

It can be seen that the relaxation time becomes independent of temperature and depends only on the viscosity of the fluid and molecular size parameters. Hildebrand adapted a theory of viscosity to self-diffusivity:

$$D_{AA} = B(V - V_{ms})/V_{ms}, \tag{9.47}$$

where $V$ is the molar volume and $V_{ms}$ is the molar volume at which fluidity is zero. The Siddiqi and Lucas correlation [17] for aqueous solutions can be written as

$$D_{Aw}^0 = 2.98\ E{-}7\ V_A^{-0.5473}\mu_w^{-1.026}T. \tag{9.48}$$

For hydrocarbon mixtures, the Haydeek–Minhas [18] correlation can be used:

$$D_{AB}^0 = 13.3\ E{-}8\ T^{1.47}\mu_B^{(10.2/V_A - 0.791)}V_A^{-.71}. \tag{9.49}$$

When electrolytes are added to a solvent, they dissociate to a certain degree. It would appear that the solution contains at least three components: solvent, anions, and cations. If the solution is to remain neutral in charge at each point, assuming the absence of any applied electric potential field, the anions and cations diffuse effectively as a single component, as with molecular diffusion. The diffusion of the anionic and cationic species in the solvent can thus be treated as a binary mixture. The theory of dilute diffusion of salts is well developed and has been experimentally verified. For dilute solutions of a single salt, the Nernst–Haskell equation is applicable:

$$D_{AB}^0 = RT/F^2\ (\ |1/n^+| + |1/n\Delta{-}|\ )/(1/\lambda_+^0 + 1/\lambda\Delta^0), \tag{9.50}$$

where $D_{AB}^0$ is diffusivity based on molarity rather than normality of dilute salt A in solvent B in cm²/s.

### 9.6.1   DIFFUSION IN CONCENTRATED SOLUTIONS

The correlations discussed earlier pertain to diffusion in dilute solutions. With increased concentration, some things are different and the considerations will thus be different. Diffusion coefficients vary with the volume fraction of the solute, often in a complex manner with extrema. Diffusion coefficients are no longer a proportionality constant, but vary with concentration and become concentration dependent. In one approach, the hydrodynamic interaction of the spheres was taken into account and the friction factor $f$ corrected for as follows (Batchelor, [19]):

$$f = 6\pi\mu R_o( 1+ 1.5\varphi_1 + \ldots),$$

where $\varphi_1$ is the volume fraction of the solute. Substituting the above in Equation (9.26):

$$-\nabla\mu_1 - (6\pi\mu R_o)(1 +1.5\varphi_1 + \ldots)\, v_1 = mdv_1/dt \tag{9.51}$$

$$k_B T\nabla c_1 + mc_1 dv_1/dt = - (6\pi\mu R_o)(1 + 1.5\varphi_1 +\ldots)c_1 v_1 \tag{9.52}$$

or

$$-(k_B T/6\pi\mu R_o)/(1 + 1.5\varphi_1 +\ldots]\, \nabla c_1 = m/6\pi\mu R_o\, /(1 + 1.5\varphi_1 +\ldots]\partial J''/\partial t + J'' \tag{9.53}$$

and

$$D = k_B T/6\pi\mu R_o\, (1 + 1.5\varphi_1 + \ldots)$$

$$\tau_r = m/6\pi\mu R_o\, /(1 + 1.5\varphi_1 +\ldots]. \tag{9.54}$$

For nonideal solutions, the chemical potential can be written as

$$\mu_1 = \mu_1^0 + k_B T\, \ln(c_1\gamma_1), \tag{9.55}$$

where $\gamma_1$ is the activity coefficient,

$$\nabla\mu_1 = k_B T/c_1\gamma_1(\gamma_1\nabla c_1 + c_1\nabla\gamma_1).$$

Substituting the above equation in Equation (9.51):

$$-D\nabla c_1\, (1 + \partial\ln\gamma_1/\partial\ln c_1) = J'' + m/(6\pi\mu R_o)\, \partial J''/\partial t. \tag{9.56}$$

The correction for diffusion coefficient given in Equation (9.56) may be attributed to a cluster of molecules in the solution.

## 9.7 DIFFUSION IN SOLIDS

### 9.7.1 MECHANISMS OF DIFFUSION

Atomic diffusion in solids is of increased interest since the phenomenal growth in VLSI (very large-scale integration) of transistors on the silicon chip. An interstitial or substitutional mechanism of diffusion is said to occur when atoms occupy specific sites in a lattice. In an interstitial mechanism of diffusion, impurity jumps from one interstitial site to the next. In the substitutional mechanism of diffusion, impurity jumps from one lattice site to the neighboring vacant lattice site. Since the concentration of vacancies is low, substitutional diffusion is much slower than interstitial diffusion. In concentrated diffusion, the atom replaces the lattice atom and moves through the interstices.

The mechanism of diffusion varies greatly depending on the crystalline structure and the nature of the solute. For crystals with lattices of cubic symmetry, the diffusivity is isotropic, but not so for noncubic crystals. In interstitial mechanisms of diffusion, small diffusing solute atoms pass through from one interstitial site to the next. The matrix atoms of the crystal lattice move apart temporarily to provide the necessary space. When there are vacancies where lattice sites are unoccupied, an atom in an adjacent site may jump into such a vacancy. This is called the *vacancy mechanism*.

NEC corporation [20] has developed an interstitial concentration simulation method. Here, a mesh is set in a simulation region of a semiconductor device. Under the condition that an area outside of the simulation region is infinite, a provisional interstitial diffusion flux at the boundary of the simulation region is calculated. Then, an interstitial diffusion rate at the boundary of the simulation is calculated by a ratio of the provisional interstitial diffusion flux to the provisional interstitial concentration. Finally, an interstitial diffusion equation is solved for each element of the mesh using the interstitial diffusion rate at the boundary.

*Crowd ion mechanism* refers to the displacement of an extra atom that can displace several atom positions, thus producing a diffusion flux. The diffusivity in a single crystal is always substantially smaller than that of a multicrystalline sample, because the latter has diffusion along the grain boundaries.

For diffusion in metals, Franklin [21] and Stark [22] gave the following expression:

$$D = a_0^2 N_f \omega, \qquad (9.57)$$

where $a_0$ is the spacing between the atoms, $N_f$ the fraction of sites vacant in the crystal, and $\omega$ the jump frequency, i.e., the number of jumps per unit time from one position to the next.

### Example 9.3 Steady Diffusion in a Hollow Cylinder

Develop the concentration profile in a hollow cylinder when a species is diffusing without any chemical reaction. Consider the concentration of the species to be held constant at the inner and outer surface of the cylinder at $C_{Ai}$ and $C_{Ao}$, respectively.

A mass balance on a thin shell of thickness $\Delta r$ at radius $r$ in the cylinder would yield

$$J''_A 2\pi r L - J''_A 2\pi (r + \Delta r)L = 0. \qquad (9.58)$$

In the limit when $\Delta r$ tends to 0,

$$-\frac{\partial (r J''_A)}{\partial r} = 0. \qquad (9.59)$$

Upon integration:

$$J_A'' = -\frac{c_1}{r} z \qquad (9.60)$$

or

$$-D_{AB}\frac{\partial C_A}{\partial r} = -\frac{c_1}{r}. \qquad (9.61)$$

Upon integration:

$$C_A = \frac{c_1 \ln(r)}{D_{AB}} + c_2. \qquad (9.62)$$

From the boundary conditions, $c_1$ and $c_2$ can be solved for:

$$r = R_o, \; C_A = C_{Ao}. \qquad (9.63)$$

$$r = R_i, \; C_A = C_{Ai}. \qquad (9.64)$$

$$c_1 = \frac{D_{AB}\left(C_{Ao} - C_{Ai}\right)}{\ln\left(\dfrac{R_o}{R_i}\right)}. \qquad (9.65)$$

$$c_2 = C_{Ao} - \ln(R_o)\frac{\left(C_{Ao} - C_{Ai}\right)}{\ln\left(\dfrac{R_o}{R_i}\right)}$$

$$c_2 = C_{Ao} - \ln(R_o)\,(C_{Ao} - C_{Ai}]/\ln(R_o/R_i). \qquad (9.66)$$

$$\frac{C_A - C_{A0}}{C_{A0} - C_{Ai}} = \frac{\ln(r) - \ln(R_o)}{\ln\left(\dfrac{R_o}{R_i}\right)}. \qquad (9.67)$$

Defining a log mean radius:

$$<R_l> = \frac{\left(R_o - R_i\right)}{\ln\left(\dfrac{R_o}{R_i}\right)}. \qquad (9.68)$$

$$\frac{C_A - C_{Ao}}{C_{Ao} - C_{Ai}} = <R_l> \frac{\ln\left(\dfrac{R_o}{r}\right)}{R_o - R_i}. \qquad (9.69)$$

## 9.7.2 Diffusion in Porous Solids

Solute movement by diffusion can be by virtue of concentration difference or by means of pressure difference. Micropores, mesopores, and macropores can be distinguished by means of pore sizes. Pore diffusion is discussed in the literature in problems such as gas–solid reactions and advanced catalysis. Knudsen diffusion may occur when the mean free path of the molecule is comparable to the pore size. When the ratio of pore size to mean free path of the molecule is about 20, molecular diffusion prevails. When $d/\lambda < 0.2$, the rate of diffusion is a function of the number of collisions of the gas molecules with the wall, and Knudsen diffusion is said to occur:

$$N_A = du_A \Delta p / 3RTl. \tag{9.70}$$

where $u_A$ is the molecular velocity of A. The Knudsen diffusion coefficient

$$D_{KA} = d/3 \, (8RT/\pi M_A)^{1/2} \tag{9.71}$$

and mean free path $\lambda$ is expressed as

$$\lambda = 3.2 \, \mu \, (RT/2\pi M). \tag{9.72}$$

In the range of $d/\lambda$ of about 0.2–20, both molecular diffusion and Knudsen diffusion are important:

$$N_A = (D_{AB,eff} \, p_t/RTz) \, \ln(N_A/N(1+ D_{AB, eff}/D_{KA,eff})$$
$$- y_{A2})/(N_A/N(1+ D_{B,eff}/D_{KA,eff}) - y_{A1})). \tag{9.73}$$

Hydrodynamic flow of gases will occur when there is a difference in absolute pressure across a porous solid. Consider a solid consisting of uniform straight capillary tubes of diameter $d_c$ and length $l$ reaching from the high-pressure to the low-pressure side. Assuming laminar flow, the Hagen–Poiseuille's law for a compressible fluid that obeys the ideal gas law can be written as

$$N_A = d^2 \, p_{t,av} \, (p_{t,1} - p_{t,2}) / 32 \mu l RT. \tag{9.74}$$

The entire pressure difference is assumed as the result of friction in the pores and ignores entrance and exit losses and kinetic energy effects.

## 9.8  DIFFUSION COEFFICIENTS IN POLYMERS

The diffusion coefficients of polymers lie between those of solids and liquids. Diffusion of high-molecular-weight assumes importance when the polymer forms a solute of a dilute solution or one component of a polymer–polymer blend. A polymer blend can be miscible or immiscible, compatible or incompatible. When two polymers are mixed to yield a product with improved properties, they are said to be a *compatible* blend. When two polymers mix at the molecular level, they are said to be a *miscible* blend. Concentrated systems in which the volume fraction of the polymer solute is large make up another category where diffusion has to be treated in a different manner compared with other systems.

A polymer molecule dissolved in a solvent can be envisioned as a necklace comprising spherical beads connected by a string [23]. The polymer molecules are separated and only interact through the solvent. The Stokes–Einstein equation for the diffusion coefficient of the polymer can be used for a Flory theta solvent. The root-mean-square radius of gyration used as a measure of the size of the polymer can be used as the radius of the solute in the Stokes–Einstein formula. These values can be measured by light scattering. For concentrated solutions, the diffusivity is given by

$$D = D_0(1 + \partial \ln \gamma_1 / \partial \ln \varphi_1).$$  (9.75)

$D_0$ includes the solute's activation energy. This must be sufficient to overcome any attractive forces that constrain near-neighboring polymer segments. This coefficient can be expected to vary with the free volume of the polymer chains. Only a fraction of the free volume, the hole free volume, will be accessed by the solute:

$$D_0 = D_0{}' \exp(-E/RT) \exp(-\omega_1 V_{10} + \omega_2 V_{20}/(\omega_1 K_1 + \omega_2 K_2).$$  (9.76)

where $E$ is the solute-polymer attractive energy, $\omega_i$ the mass fractions, $V_{i0}$ the specific critical free volumes, and $K_i$ the additional free volume parameters. These parameters are strong functions of the actual temperature minus the glass transition temperatures.

For polymer blends, the Rouse model is suggested:

$$D = k_B T / N_p \zeta,$$  (9.77)

where $N_p$ is the degree of polymerization and $\zeta$ the friction coefficient characteristic of the interaction of a bead with its surroundings.

## 9.9   TRANSIENT DIFFUSION

The transient concentration profile due to molecular diffusion can be described using Fick's second law of diffusion, and the damped wave diffusion and relaxation equation [24]. The parabolic Fick model and damped wave diffusion and relaxation model for transient mass flux at the surface, for the problem of transient diffusion in a semi-infinite medium subject to a step change in concentration at the surface, were found by Sharma [25] to be within 10% of each other for times $t > 2\tau_r$. This agrees with the Boltzmann transformation; the hyperbolic governing equation reverts to the parabolic at long times. At short times, there is a "blow-up" in the parabolic model. In the hyperbolic model, there is no singularity. This has significant implications in several industrial applications such as gel acrylamide electrophoresis, which is used in obtaining the sequence distribution of DNA and protein microstructure.

The Fick regime is valid for materials with small relaxation times, long times, and moderate-to-small mass flux rates. The wave regime and the hyperbolic model are valid for short times, high mass flux rates, and materials with long relaxation times. There were some concerns expressed in the literature that the hyperbolic mass diffusion equation violates the second law of thermodynamics. Solutions to the hyperbolic mass diffusion equation were developed. These solutions were found to be well bounded without any singularities. The hyperbolic equations were shown by Sharma [31] to yield well-bounded solutions in accordance with the second law of thermodynamics when the final condition in time was used. This condition is a more realistic representation of the transient events in molecular diffusion in practice. The physical significance of the damped wave equation needs to be borne in mind when applying it.

The solution developed by Baumeister and Hamill [32] employed Laplace transforms and was further integrated into a useful expression. A Chebyshev polynomial approximation was used to approximate the integrand with modified Bessel composite function of space and time of the first kind and first order. The error involved in the method of Chebyshev economization was estimated to be about $4.1\ 10^{-5}\ \eta\xi$. The useful expression for transient temperature was shown in Sharma [25] for a typical time of $\tau = 5$. The dimensionless temperature as a function of dimensionless distance is shown in Figure 9.3. The predictions from Baumeister and Hamill and the solution by the method of relativistic transformation are within 12% of each other on average. Close to the wave front, the error in the Chebyshev economization is expected to be small and was verified accordingly. Close to the surface, the numerical error involved in the Chebyshev economization can be expected to be significant. The method of relativistic transformation yields bounded solutions without any singularities. The transformation variable, $\psi$, is symmetric with respect to space and time. It transforms the partial differetial equation (PDE) that governs the wave temperature into a Bessel differential equation. The penetration distance beyond which there is no effect of the step change in temperature at the surface for a considered instant in time is shown in Figure 9.8. The solutions from the relativistic transformation of coordinates are an improvement over the Baumeister and Hamill solution and parabolic Fourier solution in depicting the transient heat events in a semi-infinite medium subject to a step change in boundary temperature. Four regimes in the transient temperature solution for the hyperbolic governing equation using the method of relativistic transformation of coordinates are recognized, and closed-form analytical solutions in each regime are given without any singularities. The transient temperature is also found to be in agreement with the second law of thermodynamics in all four regimes.

### 9.9.1   Fick Molecular Diffusion—Semi-Infinite Medium

Consider a semi-infinite medium at an initial concentration of a species A at $C_{A0}$ (Figure 9.2). For times greater than 0, the surface at $x = 0$ is maintained at constant surface concentration at $C_A = C_{As}$, $C_{As} > C_{A0}$. The boundary conditions and initial condition are as follows:

$$t = 0,\ C_A = C_{A0}. \tag{9.78}$$

$$x = 0,\ C_A = C_{As}. \tag{9.79}$$

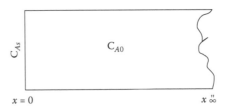

**FIGURE 9.2**   Semi-infinite slab at a initial concentration $C_{A0}$ and surface concentration $C_{As}$.

$$x = \infty, \, C_A = C_{A0}. \tag{9.80}$$

The transient concentration in the semi-infinite medium can be obtained by solving the Fick parabolic mass diffusion equations using the Boltzmann transformation, $\eta = x/\sqrt{4Dt}$, as follows. The governing equation for molecular diffusion in one dimension using Fick's second law can be written as

$$\frac{\partial C_A}{\partial t} = D_{AB} \frac{\partial^2 C_A}{\partial x^2}. \tag{9.81}$$

Equation (9.81) is a parabolic PDE of the second order in space and time.

Let

$$u = \frac{(C_A - C_{A0})}{(C_{As} - C_{A0})}; \eta = \frac{x}{\sqrt{4D_{AB}t}}. \tag{9.82}$$

Equation (9.81) becomes

$$-\frac{\eta \partial u}{2t \partial \eta} = \frac{\partial^2 u}{\partial \eta^2} \frac{D_{AB}}{4t D_{AB}} \tag{9.83}$$

or
$$-\frac{2\eta \partial u}{\partial \eta} = \frac{\partial^2 u}{\partial \eta^2}. \tag{9.84}$$

The three conditions, one in time and two in space, given by Equations (9.78–9.80), become

$$\eta = 0, u = 1. \tag{9.85}$$

$$\eta = \infty, u = 0. \tag{9.86}$$

Thus, a PDE of the second order in space and time can be transformed into an ordinary differential equation (ODE) in one variable. The transformation $\eta = x/\sqrt{4D_{AB}t}$ is called the *Boltzmann transformation*. The solution to the ODE in the transformed variable, $\eta$, can be written as

$$u = c_1 \int_0^\eta e^{-\eta^2} d\eta + c_2. \tag{9.87}$$

The integration constants, $c_1$ and $c_2$, can be solved for using the boundary conditions given by Equations (9.85–9.86). Thus:

$$u = \frac{(C_A - C_{A0})}{(C_{As} - C_{A0})} = 1 - erf\left(\frac{x}{\sqrt{4D_{AB}t}}\right). \tag{9.88}$$

The mass flux can be written as

$$J^* = \frac{J''}{(C_{As} - C_{A0})} \sqrt{\frac{\tau_r}{D_{AB}}} = \frac{\sqrt{\tau_r}}{\sqrt{4\pi t}} \exp\left(-\frac{x^2}{4D_{AB}t}\right).$$ (9.89)

The dimensionless mass flux at the surface is then given by

$$q_s^* = \frac{1}{\sqrt{\pi\tau}}.$$ (9.90)

## 9.10 DAMPED WAVE DIFFUSION AND RELAXATION

The semi-infinite medium is employed to study the spatiotemporal patterns that the solution of the non-Fick damped wave diffusion and relaxation equation exhibits. This medium has been used in the study of Fick mass diffusion. The boundary conditions can be different kinds, such as constant wall concentration, constant wall flux (CWF), pulse injection, and convective, impervious, and exponential decay. The similarity or Boltzmann transformation worked out well in the case of the parabolic PDE, where an error function solution can be obtained in the transformed variable. The conditions at infinite width and zero time are the same. The conditions at zero distance from the surface and at infinite time are the same.

Baumeister and Hamill [32] solved the hyperbolic heat conduction equation in a semi-infinite medium subjected to a step change in temperature at one of its ends using the method of Laplace transform. The space-integrated expression for the temperature in the Laplace domain had the inversion readily available within the tables. This expression was differentiated using Leibniz's rule, and the resulting temperature distribution was given for $\tau > X$ as

$$u = \frac{(C_A - C_0)}{(C_{As} - C_0)} = \exp\left(\frac{-X}{2}\right) + X\int_X^\tau \exp\left(\frac{-p}{2}\right) \frac{I_1\sqrt{p^2 - X^2}}{\sqrt{p^2 - X^2}} dp.$$ (9.91)

The method of relativistic transformation of coordinates is evaluated to obtain the exact solution for the transient temperature. Consider a semi-infinite slab at initial concentration $C_0$, imposed by a constant wall concentration $C_s$ for times greater than zero at one of the ends. The transient concentration as a function of time and space in one dimension is obtained, yielding the dimensionless variables

$$u = \frac{(C_A - C_{A0})}{(C_{AS} - C_{A0})}; \ \tau = \frac{t}{\tau_{mr}}; \ X = \frac{x}{\sqrt{D\tau_{mr}}}; \ J^* = \frac{J''}{\sqrt{\frac{D}{\tau_r}}(C_{AS} - C_{A0})}.$$ (9.92)

The mass balance on a thin spherical shell at $x$ with thickness $\Delta x$ is written in one dimension as $-\partial J^*/\partial X = \partial u/\partial \tau$. The governing equation can be obtained in terms of the mass flux after eliminating the concentration between the mass balance equation and the non-Fick expression:

$$\frac{\partial J^*}{\partial \tau} + \frac{\partial^2 J^*}{\partial \tau^2} = \frac{\partial^2 J^*}{\partial X^2}.$$

(9.93)

It can be seen that the governing equation for the dimensionless mass flux is identical in form with that of the dimensionless concentration. The initial condition is

$$\tau = 0, \ J^* = 0.$$

(9.94)

The boundary conditions are

$$X = \infty, \ J^* = 0.$$

(9.95)

$$X = 0, \ C = C_s; \ u = 1.$$

(9.96)

Let us suppose that the solution for $J^*$ is of the form $w\exp(-n\tau)$ for $\tau > 0$, where $W$ is the transient wave flux. Then, when $n = \frac{1}{2}$, Equation (9.95) becomes

$$\frac{\partial^2 w}{\partial \tau^2} - \frac{w}{4} = \frac{\partial^2 w}{\partial x^2}.$$

(9.97)

The solution to Equation (9.97) can be obtained by the following relativistic transformation of coordinates, for $\tau > X$. Let $\eta = (\tau^2 - X^2)$. Then Equation (9.97) becomes

$$\frac{\partial^2 w}{\partial \tau^2} = 4\tau^2 \frac{\partial^2 w}{\partial \eta^2} + 2\frac{\partial w}{\partial \eta}.$$

(9.98)

$$\frac{\partial^2 w}{\partial X^2} = 4X^2 \frac{\partial^2 w}{\partial \eta^2} - 2\frac{\partial w}{\partial \eta}.$$

(9.99)

Combining Equations (9.98 and 9.99) into Equation (9.97):

$$4(\tau^2 - X^2)\frac{\partial^2 w}{\partial \eta^2} + 4\frac{\partial w}{\partial \eta} - \frac{-w}{4} = 0.$$

(9.100)

$$\eta^2 \frac{\partial^2 w}{\partial \eta^2} + \eta\frac{\partial w}{\partial \eta} - \frac{\eta w}{16} = 0.$$

(9.101)

Equation (9.101) can be seen to be a special differential equation with one independent variable. The number of variables in the hyperbolic PDE has thus been reduced from two to one. Comparing Equation (9.101) with the generalized form of Bessel's equation, it can be seen that $a = 1$, $b = 0$, $c = 0$, $s = \frac{1}{2}$, $d = -1/16$. The order of the solution is calculated as 0, and the general solution is given by

$$w = c_1 I_0 \left[ \frac{\sqrt{\tau^2 - X^2}}{2} \right] + c_2 K_0 \left[ \frac{\sqrt{\tau^2 - X^2}}{2} \right] \qquad (9.102)$$

The wave flux $w$ is finite when $\eta = 0$ and, hence, $c_2$ can be seen to be zero. $c_1$ can be solved from the boundary condition given in Equation (9.93). The expression for the dimensionless mass flux for times $\tau$ greater than $X$ is thus,

$$J^* = c_1 \exp\left(\frac{-\tau}{2}\right) I_0 \left[ \frac{1}{2} \sqrt{\tau^2 - X^2} \right]. \qquad (9.103)$$

For large times, the modified Bessel's function can be given as an exponential and reciprocal of the square root of time by asymptotic expansion. Consider the surface flux, that is, when in Equation (9.102) $X$ is set to zero.

$$J^* = c_1 \exp\left(\frac{-\tau}{2}\right) \frac{\exp\left(\frac{\tau}{2}\right)}{\sqrt{\pi \tau}} = \frac{c_1}{\sqrt{\pi \tau}} \qquad (9.104)$$

For times, when $\exp(\tau)$ is much greater than the mass flux, it can be seen that the second derivative in time of the dimensionless flux in Equation (9.93) can be neglected compared with the first derivative. The resulting expression is the familiar expression for surface flux from the Fourier parabolic governing equation for constant wall concentration in a semi-infinite medium and is given by

$$J^* = \frac{1}{\sqrt{\pi \tau}}. \qquad (9.105)$$

Comparing Equations (9.104) and (9.105) it can be seen that $c_1$ is 1. Thus, the dimensionless heat flux is given by

$$J^* = \exp\left(\frac{-\tau}{2}\right) I_0 \left( \frac{\sqrt{\tau^2 - X^2}}{2} \right). \qquad (9.106)$$

The solution for $J^*$ needs to be converted to the dimensionless concentration $u$ and then the boundary conditions applied. From the mass balance:

$$-\frac{\partial J^*}{\partial X} = \frac{\partial u}{\partial \tau}. \qquad (9.107)$$

Thus, differentiating Equation (9.106) with respect to $X$, and substituting in Equation (9.107) and integrating both sides with respect to $\tau$: for $\tau > X$,

$$u = \int \exp\left(\frac{-\tau}{2}\right) \left[ \frac{I_1 \frac{1}{2}\sqrt{\tau^2 - X^2}}{\sqrt{\tau^2 - X^2}} \right] d\tau + c(X).$$ (9.108)

It can be left as an indefinite integral, and the integration constant can be expected to be a function of space. $c(X)$ can be solved for by examining what happens at the wave front. At the wave front, $\eta = 0$ and time elapsed equals the time taken for a mass disturbance to reach the location $x$ given the wave speed $\sqrt{D/\tau_{mr}}$. The governing equations for the dimensionless mass flux and dimensionless concentration are identical in form. At the wave front, Equation (9.100) reduces to

$$\frac{\partial w}{\partial \eta} = \frac{w}{16}$$

or

$$w = c' \exp\frac{\eta}{16} = c'.$$ (9.109)

$$u = c' \exp\frac{-\tau}{2} = c' \exp\frac{-X}{2}.$$ (9.110)

Thus, $c(X) = c' \exp- X/2$. From the boundary condition in Equation (9.95), it can be seen that $c' = 1$. Thus, for $\tau > X$,

$$u = \int X \exp\frac{-\tau}{2} \frac{I_1 \frac{1}{2}\sqrt{\tau^2 - X^2}}{\sqrt{\tau^2 - X^2}} d\tau + c' \exp\frac{-X}{2}.$$ (9.111)

It can be seen that the boundary conditions are satisfied by the Equation (9.111) and describe the transient concentration as a function of space and time that is governed by the hyperbolic wave diffusion and relaxation equation. The flux expression is given by Equation (9.106).

It can be seen that expressions for dimensionless mass flux and dimensionless concentration given by Equations (9.106) and (9.111) are valid only in the open interval for $\tau > X$. When $\tau = X$, the wave front condition results, and the dimensionless mass flux and concentration are identical and is

$$J^* = u = \exp\frac{-X}{2} = \exp\left(\frac{-\tau}{2}\right).$$ (9.112)

When $X > \tau$, the transformation variable can be redefined as $\eta = X^2 - \tau^2$. Equation (9.101) becomes

$$\eta^2 \frac{\partial^2 w}{\partial \eta^2} + \eta \frac{\partial w}{\partial \eta} + \eta \frac{w}{16} = 0. \tag{9.113}$$

The general solution for this Bessel equation is given by

$$w = c_1 J_0\left[\frac{\sqrt{\eta}}{2}\right] + c_2 Y_0\left[\frac{\sqrt{\eta}}{2}\right]. \tag{9.114}$$

The wave temperature, $W$, is finite when $\eta = 0$ and, hence, it can be seen that $c_2$ can be seen to be zero. $c_1$ can be solved from the boundary condition given in Equation (9.96). The expression in the open interval or the dimensionless heat flux for times $\tau$ smaller than $X$ is thus

$$J^* = c_1 \exp(\frac{-\tau}{2}) J_0\left[\frac{\sqrt{X^2 - \tau^2}}{2}\right] \tag{9.115}$$

On examining the Bessel function in Equation (9.115), it can be seen that the first zero of the Bessel occurs when the argument becomes 2.4048. Beyond that point, the Bessel function will take on negative values, indicating a reversal of heat flux. There is no good reason for the mass flux to reverse in direction at short times. Hence, Equation (9.115) is valid from the wave front down to where the first zero of the Bessel function occurs. Thus, the plane of zero transfer explains the initial condition verification from the solution.

By using the expression at the wave front for the dimensionless mass, flux $c_1$ can be solved for and is found to be 1. Equation (9.115) can also be obtained directly from Equation (9.104) by using $I_0(\eta) = J_0(i\eta)$. The expression for temperature in a similar vein for the open interval $X > \tau$ is thus

$$u = \int X \exp\left(\frac{-\tau}{2}\right) \frac{J_1\left[\frac{\sqrt{\tau^2 - X^2}}{2}\right]}{\sqrt{\tau^2 - X^2}} d\tau + \exp\left(\frac{-X}{2}\right). \tag{9.116}$$

Consider a point $X_p$ in the semi-infinite medium. Three regimes can be identified in the mass flux at this point from the surface as a function of time. The series expansion of the modified Bessel composite function of the first kind and zeroth order was used using a Microsoft Excel spreadsheet on a Pentium IV desktop microcomputer. The *four regimes* and the mass flux at the wave front are summarized as follows:

1. The first regime is a thermal inertial regime when there is no transfer.
2. The second regime is given by expression Equation (9.115) for the mass flux and

$$J^* = \exp\left(\frac{-\tau}{2}\right) J_0\left[\frac{\sqrt{X^2 - \tau^2}}{2}\right]. \tag{9.117}$$

The first zero of the zeroth-order Bessel function of the first kind occurs at 2.4048. This is when

$$2.4048 = \frac{\sqrt{X^2 - \tau^2}}{2} \quad \text{or} \quad \tau_{lag} = \sqrt{-23.132}. \tag{9.118}$$

Thus $\tau_{lag}$ is the inertial lag that will ensue before the mass flux is realized at an interior point in the semi-infinite medium at a dimensionless distance $X$ from the surface. As demonstrated, one value of $X$ is used, that is, 5. Thus, for points closer to the surface, the time lag may be zero. Only for dimensionless distances greater than 4.8096 is the time lag finite. For distances *closer than 4.8096* $\sqrt{(\alpha \tau_r)}$ the thermal lag experienced *will be zero*. For distances

$$x > 4.8096\sqrt{\alpha \tau_{mr}}. \tag{9.119}$$

the time lag experienced is given by Equation (9.118) and is $\sqrt{(X^2 - 4\beta_1^2}$, where $\beta_1$ is the first zero of the Bessel function of the first kind and zeroth order and is 2.4048. In a similar fashion, the penetration distance of the disturbance for a considered instant in time beyond which the change in initial temperature is zero can be calculated as

$$X_{pen} = \sqrt{23.132 + \tau_i^2}.$$

3. The third regime starts at the wave front and described by Equation (9.106).

$$J^* = \exp\left(\frac{-\tau}{2}\right) I_0\left(\frac{\sqrt{\tau^2 - X^2}}{2}\right). \tag{9.120}$$

4. At the wave front, $J^* = u = \exp(-X/2) = \exp(-\tau/2)$.

The expressions for transient concentration derived in the preceding section needs integration prior to use. More easily usable expressions can be developed by making suitable approximations. Realizing that for PDE, a set of functions instead of constants, as in the case of ODE, needs to be solved from the boundary condition; Equation (9.96) is allowed to vary with time. This results in an expression for transient concentration that is more readily available for direct use of the practitioner. Extension to three dimensions in space is also straightforward with this method.

In this section, the problem to find the exact solution for the transient concentration, in a semi-infinite-1D-medium is revisited since the work reported by Baumeister and Hamill [32]. An expression that does not need further integration is attempted to be derived in this section. Consider a semi-infinite slab at initial concentration $C_0$, subjected to a sudden change in concentration at one of the ends to $C_s$. The mass propagative velocity is $V_m = \sqrt{(D_{AB}/\tau_r)}$. The initial condition

$$t = 0, \; Vx, \; C = C_0 \tag{9.121}$$

$$t > 0, \; x = 0, \; C = C_s \tag{9.122}$$

$$t > 0, \; x = \infty, \; C = C_0. \tag{9.123}$$

Obtaining the dimensionless variables:

$$u = \frac{(C - C_0)}{(C_s - C_0)}; \; \tau = \frac{t}{\tau_{mr}}; \; X = \sqrt{D\tau_{mr}}. \tag{9.124}$$

The mass balance on a thin spherical shell at $x$ with thickness $\Delta x$ is written. The governing equation can be obtained from eliminating $J''$ between the mass balance equation as given in Equation (9.107) and the hyperbolic damped wave diffusion and relaxation equation as given in Equation (9.123). The governing equation can be rendered dimensionless and seen to be

$$\frac{\partial u}{\partial \tau} + \frac{\partial^2 u}{\partial \tau^2} = \frac{\partial^2 u}{\partial X^2}. \tag{9.125}$$

Suppose $u = \exp(-n\tau)\, w\,(X, \tau)$. By choosing $n = \frac{1}{2}$, the damping component of the equation is removed. Thus, for $n = \frac{1}{2}$, the governing equation becomes

$$\frac{\partial^2 w}{\partial \tau^2} - \frac{w}{4} = \frac{\partial^2 w}{\partial x^2}. \tag{9.126}$$

The solution to Equation (9.126) can be obtained by the following relativistic transformation of coordinates, for $\tau > X$. Let $\eta = (\tau^2 - X^2)$. Then, Equation (9.101) becomes

$$\frac{\partial^2 w}{\partial \tau^2} = 4\tau^2 \frac{\partial^2 w}{\partial \eta^2} + 2\frac{\partial w}{\partial \eta} \tag{9.127}$$

$$\frac{\partial^2 w}{\partial X^2} = 4X^2 \frac{\partial^2 w}{\partial \eta^2} - 2\frac{\partial w}{\partial \eta} \tag{9.128}$$

Combining Equations (9.127, 9.128) into Equation (9.126):

$$4(\tau^2 - X^2)\frac{\partial^2 w}{\partial \eta^2} + 4\frac{\partial w}{\partial \eta} - \frac{-w}{4} = 0 \tag{9.128a}$$

$$\eta^2 \frac{\partial^2 w}{\partial \eta^2} + \eta\frac{\partial w}{\partial \eta} - \frac{\eta w}{16} = 0 \tag{9.129}$$

Equation (9.129) can be seen to be a special differential equation in one independent variable. The number of variables in the hyperbolic PDE has thus been reduced from two to one. Comparing Equation (9.129) with the generalized form of Bessel's equation, it can be seen that $a = 1$, $b = 0$, $c = 0$, $s = \frac{1}{2}$, $d = -1/16$. The order of the solution is calculated as 0, and the general solution is given by

$$w = c_1 I_0\left[\frac{\sqrt{\tau^2 - X^2}}{2}\right] + c_2 K_0\left[\frac{\sqrt{\tau^2 - X^2}}{2}\right]. \tag{9.130}$$

The wave temperature, $w$, is finite when $\eta = 0$ and, hence, it can be seen that $c_2$ is zero. $c_1$ can be solved from the boundary condition given in Equation (9.122). For $X = 0$, $u$ is 1. Writing the expression for $X = 0$:

$$1 = c_1 \exp\left(\frac{-\tau}{2}\right) I_0\left(\frac{\sqrt{\eta}}{2}\right). \tag{9.131}$$

$c_1$ can be eliminated by dividing Equation (9.129) after setting $c_2 = 0$ by Equation (9.128) to yield in the open interval of $\tau > X$:

$$u = \frac{I_0\left[\dfrac{\sqrt{\tau^2 - X^2}}{2}\right]}{I_0\left[\dfrac{\tau}{2}\right]}. \tag{9.132}$$

In the open interval $X > \tau$:

$$u = \frac{J_0\left[\dfrac{\sqrt{X^2 - \tau^2}}{2}\right]}{I_0\left[\dfrac{\tau}{2}\right]}. \tag{9.133}$$

It can be inferred that an expression in time is used for $c_1$. A domain-restricted solution for short and long times may be in order. The dimensionless concentration profile as a function of dimensionless distance for different values of dimensionless times is shown in Figure 9.3.

## 9.11 PERIODIC BOUNDARY CONDITION

Consider a semi-infinite slab at an initial concentration $C_0$, imposed by a periodic concentration at one of the ends by $C_0 + C_1 \cos(wt)$. The transient concentration as a function of time and space in one dimension is obtained. Obtaining the dimensionless variables:

$$u = \frac{(C - C_0)}{C_1}; \tau = \frac{t}{\tau_{mr}}; X = \frac{x}{\sqrt{D\tau_{mr}}} \quad u = (C - C_0)/(C_1); \tau = t/\tau_r; X = x/\sqrt{(D\tau_r)}. \tag{9.134}$$

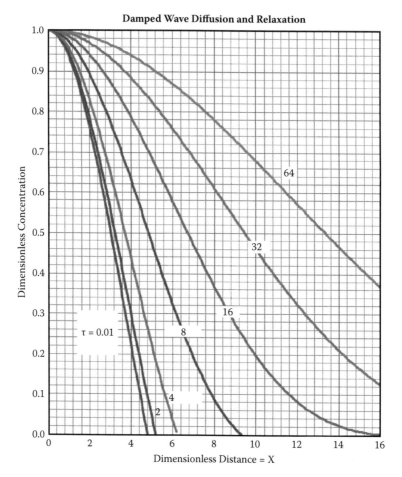

**FIGURE 9.3**  Concentration profile under damped wave diffusion and relaxation in semi-infinite medium.

The governing equation is obtained after eliminating $J$ between the mass balance equation and the derivative with respect to $x$ of the flux equation and introducing the dimensionless variables. The mass balance equation is obtained by shell balance at a distance $x$ from the origin with a slice thickness delta $x$. The initial condition is

$$t = 0, C = C_0 ; u = 0. \tag{9.135}$$

The boundary conditions are

$$X = \infty, C = C_0 ; u = 0 \tag{9.136}$$

$$X = 0, C = C_0 + C_1 \cos(\omega t); u = \cos(\omega^*\tau). \tag{9.137}$$

Let us suppose that the solution for $u$ is of the form $f(x)\exp(-i\omega^*\tau)$ for $\tau > 0$, where $\omega$ is the frequency of the concentration wave imposed on the surface and $C_1$ is the amplitude of the wave. Then

$$(-i\omega^*)\,f\exp(-i\omega^*\tau) + (i^2\omega^{*2}\,)f\exp(-i\omega\tau) = f''\exp(-i\omega^*\tau). \qquad (9.138)$$

$$i^2\,f(\omega^{*2} + i\omega^*) = f''$$

$$f(X) = c\,\exp(-iX\omega^*\sqrt{(\omega^* + i)}). \qquad (9.139)$$

$d$ can be seen to be zero as at $X = \infty$, $u = 0$.

$$u = c\exp(-iX\omega^*\sqrt{\omega^* + i}\,)\exp(-i\omega^*\tau). \qquad (9.140)$$

From the boundary condition at $X = 0$:

$$\cos(\omega^*\tau) = \text{real part } (c\exp(-i\omega^*\tau)\,) \text{ or } c = 1. \qquad (9.141)$$

$$u = \exp(-X\omega^*(A + iB)\exp(-i\omega^*\tau) = \exp(-A\omega^*X)\exp(-i(BX\omega^* + \omega^*\tau). \qquad (9.142)$$

$$\text{where } A + iB = i\sqrt{(\omega^* + i)}. \qquad (9.143)$$

Squaring both sides:

$$A^2 - B^2 + 2ABi = i^2\,(\omega^* + i) = -\omega^* - i. \qquad (9.144)$$

$$A^2 - B^2 = -\omega^*;\ 2AB = -1 \text{ or } B = -1/2A$$

$$\text{Or } A^2 - 1/4A^2 = -\omega^*. \qquad (9.145)$$

$$A^2 = (-\omega^* \pm \sqrt{(\omega^{*2} + 1)}\,)/2;\ B = -1/2A. \qquad (9.146)$$

Obtaining the real part:

$$u = \exp(-A\omega^*X)\cos(\omega^*(BX + \tau)). \qquad (9.147)$$

The time lag in the propagation of the periodic disturbance at the surface is captured by the foregoing relationship. Thus, the boundary conditions can be seen to be satisfied by Equation (9.145). In a similar vein to the supposition of $f(x)\exp(-i\omega^*\tau)$, the mass flux $J''$ can be supposed to be of the form $J^* = g(x)\exp(-i\omega^*\tau)$. Thus,

$$g = \frac{f'}{(1 - i\omega^*)}. \qquad (9.148)$$

Combining $f$ from Equation (9.139) into Equation (9.148):

$$J^* = -\omega^*(A + iB)\,\exp(-X\omega^*(A + iB)\exp(-i\omega^*\tau))$$
$$= -\omega^*(A + iB)\exp(-A\omega^*X)\exp(-i(BX\omega^* + \omega^*\tau)) \qquad (9.149)$$

$$= -\omega^*(A + iB)\exp(-A\omega^*X)(\cos(BX\omega^* + \omega^*\tau) + i\sin(BX\omega^* + \omega^*\tau)\,).$$

Obtaining the real part:

$$J'' = \sqrt{\frac{D}{\tau_{mr}}}\ \omega^*\,\exp(-A\omega^*X)(B\,\sin(\omega^*(BX + \tau)) - A\cos(\omega^*(BX + \tau))). \qquad (9.150)$$

## 9.12  SUMMARY

The Flory–Huggins interaction parameter, $\chi$, was related to the binary diffusion coefficient.

Diffusion is a phenomenon in which a species migrates from a region of higher concentration to one of lower concentration. The driving force for motion is the concentration gradient. The Skylab demonstration experiments by Fascimire document the diffusion of tea in water under reduced-gravity conditions. The lowest achievable concentration is 0 mol m$^{-3}$, by law. Fick's first and second laws of diffusion can be written as

$$J = -D_{ij} A \frac{\partial C_i}{\partial x}$$

$$D_{ij} \frac{\partial^2 C_i}{\partial x^2} = \frac{\partial C_i}{\partial t}.$$

The $N$ and $J$ fluxes are different: $J$ describes molecular diffusion and $N$ the migration due to bulk motion. The diffusion coefficient varies with temperature. In gases, the correlations of Chapman and Cowling, Wilkee-Lee, and Mathur and Todos, and Catchpole and King are presented. Binary diffusion coefficient values for commonly available gases are given in Table 9.1. For liquids, the Stokes–Einstein relation for diffusion coefficients was derived. During the derivation, when accounting for the acceleration regime of the solute molecule, the generalized Fick's laws of diffusion were derived:

$$J'' + \frac{M_i D_{ij}}{RT} \frac{\partial J''}{\partial t} = -\left( \frac{RT}{6\pi\mu R_o} \right) \frac{\partial C_i}{\partial x}$$

Correlations of Nernst–Haskell [9] for electrolytes were mentioned. The effect of concentration, that is, dilute versus concentrated solutions were separately discussed. Correlations of Wilkee–Chang, Siddiqi–Lucas, and Haydeek–Minhas were described. The diffusion mechanism in solids was discussed. The various mechanisms of diffusion such as vacancy mechanism, interstitial mechanism, substitutional mechanism, and crowd ion mechanism were outlined. Knudsen diffusion, when the mean free path of the molecule is greater than the diffusion path, as in pore diffusion, was discussed. Diffusion in polymers and the Arrhenius dependence of the diffusion coefficient with temperature were discussed.

Transient diffusion in a semi-infinite medium was studied under a constant wall concentration boundary condition using Fick's second law of diffusion and the damped wave diffusion and relaxation equation. The latter can account for finite speed of propagation of mass. A new procedure called the method of relativistic transformation was developed to obtain bounded and physically realistic solutions. These solutions were compared with the solution from Fick's second law of diffusion obtained using the Boltzmann transformation and the solution presented in the literature by Baumesiter and Hamill [32]. Four different regimes of the solution were recognized: (1) an inertial regime with zero transfer; (2) a second regime characterized by Bessel composite function of space and time of the zeroth order and first kind; (3) a third regime characterized by modified Bessel composite function of space and time of the zeroth order and the first kind and; (4) a wave front regime. The characteristics of the solution to the damped wave diffusion and relaxation equation subject to the periodic boundary condition by the method of complex concentration were discussed. The transient concentration profile from the relativistic transformation method was presented in Figure 9.3. The profile has a point of inflection, zero curvature at $X = 0$. Mathematical expressions for penetration length and inertial lag time were derived.

## PROBLEMS

1. *Estimate of Diffusion Coefficient of Argon in Hydrogen*

   Calculate the diffusion coefficient of argon in hydrogen at 2.5 atm and 210K.

   Compare this with the experimental values reported in the literature.

2. *Parabolic Law of Oxidation*

   During the corrosion of metals, an oxide layer is formed on the metal. Assuming that the oxygen diffuses through the oxide layer, show that the thickness of the oxide layer, $\delta$, can be give by $(C_{bulk} D_{AB} t/\rho_m)^{1/2}$ using Fick's law of diffusion.

   A gentle breeze is blowing at a constant velocity of $U$ over the corroded layer. Is this going to increase the rate of corrosion due to the convection contribution?

3. *Sacred Pond*

   Evaporation from ponds is retarded by the introduction of lotus leaves into the sacred ponds in temples. Assume that in a pond of area 11 m × 9 m, 4130 leaves, each with a diameter of 2 in., were placed. Calculate the reduction in diffusion rate on account of the reduction in area in the path of evaporation.

4. *Diffusion of Oxygen through Spiracles*

   Many insects breathe through spiracles. Spiracles are open tubes that extend into the insect's body. Oxygen diffuses from the surrounding air and gas exchange takes place through the walls. For every mole of oxygen diffusing in, one mole of $CO_2$ diffuses out. The walls of the spiracle are coated with a cuticle of 10 µm thickness, which reduces water losses. The oxygen concentration outside the cuticle is constant and is 5% of the equilibrium concentration. What is the local oxygen flux in the spiracle to the tissue? Derive an oxygen concentration profile within the tissue. Is the spiracle an efficient method of respiration?

   Spiracle radius—200 µm oxygen solubility

   Spiracle length—8 mm in tissue $C_t = 0.1$ mmol/L

   $D_{o,cuticle}$—2 E−5 cm²/s

   $D_{o,air}$—0.12 cm²/s

5. *Scrubbing of $SO_2$*

   During coal combustion, the emission of sulfur dioxide from power plants can be reduced by using CaO scrubbers. In the scrubber,

   $$2CaO + 2SO_2 + O_2 + 2CaSO_4.$$

Consider the diffusion of $SO_2$ into a spherical particle of CaO. Show that governing the equation can be derived from the shell balance as follows:

$$D_{AB} \frac{\partial}{r^2 \partial r}\left(r^2 \frac{\partial C_A}{\partial r}\right) = k''' C_A.$$

Show that the concentration profile of $SO_2$ in the spherical lime particle can be written as

$$\frac{C_A}{C_{As}} = \frac{I_{1/2}\left(r\sqrt{\frac{k'''}{D_{AB}}}\right)}{XI_{1/2}\left(R\sqrt{\frac{k'''}{D_{AB}}}\right)}.$$

The Thiele modulus is $\phi = R\sqrt{\frac{k'''}{D_{AB}}}$.

6. *Coextrusion*

   In the manufacture of the casings of solid rocket motor (SRM), the material requirements are bifunctional. They have to have high hoop strength on one side and high ablation resistance on the other. In order to prepare such materials, the technology of coextrusion is utilized. In a twin-screw extruder, both the materials are coextruded together. During the residence time of the polymers in the extruder, the interdiffusion of either material in the other occurs. Calculate the interlayer thickness as a function of the extruder residence time and diffusivities of the two materials.

7. *Diffusion Coefficient of Milk in the Refrigerator*

   Estimate the diffusion coefficient of lactic acid in the refrigerator. Compare this with the value at room temperature and that of the milk through the plastic container.

8. *Restriction Mapping*

   Endonucleases or restriction enzymes can be used to "cut" the unmethylated DNA at several sites and restrict their activity. About 300 restriction enzymes are known, and they act upon 100 distinct restriction sites that are palindromes. Some cut leaves' blunt ends, and others leave them sticky. The restriction fragment lengths can be measured by using the technique of gel electrophoresis. The solid matrix is the gel, usually agarose or polyacrylamide that is permeated with liquid buffer. As DNA is a negatively charged molecule, when placed in an electric field, the DNA migrates toward the positive pole. DNA migration is a function of its size. Calibration is used to relate the migration distance to size. The migration distance of DNA under a field for a set time is measured. The DNA molecule is made to fluoresce and made visible under ultraviolet light by staining the gel with ethidium bromide. A second method is to tag the DNA with a radioactive label and then expose the x-ray film to the gel. Show that the migration under gel electrophoresis can be given by

$$J_{frag} = -\left(z_A u_A F\right)\frac{\partial E}{\partial x} - D_{frag}\frac{\partial C_A}{\partial x}.$$

Show that the governing equation can be written in one dimension as

$$0 = D_{frag}\frac{\partial^2 C_A}{\partial x^2} - \left(z_A u_A F\right)\frac{\partial^2 E}{\partial x^2}.$$

9. *Pheromone and Insect Control*

During insect control, controlled release of pheromones is used. Pheromones are sex attractants released by insects. When mixed with an insecticide and used, they annihilate all of one sex of a particular insect pest. The pheromone sublimation rate in the impermeable holder can is given as

$$S_0 = 9\,E{-}16\,(\,1 - 1E6\,C_1),$$

where $C_1$ is the concentration in the vapor. The diffusivity through the polymer is 1.2 E–11 cm²/s. It can be assumed that the pheromone outside the chamber is 0. If the polymeric diffusion barrier is 600 microns thick and has an area of 1.6 cm², what is the concentration of pheromone in the vapor? How fast is the pheromone released by the device?

10. *Oxygen Transport in the Eye*

The cornea is a unique, living tissue and is a transparent window through which light enters the eye to be focused on the retina, thus forming the images of our surroundings and enabling sight. When the eye is open, it receives all of its oxygen requirements from the surrounding air. Other nutrients are likely delivered via the tear duct fluid that bathes the outer surface of the cornea or the aqueous humor that fills the chamber behind the cornea and in front of the lens. Some oxygen may enter the aqueous humor from vasculature in the muscle around the periphery of the lens. When the eye is closed, it is cut off from the $O_2$ source in the air. There is a rich microvascular bed well perfused with high vascular density on the inner surface of the eyelid, which supplies the cornea with oxygen (and possibly other nutrients). What is the $pO_2$ at the surface of the cornea when the eye is closed?

Table of Model Parameters

| Layer | Thickness (µm) | Diffusion coefficient (cm²/s) | VO₂ mL O₂·mL tissue⁻¹s⁻¹ |
|---|---|---|---|
| Epithelium | 30 | 2.8 E–10 | 3.0 E–4 |
| Stroma | 440 | 2.8 E–10 | 2.0 E–5 |
| Endothelium | 20 | 2.8 E–4 | 1.0 E–4 |

11. *Loss from Beverage Containers*

Coca-Cola bottles are made of plastic. The contents diffuse at a slow rate through the walls of the container and out into the air, resulting in some losses. It has been suggested to coat the inner wall of the container to reduce the losses. With a coating thickness of 15 μm and a diffusion coefficient in the coating of 1 E−9 m²/s, what would be the benefit to the manufacturer? Assume a thickness of 2.5 mm for the plastic container and a diffusion coefficient of the contents in the plastic container as 2 E−6 m²/s.

12. *Reaction and Diffusion in a Nuclear Fuel Rod*

In autocatalytic reactions, such as those occurring during nuclear fission, the neutrons can be studied by a first-order reaction. The mass balance in a long cylindrical rod with a first-order autocatalytic reaction can be written at steady state as

$$\frac{1}{r}\frac{\partial(rJ_r)}{\partial r} + k'''C = 0.$$

The long cylindrical rod is at zero initial concentration of autocatalytic reactant A. The surface of the rod is maintained at a constant concentration $C_s$ for times greater than zero. The boundary conditions are

$$r = 0, \quad \frac{\partial C}{\partial r} = 0$$

$$r = R, \ C = C_s.$$

Show that the steady-state solution can be obtained as follows after redefining $u^s = C/C_s$:

$$\frac{\partial^2 u^s}{\partial X^2} + \frac{1}{X}\frac{\partial u^s}{\partial X} + k*u^s = 0$$

$$X^2\frac{\partial^2 u^s}{\partial X^2} + X\frac{\partial u^s}{\partial X} + X^2 k*u^s = 0,$$

The foregoing equation can be recognized as the Bessel equation. The solution is

$$u^s = c_1 J_0\left(\sqrt{k*X^2}\right) + c_2 Y_0\left(\sqrt{k*X^2}\right),$$

It can be seen that $c_2 = 0$ as the concentration is finite at $X = 0$. The boundary condition for surface concentration is used to obtain $c_1$. Thus

$$c_1 = \frac{1}{J_0\left(R\sqrt{\dfrac{k*}{D\tau_r}}\right)}$$

Thus,

$$u^s = \frac{J_0(\sqrt{k^* X^2})}{J_0\left(R\sqrt{\dfrac{k^*}{D\tau_r}}\right)}.$$

13. *Grooming Hair with Oil*

In order to keep the hair on the human skull from becoming dehydrated, some people apply oil or hair cream daily. During the course of the day, estimate the loss of oil from the human hair by diffusion. Show that there are two contributions. One is from molecular diffusion from the head to the atmosphere in the vertical direction, and the other is by convection from a wind blowing in the horizontal direction. Show that the governing equation can be given by

$$\frac{\partial^2 u}{\partial z^2} = \frac{U d_{hair}}{D} \frac{\partial u}{\partial x}.$$

Show that the solution for the concentration profile of the oil in the surrounding region of the human skull at steady state can be given by

$$u = 1 - erf\left(Z\sqrt{\frac{Pe_m}{4X}}\right).$$

Assuming that the diameter of the hair is 2 microns, the velocity of air is 1 m/s, and the diffusivity is 1 E–7 m²/s, estimate the time taken for a layer of cream of 1 micron to be replaced. Make suitable assumptions such as a cranial area of 1500 cm² and hair length of 5 cm.

14. *Dyeing of Wool*

A dye bath at a concentration of $C_0$ and a volume $V$ is used to dye wool that is bathed in it. The dye diffuses into the wool. Measuring the concentration of the dye in the wool as a function of time (a) Can you estimate the diffusion coefficient of the dye? If so, how? (b) Can you estimate the relaxation time?

15. *Dopant Profile by Ion Implantation*

Ion implantation is used to introduce dopant atoms into the semiconductor material to alter its electrical conductivity. During ion implantation, a beam of ions containing the dopant is directed at the semiconductor surface. For example, boron atoms are implanted into silicon wafers by Lucent Technologies, Murray Hill, New Jersey. Assume that the transfer of boron into the silicon surface is on account of both the convection and diffusion contributions at steady state. Show that the governing equation for the transfer of boron at the gas–solid interface is given by

$$-\frac{\partial C_A}{\partial z} = D_{AB}\frac{\partial^2 C_A}{\partial z^2}.$$

Given a characteristic length l, show that the equation can be reduced to

$$-Pe_m \frac{\partial u}{\partial Z} = \frac{\partial^2 u}{\partial Z^2}$$

and the solution is

$$u = 1 - \frac{J_{ss}^*}{Pe_m} e^{-Pe_m Z}.$$

16. *Soot from the Steam Engine*

The steam engine that powers the train that takes you from Chennai to New Delhi in 31 h discharges coal dust at steady rate of 68 kg-mol/h. The train moves at a velocity of 90 km/h. Estimate the thickness of soot that will deposit on a passenger sitting near the window of S6 during the entire journey. S6 is about 200 feet from the engine. Assume that the diffusion coefficient of the soot in air is 1 E−6 m²/s. Repeat the analysis for a wind speed of 10 km/h. (*Hint*: Bulk concentration of soot in surrounding air can be calculated by considering the basis of time as that taken for the passenger to move 600 ft to the discharge point in fixed space, and in that time the discharge amount calculated from the discharge rate and the dispersed region from the penetration length in all three directions).

17. *Steady Diffusion in a Hollow Sphere*

Develop the concentration profile in a hollow sphere when a species is diffusing without any chemical reaction. Consider the concentration of the species to be held constant at the inner and outer surface of the cylinder at $C_{Ai}$ and $C_{Ao}$, respectively. Show that

$$\frac{C_A - C_{Ai}}{C_{Ai} - C_{Ao}} = \frac{\left(1 - \dfrac{R_i}{r}\right)}{\left(1 - \dfrac{R_i}{R_o}\right)}.$$

18. *Determination of Diffusivity*

Unimolar diffusion can be used to estimate the binary diffusivity of a binary gas pair. Consider the evaporation of $CCl_4$, carbon tetrachloride, into a tube containing oxygen. The distance between the $CCl_4$ level and the top of the tube is 16.5 cm. The total pressure in the system is 760 mm Hg and the temperature −5°C. The vapor pressure of $CCl_4$ at that temperature is 29.5 mm Hg. The area of the diffusion path in the diffusion tube may be taken as 0.80 cm². Determine the binary diffusivity of $O_2$–$CCl_4$ when in an 11 h period after steady state, 0.026 cm³ of $CCl_4$ is evaporated.

19. *Helium Separation from Natural Gas*

A novel method to separate helium from natural gas was proposed. It was noted that Pyrex glass is almost impermeable to all gases but helium. The diffusion coefficient of helium is 25 times the diffusion coefficient of hydrogen. Consider a Pyrex tubing of length $L$ and inner and outer radii, $R_i$ and $R_0$. Show that the rate at which helium will diffuse through the Pyrex can be given by

$$J_{He} = \frac{2\pi LD_{He,pyrex}\left(C_{He,1} - C_{He,2}\right)}{\ln\left(\dfrac{R_o}{R_i}\right)}.$$

20. *Solid Dissolution into a Falling Film*

A liquid is flowing in laminar motion down a vertical wall. The wall consists of a species that is slightly soluble in the liquid. Show that the governing equation for species diffusing into the liquid from the wall can be written as

$$\frac{\partial^2 u}{\partial z^2} = \frac{UL}{D}\frac{\partial u}{\partial x}.$$

Show that an error function solution results for the foregoing PDE.

21. *Carburizing Steel*

Low-carbon steel can be hardened in order to improve the wear resistance by carburizing. Steel is carburized by exposing it to gas, liquid, or solid, which provides a high carbon concentration at the surface. Given the percentage carbon versus depth graphs for various times at 930°C, how can the diffusion coefficient be estimated from the graphs?

22. *Electrophoretic Term*

$$-j_A = D\frac{\partial C_A}{\partial z} - \left(\frac{zFm}{RT}\right)C_A + \tau_{mr}\frac{\partial j_A}{\partial t}.$$

For some systems, there is a minus sign in the electrophoretic term as shown in the preceding equation. What are the implications of the minus sign in this equation? How will this manifest in applications?

## REVIEW QUESTIONS

1. What is the difference between self-, binary, and ternary diffusion coefficients?
2. During Brownian motion, the molecules follow a random zigzag path and sometimes move in the opposite direction compared with the imposed concentration difference driving the diffusion. Is this a violation of the second law of thermodynamics?
3. What are the differences between multicomponent diffusion and binary diffusion?
4. What happens to the formula for total flux during equimolar counter diffusion compared with that for molecular diffusion?
5. Correlations for diffusion in gases, liquids, and solids were discussed. Which would be appropriate for liquid diffusing in a solid or gases diffusing in a liquid?

6. Discuss the units of each term in the equation $P = DS$.
7. Explain the effect of temperature on the mass propagation velocity. What happens to the diffusion coefficient and relaxation time at high pressure?
8. Why are insects larger in size in the tropics compared with insects in the Arctic region?
9. Are the forces of gravity taken into account in the derivation of the Stokes–Einstein relationship for diffusivity coefficients?
10. Can you expect a plane of zero concentration or null transfer during drug delivery in the tissue region? If so, how?
11. The diffusion coefficient is a proportionality constant in Fick's first law of diffusion, independent of concentration. For concentrated solutions, it is said to vary with concentration. How can this be interpreted?
12. State the Onsager reciprocal relations. Show that $D_{12} = D_{21}$.
13. What was Landau's observation of infinite speed of propagation?
14. What is penetration length?
15. What is inertial lag time?
16. What is the first zero of the Bessel function of the first order? How is this used in the derivation of the penetration length and inertial lag time in a three-dimensional medium?
17. Examine $I_0(\tau/2)\exp(-\tau/2)$ in terms of extrema, asymptotic limits. Under what conditions can $I_0(\tau/2)$ be reduced to a simpler expression?
18. What is the meaning of a negative mass flux? What happens to the ratio of the accumulation and diffusion terms?

## REFERENCES

1. R. A. L. Jones, J. Klein, and A. M. McDonald, Mutual diffusion in a miscible polymer blend, *Nature*, Vol. 321, 161–162, May 8, 1986.
1a. E. L. Cussler, *Diffusion Mass Transfer in Fluid Systems,* Cambridge University Press, U.K., 1997.
2. A. Einstein, *Ann. der Phys.*, 7, 549, 1905.
3. A. E. Fick, *Poggendorff's Ann. der Phys.*, 94, 59, 1855.
4. A. E. Fick, *Philosophical Magazine*, 10, 30, 1855.
5. J. B. Fourier, *Theorie analytique de la chaleur*, English Translation by A. Freeman, 1955, Dover Publications, New York, 1955.
6. B. Fascimire, NASA Marshall Flight Center, AL, 1973.
7. E. R. Gilland, Ind. Eng. Chem., 26, 681, 1934.
8. S. Chapman and T. G. Cowling, *The Mathematical Theory of Non-Uniform Gases*, Cambridge University Press, U.K., 1970.
9. R. C. Reid, T. K. Sherwood, and J. M. Prausnitz, *Properties of Gases and Liquids,* McGraw Hill, New York, 1977.
10. C. R. Wilke and C. Y. Lee, *Ind. Eng. Chem.*, 47, 1253, 1955.
11. R. S. Brokaw, *Ind. Eng. Chem. Process. Des. Dev.*, 8, 2, 240, 1969.
12. G. P. Mathur and G. Thodos, 1965, *AIChE*, 11, 613.
13. O. J. Catchpole and M. B. King, *Ind. Eng. Chem. Process. Des. Dev.*, 33, 1828, 1994.
14. R. H. Stokes, *J. Amer. Chem. Soc.*, 72, 763, 2243, 1950.
15. R. Resnick, and D. Halliday, *Physics*, Part I, 38th Wiley Eastern Reprint, New Delhi, 1991.
16. C. R. Wilke and P. C. Chang, *AIChE J*, 1264, 1955.
17. A. A. Siddiqi and S. M. Lucas, *Can. J. Chem. Eng.*, 64, 839, 1986.
18. W. Haydeek and B. Minhas, *Can. J. Chem. Eng.*, 60, 195, 1982.
19. G. K. Batchelor, *J. Fluid Mechanics*, 52, 245, 71, 1, 1972.

20. NEC Corporation, US Patent 5,784, 300, 1998.
21. W. M. Franklin, in *Diffusion in Solids: Recent Developments*, Academic Press, New York, 1975.
22. J. P. Stark, *Solid State Diffusion*, John Wiley & Sons, New York, 1976.
23. J. S. Vrentas and J. L. Duda, *J. Appl. Polym. Sci.*, 25, 1297, 1980.
24. K. R. Sharma, *Damped Wave Transport and Relaxation*, Elsevier, Amsterdam, 2005.
25. K. R. Sharma, On the solution of damped wave conduction and relaxation equation in a semi infinite medium subject to constant wall flux, *Int. J. Heat Mass Transfer*, 51, 25–26, 6024–6031, 2008.
26. K. R. Sharma, Damped wave conduction and relaxation in a finite sphere and cylinder, *J. Thermophys. Heat Transfer*, 22, 4, 783–786, 2008.
27. K. R. Sharma, Damped wave conduction and relaxation in cylindrical and spherical coordinates, *J. Thermophys. Heat Transfer*, 21, 4, 688–693, 2007.
28. K. R. Sharma, Manifestation of acceleration during transient heat conduction, *J. Thermophys. Heat Transfer*, 20, 4, 799–808, 2006.
29. K. R. Sharma, A Fourth Mode of Heat Transfer Called Damped Wave Conduction, 42nd Annual Convention of Chemists Meeting, Santiniketan, India, February 2006.
30. K. R. Sharma, Solution Methods and Applications for Generalized Fick's Law of Diffusion, Invited Lecture, 43rd Annual Convention of Chemists, Aurangabad, December 2006.
31. Sharma, K. R., On the Second Law Violation in Fourier Conduction, 231st ACS National Meeting, Atlanta, GA, March 2006.
32. K. J. Baumeister and T. D. Hamill, Hyperbolic heat conduction equation—A solution for the semi-infinite body problem, *ASME J. Heat Transfer*, 93, 126–128, 1971.

# 10 Copolymer Composition

## LEARNING OBJECTIVES

- Random, graft, and alternating copolymerization
- Monomer polymer composition—copolymers
- Monomer polymer composition—terpolymers
- Monomer polymer composition—multi-component terpolymers
- Free radical concentration
- Reactivity ratios
- Linear algebra—equations in matrix form
- Eigenvalues
- Subcritical damped oscillations
- Illustrations of copolymer composition equation

## 10.1   INTRODUCTION

*Copolymerization* is a process by which two or more monomer repeat units enter the backbone chain of the polymer. This is different from the products obtained from condensation polymerization such as PET (polyethylene terepthalate) and PA (poly-amide). PET is made by reacting monoethylene glycol and terepthalic acid and by removal of water molecules. The polymer backbone chain has two different repeat units. In a similar manner, hexamethylene diamine and adipic acid can be reacted together, followed by removal of water molecules to synthesize polyhexamethyl-ene adipamide or nylon. Chain polymerizations can be carried out by allowing one monomer to polymerize to a polymer. By a free radical mechanism, the initiation, propagation, and termination reactions lead to the building of molecular weight and formation of a homopolymer with repeat units of the single monomer used as start-ing material. Rather than using one monomer, two or more monomers can be used as starting materials. The free radical reactions are carried out. The resulting product can be referred to as a *copolymer* containing more than one monomer repeat units. When more than two monomers are used, the term *multicomponent copolymer* can be used. The following terms can be used:

- **Homopolymer:** Polymer made from one monomer with one repeat unit throughout the entire backbone chain
- **Copolymer:** Polymer made from two monomers with two different repeat units throughout the entire backbone chain
- **Terpolymer:** Polymer made from three monomers with three different repeat units throughout the entire backbone chain

- **Tetrapolymer:** Polymer made from four monomers with four different repeat units throughout the entire backbone chain
- **Multicomponent copolymer:** Polymer made from more than two monomers with different repeat units throughout the entire backbone chain

An example of a homopolymer is PMMA (polymethylmethacrylate); that of copolymer is SAN (styrene-acrylonitrile); that of a terpolymer is styrene-acrylonitrile-alphamethyl-styrene; and that of a tetrapolymer is polynucleotide, consisting of adenine, guanine, cytosine, and thymine.

The copolymer can assume different microstructures:

i. **Alternating copolymer:** For example, a copolymer with two repeat units in the backbone chain from monomers A and B can have a microstructure that is alternating:

$-M_A M_B M_A M_B \quad M_A \quad M_B M_A M_B \quad M_A \quad M_B M_A M_B \quad M_A M_B M_A M_B$
$M_A M_B M_A M_B \; M_A M_B M_A M_B-$

ii. **Random copolymer:** Here, $M_A$ and $M_B$ are interspersed randomly with a chain sequence distribution. For example, a copolymer with two repeat units in the backbone chain from monomers A and B can have a microstructure that has the following chain sequence distribution:

$-M_A M_A M_B \qquad M_B M_B M_A M_B M_A \qquad M_B M_B M_B M_A M_A M_A M_A$
$M_A M_B M_B M_A M_A M_B M_A-$

iii. **Block copolymer:** Copolymers comprising repeat units from monomers A and B can have the following block microstructure:

$-M_A M_A M_A M_A M_A \qquad M_A M_A M_A M_A M \qquad M_A M_A M_A M_A M_A$
$M_A M_A M_A M_A M_A M_B M_B M_B M_B-$

iv. **Graft copolymer:** Copolymers comprising repeat units from monomers A and B can assume a graft microstructure, as shown in Figure 10.1.

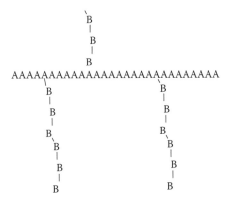

**FIGURE 10.1**   Graft microstructure of rubber-modified polystyrene.

## 10.2   COMPOSITION FOR RANDOM COPOLYMERS

Copolymer composition can be calculated as a function of monomer composition when the polymer is formed by free radical polymerization in a CSTR (continuous stirred tank reactor). Consider two monomers 1 and 2 as starting materials for forming a copolymer with repeat units of 1 and 2. The initiation can be effected by thermal means or by using a peroxy initiator.

The propagation reactions can be of four types:

$$M_1^* + M_1 \xrightarrow{k_{11}} M_1 M_1^*$$

$$M_1^* + M_2 \xrightarrow{k_{12}} M_1 M_2^*$$

$$M_2^* + M_1 \xrightarrow{k_{21}} M_2 M_1^*$$ 

$$M_2^* + M_2 \xrightarrow{k_{22}} M_2 M_2^*.$$

$$(10.1)$$

The rate of irreversible reactions can be written as

$$\frac{d[M_1]}{dt} = -k_{11}[M_1][M_1^*] - k_{21}[M_1][M_2^*]$$

$$\frac{d[M_2]}{dt} = -k_{22}[M_2][M_2^*] - k_{12}[M_2][M_1^*].$$

$$(10.2)$$

The free radical species formed may be assumed to be highly reactive and can be assumed to be consumed as rapidly as they are formed. This is referred to as the quasi-steady-state assumption (QSSA). Thus:

$$\frac{d[M_1^*]}{dt} = -k_{11}[M_1][M_1^*] - k_{12}[M_1^*][M_2]$$

$$\frac{d[M_2^*]}{dt} = -k_{22}[M_2][M_2^*] - k_{21}[M_2^*][M_1].$$

$$(10.3)$$

$$\frac{d[M_1 M_1^*]}{dt} = k_{11}[M_1][M_1^*]$$

$$\frac{d[M_2 M_2^*]}{dt} = k_{22}[M_2][M_2^*]$$

$$\frac{d[M_1 M_2^*]}{dt} = k_{12}[M_2][M_1^*]$$

$$\frac{d[M_2 M_1^*]}{dt} = k_{21}[M_1][M_2^*].$$

$$(10.4)$$

By QSSA:

$$\frac{d[M_1^*]}{dt} + \frac{d[M_1 M_1^*]}{dt} + \frac{d[M_2 M_1^*]}{dt} = 0$$

$$\frac{d[M_2^*]}{dt} + \frac{d[M_2 M_2^*]}{dt} + \frac{d[M_1 M_2^*]}{dt} = 0$$

$(10.5)$

Thus:

$$k_{12}[M_1^*][M_2] = k_{21}[M_1][M_2^*]$$

or

$$\frac{[M_2] k_{12}}{[M_1] k_{21}} = \frac{[M_2^*]}{[M_1^*]}.$$

$(10.6)$

The *reactivity ratios* are defined as follows:

$$r_{12} = \frac{k_{11}}{k_{12}}; r_{21} = \frac{k_{22}}{k_{21}}.$$

$(10.7)$

Combining Equations (10.2) and (10.6):

$$\frac{d[M_1]}{d[M_2]} = \frac{-k_{11}[M_1][M_1^*] - k_{21}[M_1][M_2^*]}{-k_{22}[M_2][M_2^*] - k_{21}[M_1][M_2^*]} = \frac{[M_1](r_{12}[M_1] + [M_2])}{[M_2](r_{21}[M_2] + [M_1])}.$$

$(10.8)$

Equation (10.8) is called the *copolymerization composition equation*. In a CSTR, the effluent concentration and the reactor concentration are the same. The reactor concentrations of monomers 1 and 2 are $[M_1]$ and $[M_2]$, respectively. The polymer composition of monomer 1 repeat unit in the polymer $F_1$ can be seen to be

$$F_1 = \frac{d[M_1]}{d[M_1] + d[M_2]}.$$

$(10.9)$

Equation (10.9) gives the rate at which the monomer repeat unit 1 enters the copolymer compared with the rates at which both the monomer repeat units or rate at which all the monomer repeat units enter the backbone chain of the polymer. Equations (10.9) and (10.8) can be combined as

$$F_1 = \frac{1}{1 + \dfrac{d[M_2]}{d[M_1]}} = \frac{1}{1 + \dfrac{[M_2](r_{21}[M_2] + [M_1])}{[M_1](r_{12}[M_1] + [M_2])}}.$$

$(10.10)$

Defining the CSTR concentration of monomer 1 as $f_1$:

$$f_1 = \frac{[M_1]}{[M_1] + [M_2]} = \frac{1}{1 + \dfrac{[M_2]}{[M_1]}}$$

$$or \frac{[M_2]}{[M_1]} = \frac{1}{f_1} - 1 = \frac{f_2}{f_1}.$$

$(10.11)$

Combining Equations (10.11) and (10.10):

$$F_1 = \frac{f_1^2 r_{12} + f_1 f_2}{\left( f_1^2 r_{12} + 2 f_1 f_2 + r_{21} f_2^2 \right)}.$$

(10.12)

The copolymer composition in the polymer as a function of reactor monomer concentrations in CSTR is given by Equation (10.12) [1]. Equation (10.12) is plotted for different values of reactivity ratios in Figure 10.2. It can be seen that when one of the two reactivity ratios $r_{12} = r_{21} = 0$ the copolymer is expected to assume the alternating microstructure as shown in Section 10.1. When one of the reactivity ratios is much greater than 1, block architecture may be expected, as shown in Section 10.1. The copolymerization is said to be ideal when the product of the two reactivity ratios $r_{12} r_{21} = 1$. Here, the monomer composition and polymer composition are equal to each other. The copolymer is said to be formed at the *azeotropic* composition, when the monomer and polymer compositions are equal to each other.

The copolymer compositions versus the monomer concentration in a CSTR for SAN (styrene acrylonitrile) and AMS–AN (alphamethyl-styrene–acrylonitrile) copolymers are shown in Figure 10.2. The copolymer composition equation changes for AMS–AN copolymers, as explained in detail in Chapter 12, due to the lack of AMS–AMS propagation reactions. The azeotropic compositions for SAN and AMS–AN copolymers with and without the ceiling temperature effects are shown

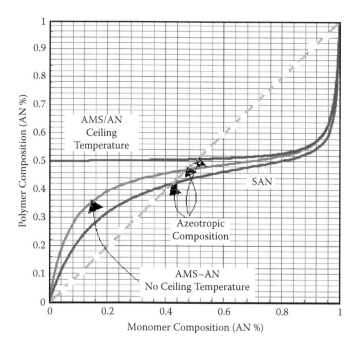

**FIGURE 10.2** Copolymer composition as a function of monomer composition for SAN and AMS–AN copolymers.

in Figure 10.2. It can be seen that the copolymer composition dependence on the monomer composition is sensitive to the reactivity ratios. Even without the ceiling temperature effect, the polymer composition for AMS–AN copolymers only changes marginally for a large range of monomer AN compositions. With the ceiling temperature effect, it can be seen that the sequence formation of AN–AN dyads, triads, tetrads, etc., is pronounced in AMS–AN copolymers. This topic is further discussed separately in Section 11.1 under sequence distribution of copolymers.

In a PFR (plug flow reactor), the polymer composition can be found by integration of the incremental addition to the polymer across the length of the reactor. During the progression of the monomers in a PFR, the composition varies with conversion:

$$\int_{M_0}^{M} \frac{dM}{M} = \int \frac{df_1}{(F_1 - f_1)}. \tag{10.13}$$

## 10.3   COMPOSITION OF RANDOM TERPOLYMERS

Synthesis of terpolymers has increased in commercial significance over the past few decades. Consider three monomer repeat units entering a polymer backbone chain in a CSTR. There are nine propagation reactions now:

$$M_1^* + M_1 \xrightarrow{k_{11}} M_1 M_1^*$$

$$M_1^* + M_2 \xrightarrow{k_{12}} M_1 M_2^*$$

$$M_1^* + M_3 \xrightarrow{k_{13}} M_1 M_3^*$$

$$M_2^* + M_1 \xrightarrow{k_{21}} M_2 M_1^*$$

$$M_2^* + M_2 \xrightarrow{k_{22}} M_2 M_2^* \tag{10.14}$$

$$M_2^* + M_3 \xrightarrow{k_{23}} M_2 M_3^*$$

$$M_3^* + M_1 \xrightarrow{k_{31}} M_3 M_1^*$$

$$M_3^* + M_2 \xrightarrow{k_{32}} M_3 M_2^*$$

$$M_3^* + M_3 \xrightarrow{k_{33}} M_3 M_3^*.$$

The rate of irreversible reactions can be written as

$$\frac{d[M_1]}{dt} = -k_{11}[M_1][M_1^*] - k_{21}[M_1][M_2^*] - k_{31}[M_1][M_3^*]$$

$$\frac{d[M_2]}{dt} = -k_{22}[M_2][M_2^*] - k_{12}[M_2][M_1^*] - k_{32}[M_2][M_3^*] \qquad (10.15)$$

$$\frac{d[M_3]}{dt} = -k_{33}[M_3][M_3^*] - k_{13}[M_3][M_1^*] - k_{23}[M_3][M_2^*].$$

The free radical species formed may be assumed to be highly reactive and can be assumed to be consumed as rapidly as formed. This is referred to as the quasi-steady-state assumption, QSSA. Thus,

$$\frac{d[M_1^*]}{dt} = -k_{11}[M_1][M_1^*] - k_{12}[M_1^*][M_2] - k_{13}[M_1^*][M_3]$$

$$\frac{d[M_2^*]}{dt} = -k_{22}[M_2][M_2^*] - k_{21}[M_2^*][M_1] - k_{23}[M_2^*][M_3] \qquad (10.16)$$

$$\frac{d[M_3^*]}{dt} = -k_{33}[M_3][M_3^*] - k_{32}[M_3^*][M_2] - k_{31}[M_3][M_1^*].$$

$$\frac{d[M_1 M_1^*]}{dt} = k_{11}[M_1][M_1^*]$$

$$\frac{d[M_2 M_2^*]}{dt} = k_{22}[M_2][M_2^*]$$

$$\frac{d[M_3 M_3^*]}{dt} = k_{33}[M_3][M_3^*]$$

$$\frac{d[M_1 M_2^*]}{dt} = k_{12}[M_2][M_1^*]$$

$$\frac{d[M_2 M_1^*]}{dt} = k_{21}[M_1][M_2^*] \qquad (10.17)$$

$$\frac{d[M_3 M_1^*]}{dt} = k_{31}[M_1][M_3^*]$$

$$\frac{d[M_1 M_3^*]}{dt} = k_{13}[M_1][M_3^*]$$

$$\frac{d[M_2 M_3^*]}{dt} = k_{23}[M_3][M_2^*]$$

$$\frac{d[M_3 M_2^*]}{dt} = k_{32}[M_2][M_3^*].$$

By QSSA

$$\frac{d[M_1^*]}{dt} + \frac{d[M_1M_1^*]}{dt} + \frac{d[M_2M_1^*]}{dt} + \frac{d[M_3M_1^*]}{dt} = 0$$

$$\frac{d[M_2^*]}{dt} + \frac{d[M_2M_2^*]}{dt} + \frac{d[M_1M_2^*]}{dt} + \frac{d[M_3M_2^*]}{dt} = 0 \qquad (10.18)$$

$$\frac{d[M_3^*]}{dt} + \frac{d[M_3M_3^*]}{dt} + \frac{d[M_2M_3^*]}{dt} + \frac{d[M_1M_3^*]}{dt} = 0.$$

Thus,

$$k_{12}[M_1^*][M_2] + k_{13}[M_1^*][M_3] = k_{21}[M_1][M_2^*] + k_{31}[M_1][M_3^*]$$

$$or \; \frac{[M_2]}{[M_1]}\frac{1}{r_{12}} + \frac{1}{r_{13}}\frac{[M_3]}{[M_1]} = \frac{k_{22}}{r_{21}k_{11}}\frac{[M_2^*]}{[M_1^*]} + \frac{k_{33}}{k_{11}r_{31}}\frac{[M_3^*]}{[M_1^*]}. \qquad (10.19)$$

$$\frac{[M_1]}{[M_2]r_{21}} + \frac{[M_3]}{r_{23}[M_2]} = \frac{k_{11}[M_1^*]}{r_{12}k_{22}[M_2^*]} + \frac{k_{33}[M_3^*]}{r_{32}k_{22}[M_2^*]}. \qquad (10.20)$$

$$\frac{[M_1]}{[M_3]r_{31}} + \frac{[M_2]}{r_{32}[M_3]} = \frac{k_{11}[M_1^*]}{r_{13}k_{33}[M_3^*]} + \frac{k_{22}[M_2^*]}{r_{23}k_{33}[M_3^*]}. \qquad (10.21)$$

The polymer composition of monomer 1 in the terpolymer is given by

$$F_1 = \frac{\dfrac{d[M_1]}{dt}}{\dfrac{d[M_1]}{dt} + \dfrac{d[M_2]}{dt} + \dfrac{d[M_3]}{dt}}. \qquad (10.22)$$

Combining Equation (10.15) and the results from QSSA given by Equations (10.19–10.21), it can be seen that

$$\frac{F_1}{F_2} = \frac{[M_1]}{[M_2]}\frac{\left([M_1] + \dfrac{[M_2]}{r_{12}} + \dfrac{[M_3]}{r_{13}}\right)}{\dfrac{r_{21}}{r_{12}}\left(\dfrac{[M_1]}{r_{21}} + [M_2] + \dfrac{[M_3]}{r_{23}}\right)} = \frac{(r_{12}r_{23}(r_{13}-1))f_1^2 + (r_{13}-r_{12})r_{23}f_1f_2 + r_{12}r_{23}f_1}{r_{13}(r_{23}-r_{21})f_1f_2 + r_{13}r_{21}(r_{23}-1)f_2^2 + r_{13}r_{21}f_2}. \qquad (10.23)$$

In a similar manner,

$$\frac{F_1}{F_3} = \frac{(r_{13}r_{32}(r_{12}-1))f_1^2 + (r_{12}-r_{13})r_{32}f_1f_3 + r_{13}r_{32}f_1}{r_{12}(r_{32}-r_{31})f_1f_3 + r_{12}r_{31}(r_{32}-1)f_3^2 + r_{12}r_{31}f_3}. \qquad (10.24)$$

$$\frac{F_2}{F_3} = \frac{(r_{23}r_{31}(r_{21}-1))f_2^2 + (r_{31}r_{21}-r_{23}r_{31})f_2f_3 + r_{23}r_{31}f_2}{r_{21}(r_{31}-r_{32})f_2f_3 + r_{21}r_{32}(r_{31}-1)f_3^2 + r_{21}r_{32}f_3}. \tag{10.25}$$

## 10.4 REACTIVITY RATIOS

It can be seen that the monomer–polymer composition for terpolymerization in a CSTR will depend on the values of reactivity ratios. Consider the expression derived for monomer–polymer composition during copolymerization in a CSTR. Equation (10.12) rewritten in terms of monomer 1 composition only is

$$F_1 = \frac{f_1^2(r_{12}-1)+f_1}{(f_1^2(r_{12}+r_{21}-2)+2f_1(1-r_{21})+r_{21})}. \tag{10.26}$$

For the special case when the reactivity ratio $r_{12} = 1$, Equation (10.26) becomes

$$F_1 = \frac{f_1}{(1-f_1)^2(r_{21}-1)+1}. \tag{10.27}$$

Equation (10.27) can be used to calculate the copolymer composition given the monomer compositions. The inverse problem of finding the monomer composition for a desired polymer composition would require rearranging Equation (10.26) and expressing $f_1$ in terms of $F_1$. It appears that the expression would be quadratic in $f_1$. Does this mean that there would be multiple or two roots to a desired polymer composition for some set of reactivity ratios? The monomer and copolymer composition for the system of diethyl fumarate and acrylonitrile is shown in Figure 10.3.

It can be seen from Figure 10.3 that for a polymer composition of 30 mole% diethyl fumarate, two monomer compositions of DEF are obtained. This is referred to as *multiplicity* of compositions. It is not desirable to operate the industrial reactors at this composition. Mathematically, a quadratic equation can lead to two roots, real and imaginary. This leads to an interesting criterion for copolymer compositional stability. It is conceivable that for some set of reactivity ratios, the composition may not be stable. Equation (10.26) rearranged to provide $f_1$ the composition of the monomers in a CSTR for a desired copolymer composition $F_1$ would be

$$f_1^2(F_1(r_{12}+r_{21}-2)+1-r_{12}) + f_1(2F_1(1-r_{21})-1)+r_{21}F_1 = 0. \tag{10.28}$$

The functionality of $F_1$ versus $f_1$ can be seen from Figure 10.2 and Equation (10.28) to change from *concave* to *convex* curvature as the composition of monomer 1 is increased. The curve *inflects*. The dependence of polymer composition in the copolymer on the monomer composition in the CSTR depends on the reactivity ratios of the monomers. The reactivity ratios of some commonly used monomers in commercial industrial practice are given in Table 10.1.

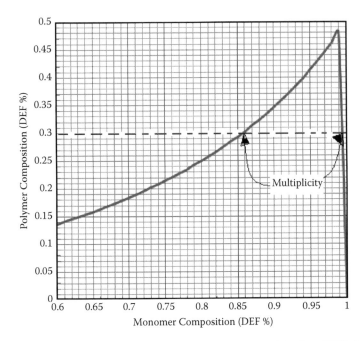

**FIGURE 10.3** Copolymer composition versus monomer composition in a CSTR for diethyl fumarate acrylonitrile system.

## Example 10.1   Discuss the Monomer–Copolymer Composition Curve for the System Methacrylonitrile–Styrene Formed in a CSTR

Let the monomer 1 be methacrylonitrile (MAN) and monomer 2 be styrene (STY). It can be seen from Table 10.1 that the reactivity ratios $r_{12}$ and $r_{21}$ are equal to each other and are 0.25. The monomer and copolymer composition curve is shown in Figure 10.4.

It can be seen from Figure 10.4 that the copolymer composition curve is *S shaped or sigmoidal*. Below the azeotropic composition, MAN is favored into the polymer and is more selective. Above the azeotropic composition of 0.5 mole%, the styrene is more selectively added into the copolymer chain.

## Example 10.2   Discuss the Terpolymer Composition Curves for the Termonomer Systems Acrylonitrile–Styrene–Alphamethyl-Styrene by Free Radical Polymerization in a CSTR

Equations (10.24) and (10.25) were used to simulate the terpolymer composition curves for different termonomer compositions using an MS Office Excel 2007 spreadsheet. The reactivity ratios from Table 10.1 were used. It can be seen that five of the six reactivity ratios are less than 1, and one of them, $r_{23}$, is greater than 1. The terpolymer composition of AN at two different compositions of AMS in the monomer phase in a CSTR was plotted in Figure 10.5. The curves were found to

**TABLE 10.1**
**Reactivity Ratios of Monomers Used in Industrial Practice**

| # | Monomer 1 | Monomer 2 | $r_{12}$ | $r_{21}$ |
|---|-----------|-----------|----------|----------|
| 1 | Acrylonitrile | Styrene | 0.02 | 0.29 |
|   |           | Alphamethyl-styrene | 0.03 | 0.14 |
|   |           | Methacrylonitrile | 1.5 | 0.84 |
|   |           | Methyl methacrylate | 0.4 | 1.3 |
|   |           | Vinyl chloride | 3.6 | 0.044 |
|   |           | 1,3-Butadiene | 0.046 | 0.36 |
|   |           | Vinyl acetate | 5.5 | 0.060 |
|   |           | Methyl acrylate | 1.5 | 0.84 |
|   |           | Diethyl fumarate | 8.0 | 0.0 |
|   |           | Acrylamide | 0.86 | 0.81 |
|   |           | 2-Vinyl pyridine | 0.020 | 0.43 |
|   |           | Maleic anhydride | 6.7 | 0 |
|   |           | Methacrylic acid | 0.092 | 2.4 |
| 2 | Styrene | Alphamethyl-styrene | 1.2 | 0.14 |
|   |         | Methacrylonitrile | 0.25 | 0.25 |
|   |         | Vinyl chloride | 15 | 0.010 |
|   |         | Acrylic acid | 0.15 | 0.25 |
|   |         | Allyl acetate | 90 | 0 |
|   |         | Diethyl fumarate | 0.30 | 0.070 |
|   |         | Maleic anhydride | 0.29 | 0.006 |
|   |         | Methacrylic acid | 0.12 | 0.60 |
|   |         | Methyl acrylate | 0.19 | 0.80 |
|   |         | Methyl methacrylate | 0.52 | 0.46 |
|   |         | Ethyl vinyl ether | 90 | 0 |

be sigmoidal. The AMS composition did not affect the polymer AN composition for monomer compositions less than 0.45%. For values greater than this, AN was found to be more selectively added to the polymer as the AN composition in the monomer phase was increased. A table of values from the spreadsheet used is shown as Table 10.2

*Multiplicity* in termonomer compositions for a given terpolymer composition can be expected at large values of reactivity ratios and compositions.

## 10.5  MULTICOMPONENT COPOLYMERIZATION—*n* MONOMERS

The analysis of copolymerization and terpolymerization resulting in copolymer composition versus monomer composition relations in the prior sections can be generalized to *n* monomers. Consider *n* monomer repeat units entering a polymer backbone chain in a CSTR. There are $n^2$ propagation reactions that can be involved in the

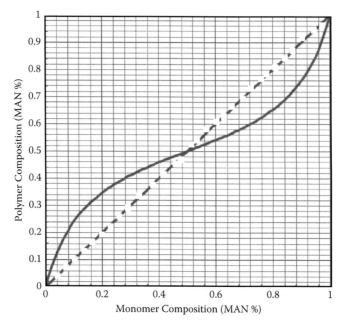

**FIGURE 10.4** Copolymer composition as a function of monomer composition in a CSTR for methacrylonitrile and styrene.

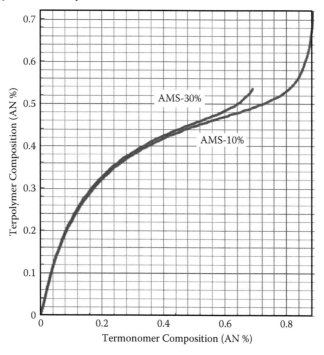

**FIGURE 10.5** Terpolymer composition versus monomer composition for STY–AN–AMS system for two different compositions of AMS.

**TABLE 10.2**

**Terpolymer versus Termonomer Composition Results for Example 10.2**

| | | | | | | | |
|---|---|---|---|---|---|---|---|
| $r_{12}$ | 0.02 | 0.02 | $f_3$ | 0.1 | | | |
| $r_{21}$ | 0.29 | 0.29 | $f_1$ | $f_2$ | $F_1/F_2$ | $F_1/F_3$ | $F_1$ |
| $r_{13}$ | 0.14 | 0.14 | 0 | 0.9 | 0 | 0 | 0 |
| $r_{31}$ | 0.03 | 0.03 | 0.1 | 0.8 | 0.286483 | 23.00781 | 0.220552 |
| $r_{23}$ | 1.2 | 1.2 | 0.2 | 0.7 | 0.480432 | 31.88669 | 0.321252 |
| $r_{32}$ | 0.14 | 0.14 | 0.3 | 0.6 | 0.622568 | 34.35353 | 0.379455 |
| | | | 0.4 | 0.5 | 0.734106 | 33.37104 | 0.418031 |
| | | | 0.5 | 0.4 | 0.828467 | 30.32031 | 0.446423 |
| | | | 0.6 | 0.3 | 0.917607 | 25.93192 | 0.469847 |
| | | | 0.7 | 0.2 | 1.021521 | 20.62862 | 0.493241 |
| | | | 0.8 | 0.1 | 1.221667 | 14.67198 | 0.530023 |
| | | | 0.89 | 0.01 | 4.084358 | 8.894603 | 0.73677 |
| $f_3$ | 0.3 | | | | | | |
| | | | | 0.7 | 0 | 0 | 0 |
| | | | | 0.6 | 0.310177 | 5.247965 | 0.226525 |
| | | | | 0.5 | 0.523935 | 6.986312 | 0.327679 |
| | | | | 0.4 | 0.689136 | 7.078854 | 0.385749 |
| | | | | 0.3 | 0.836121 | 6.268162 | 0.424532 |
| | | | | 0.2 | 1.002606 | 4.908486 | 0.454312 |
| | | | | 0.1 | 1.324509 | 3.190246 | 0.483453 |

multicomponent copolymerization. Thus, for tetrapolymerization there would be 16 propagation reactions involved. These reactions are listed as follows:

$$M_1^* + M_1 \xrightarrow{k_{11}} M_1 M_1^*$$

$$M_1^* + M_2 \xrightarrow{k_{12}} M_1 M_2^*$$

$$M_1^* + M_3 \xrightarrow{k_{13}} M_1 M_3^*$$

$$\cdots\cdots\cdots\cdots\cdots\cdots\cdots$$

$$M_1^* + M_n \xrightarrow{k_{1n}} M_1 M_n^*$$

$$M_2^* + M_1 \xrightarrow{k_{21}} M_2 M_1^*$$

$$M_2^* + M_2 \xrightarrow{k_{22}} M_2 M_2^*$$

$$M_2^* + M_3 \xrightarrow{k_{23}} M_2 M_3^*$$

$$\cdots\cdots\cdots\cdots\cdots\cdots\cdots$$

$$M_2^* + M_n \xrightarrow{k_{2n}} M_2 M_n^*$$

$$M_3^* + M_1 \xrightarrow{k_{31}} M_3 M_1^*$$

$$M_3^* + M_2 \xrightarrow{k_{32}} M_3 M_2^*$$

$$M_3^* + M_3 \xrightarrow{k_{33}} M_3 M_3^*$$

$$\dots\dots\dots\dots\dots\dots$$

$$M_3^* + M_n \xrightarrow{k_{3n}} M_3 M_n^*$$                                    (10.29)

$$\dots\dots\dots\dots\dots\dots$$

$$M_n^* + M_1 \xrightarrow{k_{n1}} M_n M_1^*$$

$$M_n^* + M_2 \xrightarrow{k_{n2}} M_n M_2^*$$

$$M_n^* + M_3 \xrightarrow{k_{n3}} M_n M_3^*$$

$$\dots\dots\dots\dots\dots\dots$$

$$M_n^* + M_n \xrightarrow{k_{nn}} M_n M_n^*$$

The rate of irreversible reactions can be written as

$$\frac{d[M_1]}{dt} = -k_{11}[M_1][M_1^*] - k_{21}[M_1][M_2^*] - k_{31}[M_1][M_3^*] - \dots - k_{n1}[M_1][M_n^*]$$

$$\frac{d[M_2]}{dt} = -k_{22}[M_2][M_2^*] - k_{12}[M_2][M_1^*] - k_{32}[M_2][M_3^*] - \dots - k_{n2}[M_2][M_n^*]$$

$$\frac{d[M_3]}{dt} = -k_{33}[M_3][M_3^*] - k_{13}[M_3][M_1^*] - k_{23}[M_3][M_2^*] - \dots - k_{n3}[M_3][M_n^*]$$

$$\dots\dots\dots\dots\dots\dots\dots\dots\dots\dots\dots\dots\dots\dots\dots\dots\dots\dots\dots\dots\dots$$

$$\frac{d[M_n]}{dt} = -k_{nn}[M_n][M_n^*] - k_{1n}[M_n][M_1^*] - k_{2n}[M_n][M_2^*] - \dots - k_{n-1n}[M_n^*][M_{n-1}^*]$$

(10.30)

The free radical species formed may be assumed to be highly reactive and can be assumed to be consumed as rapidly as they are formed. This is referred to as the QSSA. Thus:

$$\frac{d[M_1^*]}{dt} = -k_{11}[M_1][M_1^*] - k_{12}[M_1^*][M_2] - k_{13}[M_1^*][M_3] - \ldots - .k_{1n}[M_1^*][M_n]$$

$$\frac{d[M_2^*]}{dt} = -k_{22}[M_2][M_2^*] - k_{21}[M_2^*][M_1] - k_{23}[M_2^*][M_3] - \ldots - k_{2n}[M_2^*][M_n]$$

$$\frac{d[M_3^*]}{dt} = -k_{33}[M_3][M_3^*] - k_{32}[M_3^*][M_2] - k_{31}[M_3][M_1^*] - \ldots - k_{3n}[M_3^*][M_n]$$

$$\frac{d[M_n^*]}{dt} = -k_{nn}[M_n][M_n^*] - k_{nn-1}[M_n^*][M_{n-1}] - \ldots - k_{n1}[M_n][M_1^*]$$

$$(10.31)$$

$$\frac{d[M_1 M_1^*]}{dt} = k_{11}[M_1][M_1^*]$$

$$\frac{d[M_1 M_2^*]}{dt} = k_{12}[M_2][M_1^*]$$

$$\frac{d[M_1 M_3^*]}{dt} = k_{13}[M_3][M_1^*]$$

$$\frac{d[M_1 M_n^*]}{dt} = k_{1n}[M_n][M_1^*]$$

$$(10.32)$$

$$\frac{d[M_2 M_1^*]}{dt} = k_{21}[M_1][M_2^*]$$

$$\frac{d[M_2 M_2^*]}{dt} = k_{22}[M_2][M_2^*]$$

$$\frac{d[M_2 M_3^*]}{dt} = k_{23}[M_3][M_2^*]$$

$$\frac{d[M_2 M_n^*]}{dt} = k_{2n}[M_n][M_2^*]$$

$$\frac{d[M_3 M_1^*]}{dt} = k_{31}[M_1][M_3^*]$$

$$\frac{d[M_3 M_2^*]}{dt} = k_{32}[M_2][M_3^*]$$

$$\frac{d[M_3 M_3^*]}{dt} = k_{33}[M_3][M_3^*]$$

............................................

$$\frac{d[M_3 M_n^*]}{dt} = k_{3n}[M_n][M_3^*]$$

............................................

$$\frac{d[M_n M_1^*]}{dt} = k_{n1}[M_1][M_n^*]$$

$$\frac{d[M_n M_2^*]}{dt} = k_{n2}[M_2][M_n^*]$$

$$\frac{d[M_n M_3^*]}{dt} = k_{n3}[M_3][M_n^*]$$

............................................

$$\frac{d[M_n M_n^*]}{dt} = k_{nn}[M_n][M_n^*]$$

By QSSA:

$$\frac{d[M_1^*]}{dt} + \frac{d[M_1 M_1^*]}{dt} + \frac{d[M_2 M_1^*]}{dt} + \frac{d[M_3 M_1^*]}{dt} + \cdots + \frac{d[M_n M_1^*]}{dt} = 0$$

$$\frac{d[M_2^*]}{dt} + \frac{d[M_1 M_2^*]}{dt} + \frac{d[M_2 M_2^*]}{dt} + \frac{d[M_3 M_2^*]}{dt} + \cdots + \frac{d[M_n M_2^*]}{dt} = 0$$

$$\frac{d[M_3^*]}{dt} + \frac{d[M_1 M_3^*]}{dt} + \frac{d[M_2 M_3^*]}{dt} + \frac{d[M_3 M_3^*]}{dt} + \cdots + \frac{d[M_n M_3^*]}{dt} = 0$$

..............................................................................................

$$\frac{d[M_n^*]}{dt} + \frac{d[M_1 M_n^*]}{dt} + \frac{d[M_2 M_n^*]}{dt} + \frac{d[M_3 M_n^*]}{dt} + \cdots + \frac{d[M_n M_n^*]}{dt} = 0 \tag{10.33}$$

Let the $n$ monomer concentrations be represented by the vector M and $n$ free radical concentrations by represented by vector M* as follows:

$$
M = \begin{pmatrix} M_1 \\ M_2 \\ M_3 \\ ..... \\ M_n \end{pmatrix}
\qquad
M^* = \begin{pmatrix} M_1^* \\ M_2^* \\ M_3^* \\ ..... \\ M_n^* \end{pmatrix}
\tag{10.34}
$$

The multicomponent copolymerization rate constants can be represented by a rate matrix as follows:

$$
K = \begin{pmatrix}
k_{11} & k_{21} & k_{31} & \cdots & k_{n1} \\
k_{12} & k_{22} & k_{32} & \cdots & k_{n2} \\
k_{13} & k_{23} & k_{33} & .. & k_{n3} \\
..... & ... & .. & ... & .. \\
k_{1n} & k_{2n} & k_{3n} & \cdots & k_{nn}
\end{pmatrix}
=
\begin{pmatrix}
k_{11} & \dfrac{k_{22}}{r_{21}} & \dfrac{k_{33}}{r_{31}} & \cdots & \dfrac{k_{nn}}{r_{n1}} \\
\dfrac{k_{11}}{r_{12}} & k_{22} & \dfrac{k_{33}}{r_{32}} & \cdots & \dfrac{k_{22}}{r_{21}} \\
\dfrac{k_{11}}{r_{13}} & \dfrac{k_{22}}{r_{23}} & k_{33} & \cdots & \dfrac{k_{33}}{r_{n3}} \\
... & ... & .. & ... & ... \\
\dfrac{k_{11}}{r_{1n}} & \dfrac{k_{22}}{r_{2n}} & \dfrac{k_{33}}{r_{3n}} & \cdots & k_{nn}
\end{pmatrix}
\tag{10.35}
$$

The rate matrix for multicomponent copolymerization can thus also be expressed in terms of the homopolymerization propagation rate constants of the $n$ monomers, and reactivity ratios as defined in a similar manner to Equation (10.7). The set of copolymerization rate equations given by Equation (10.30) can be given in matrix form as follows:

$$
\frac{d\tilde{M}}{dt} = -K(M * M^T),
\tag{10.36}
$$

where

$$
\begin{pmatrix}
M_1 & 0 & 0 & 0 & 0 \\
0 & M_2 & 0 & 0 & 0 \\
0 & 0 & M_3 & 0 & 0 \\
.. & .. & .. & .. & .. \\
0 & 0 & 0 & 0 & M_n
\end{pmatrix} = \tilde{M}
$$

Only the diagonal elements of the resulting matrix from the right-hand side of Equation (10.36) are of interest. The rate equations that represent the radical generation and consumption during propagation can be given in the matrix form as

$$\frac{d\tilde{M}^*}{dt} = -K^T(MM^{*T}) + K(M^*M^T) \tag{10.37}$$

Equations (10.36) and (10.37) form a set of autonomous systems. For certain values of the rate propagation matrix, using eigenvectors and stability analysis it can be shown that *periodic solutions* may result. The initial conditions of the monomers can be represented by

$$M\,(t{=}0) = M0. \tag{10.38}$$

By the QSSA, that is, when the radical species are highly reactive, Equation (10.27) can be set to zero. Or,

$$K^T(MM^{*T}) + K(M^*M^T) \tag{10.39}$$

For a given set of reactivity ratios, homopolymer propagation constants and initial condition of multicomponent comonomers, $M^*$ can be solved for using Equation (10.39) and then substituting in Equation (10.37). Then the monomer concentrations with time as well as the polymer composition can be calculated from Equation (10.37).

Observe that Equation (10.37) can be solved for by the method of *eigenvectors* [2]. It can be seen that nontrivial equations to the set of ordinary differential equations with constant coefficients represented by Equation (10.36) exist only for certain specific values of $\lambda$ called *eigenvalues*. These are the solutions to

$$|K - \lambda I| = 0. \tag{10.40}$$

Equation (10.40) upon expansion may lead to a polynomial degree of $n$ in $\lambda$;

$$P_n(\lambda) = 0. \tag{10.41}$$

In order to obtain numerical values for $\lambda$, the expansion of Equation (10.40) is obtained by Laplace development of the determinant;

$$P_n(\lambda) = (-1)^n\,\lambda^n + s_1\,\lambda^{n-1} + s_2\,\lambda^{n-2} + s_{n-1}\lambda + {}^{n-2}\,|K| = 0 \tag{10.42}$$

where

$$s_j = (-1)^{n-j}(\text{sum of all principal minors of order } j \text{ of } \kappa] \tag{10.43}$$

where $j = 1, 2, 3, \ldots , (n-1)$.

The copolymer composition can be obtained using the trace of the matrix defined in Equation (10.36). Trace is a sum of all the diagonal elements of a matrix. Thus:

$$
\tilde{F} = \begin{pmatrix} F_1 & 0 & 0 & 0 & 0 \\ 0 & F_2 & 0 & 0 & 0 \\ 0 & 0 & F_3 & 0 & 0 \\ \cdots & \cdots & \cdots & \cdots & \cdots \\ 0 & 0 & 0 & 0 & F_n \end{pmatrix} = \dfrac{\dfrac{d\tilde{M}}{dt}}{tr\left(\dfrac{d\tilde{M}}{dt}\right)}. \tag{10.44}
$$

The corresponding monomer compositions in the CSTR for $n$ comonomers entering the copolymer backbone chain can be written as

$$
\tilde{F} = \begin{pmatrix} f_1 & 0 & 0 & 0 & 0 \\ 0 & f_2 & 0 & 0 & 0 \\ 0 & 0 & f_3 & 0 & 0 \\ \cdots & \cdots & \cdots & \cdots & \cdots \\ 0 & 0 & 0 & 0 & f_n \end{pmatrix} = \dfrac{\tilde{M}}{tr\left(\tilde{M}\right)}. \tag{10.45}
$$

Equation (10.45) can be differentiated with respect to time and combined with Equation (10.44) to obtain the relationship between the multicomponent copolymer composition and CSTR monomer compositions. It can be seen that trace of a square matrix is a scalar quantity.

Equation (10.36) is in terms of $M^T$ and diagonal matrix $\tilde{M}$. During copolymerization from the QSSA as given by Equation (10.6), it can be seen that

$$
k_{12}[M_1^*][M_2] = k_{21}[M_1][M_2^*]. \tag{10.46}
$$

This is because the bond $M_1{}^*M_2$ and $M_1M_2{}^*$ are formed at the same rate. The order of addition of the radical does not matter.

Equation (10.46) can be derived [3–19] by evaluation of

$$
\dfrac{dM^*}{dt} = -(K\tilde{M}^*)^T M(K\tilde{M}^*)M. \tag{10.46a}
$$

From QSSA, it can be seen that Equation (10.46a) ought to be set to 0. Differentiating the resulting equation with respect to time:

$$
(K\tilde{M}^*)^T \dfrac{dM}{dt} =
$$
$$
K^T M^{*T} \dfrac{dM}{dt} = KM^* \dfrac{dM^T}{dt}. \tag{10.46b}
$$

It can be seen from Equation (10.46b) that

$$
(K\tilde{M}^*)^T = (K\tilde{M}^*). \tag{10.46c}
$$

Equation (10.46) would fall out from Equation (10.46c) when the order of the matrix is 2.

Substituting Equation (10.46) in the set of copolymerization rate equations, the rate matrix can be rewritten as

$$\frac{dM}{dt} = -(K\tilde{M}^*)^T M, \tag{10.47}$$

i.e.,

$$\frac{d}{dt}\begin{pmatrix} M_1 \\ M_2 \end{pmatrix} = \left( \begin{pmatrix} k_{11} & k_{21} \\ k_{12} & k_{22} \end{pmatrix} \begin{pmatrix} M_1^* & 0 \\ 0 & M_2^* \end{pmatrix} \right)^T \begin{pmatrix} M_1 \\ M_2 \end{pmatrix}.$$

Equation (10.46) is a reasonable assumption for multicomponent copolymerization for the general case of $n$ monomers as well. This implies that the rate of copolymerization propagation of $M_iM_j^*$ equals the formation of $M_jM_i^*$ radical. When this is applied to all possible pairs of monomers in $n$ comonomers, the QSSA holds as the net production of radicals equals the consumption during the copolymerization propagation itself. Nothing about the termination reactions is brought forward into this analysis. Thus, for $n$ comonomers:

$$\frac{d}{dt}\begin{pmatrix} M_1 \\ M_2 \\ M_3 \\ \cdots \\ M_n \end{pmatrix} = \left( \begin{pmatrix} k_{11} & k_{21} & k_{31} & \cdots & k_{n1} \\ k_{12} & k_{22} & k_{32} & \cdots & k_{n2} \\ k_{13} & k_{23} & k_{33} & \cdots & k_{n3} \\ \cdots & \cdots & \cdots & \cdots & \cdots \\ k_{1n} & k_{2n} & k_{3n} & \cdots & k_{nn} \end{pmatrix} \begin{pmatrix} M_1^* & 0 & 0 & 0 & 0 \\ 0 & M_2^* & 0 & 0 & 0 \\ 0 & 0 & M_3^* & 0 & 0 \\ 0 & 0 & 0 & \cdots & 0 \\ 0 & 0 & 0 & 0 & M_n^* \end{pmatrix} \right)^T \begin{pmatrix} M_1 \\ M_2 \\ M_3 \\ \cdots \\ M_n \end{pmatrix}$$

(10.48)

The solution to Equation (10.47) can be obtained by the method of eigenvectors. Assume that the solution to Equation (10.47) has the form

$$M = Ze^{\lambda t}, \tag{10.49}$$

where $Z$ is the unknown vector of constants and $\lambda$ are the eigenvalues. Substituting Equation (10.49) in Equation (10.47):

$$\lambda M = (K\tilde{M}^*)^T Z. \tag{10.50}$$

Equation (10.50) requires that $e^{\lambda t} \neq 0$. Equation (10.49) has the form of the solution to the set of ordinary differential equations with constant coefficients requires that $Z \neq 0$. It can be seen from Equation (10.50) that $\lambda$ is an eigenvalue of the rate matrix modified by the free radical concentrations. The free radical concentrations can be obtained from Equation (10.39). The modified rate matrix transposed, is a square matrix of $n \times n$. Therefore, $n$ eigenvalues can be expected. The eigenvalues must obey

$$|(K\tilde{M}^*)^T - \lambda_j I| = 0$$

$$j = 1, 2, 3, \dots, n \tag{10.51}$$

Corresponding to each eigenvalue exists an eigenvector $Z_j$. There are $n$ solutions of the form of Equation (10.49). The general solution of the *homogenous equation* Equation (10.47) is given by a linear combination of these solutions as

$$M = \sum_{j=1}^{n} c_j Z_j e^{\lambda_j t}, \tag{10.52}$$

where $c_j$ are integration constants that can be solved for from the initial condition given by Equation (10.38). When the eigenvalues when solved for from the polynomial equation given by Equation (10.42) become complex, the solution to the monomer concentration with time becomes *subcritical damped oscillatory* [3–10].

## 10.6   SUMMARY

Copolymerization is a process by which two or more monomer repeat units enter the backbone chain of the polymer. The terms *homopolymer, copolymer, terpolymer, tetrapolymer*, and *multicomponent copolymer* were introduced and distinguished from each other. Alternating, random, block, and graft copolymers have different microstructures. The composition of a copolymer as a function of comonomer composition, reactivity ratios, and reactor choice was derived from the kinetics of free radical initiation, propagation, and termination reactions. The copolymerization equation obtained using the QSSA, quasi-steady-state approximation, is given by Equation (10.8). As an example, AN composition in SAN copolymer made in a CSTR as a function of AN monomer composition in the mixture of comonomers is illustrated in Figure 10.2 made in CSTR. The equation would depend on whether it is made in CSTR or PFR. In a similar manner, the composition equation for terpolymer was also developed. The reactivity ratios for commonly used comonomers are presented in Table 10.1. The copolymer composition is sensitive to reactivity ratios to a considerable extent. The copolymer composition equation for a multicomponent copolymer with $n$ monomers was derived. This methodology was generalized for $n$ monomers. The general form of the equation was represented in the matrix form using linear algebra. The rate matrix and rate equation is given in Equation (10.47). The QSSA in the vector notation becomes

$$K^T(MM^{*T}) = (M^*M^T)$$

When the eigenvalues of the rate equation become imaginary, the monomer concentration can be expected to undergo subcritical damped oscillations as a function of time.

## PROBLEMS

1. Develop the copolymer composition curve for the methacrylonitrile–vinyl acetate system. The reactivity ratios are $r_{12} = 12$ and $r_{21} = 0.01$.
2. Develop the terpolymer composition curve for 1-3-butadiene, acrylonitrile, styrene. The reactivity ratios are given as $r_{12} = 0.36$; $r_{21} = 0.046$; $r_{13} = 1.4$; $r_{31} = 0.58$; $r_{23} = 0.02$; $r_{32} = 0.29$.

3. Similar to Equation (10.12), which gives the copolymer composition as a function of the comonomer compositions and reactivity ratios, develop an expression for comonomer composition given the copolymer composition.

4. Similar to Equations (10.23–10.25), which expresses the terpolymer composition as a function of comonomer compositions and reactivity ratios, develop an expression for comonomer compositions given the terpolymer composition.

5. Repeat the analysis in Problem 1 for a PFR at 15% conversion.

6. Repeat the analysis in Problem 2 for a PFR at 35% conversion.

7. Develop the copolymer composition curve for the styrene and ethyl vinyl ether system. The reactivity ratios are $r_{12} = 90$ and $r_{21} = 0$.

8. Develop the copolymerization composition curve for methacrylic acid and 2-vinyl pyridine. The reactivity ratios are $r_{12} = 0.58$ and $r_{21} = 1.7$.

9. Calculate the copolymer composition curve for alphamethyl-styrene and maleic anhydride. The reactivity ratios are $r_{12} = 0.14$ and $r_{21} = 0.03$.

10. Develop the terpolymer composition curve for alphamethyl-styrene, styrene, and maleic anhydride.

11. Similar to the *multiplicity*, as shown in Figure 10.3, what problems can be anticipated in a PFR for certain comonomer systems?

## REVIEW QUESTIONS

1. What is meant by an ABC triblock terpolymer?

2. What is the difference between a random tetrapolymer and an alternating tetrapolymer?

3. Can a polymer chain become circular when the end groups are bonded to each other?

4. What is the difference between an alternating copolymer and a condensation homopolymer?

5. What is the difference between graft and block copolymers?

6. How sensitive are the model equations to the QSSA?

7. Is there a limit to the number of monomers that can enter a copolymer backbone chain?

8. When one of the reactivity ratios is zero, what kind of microstructure of the copolymer can be expected?

9. When the product of two reactivity ratios is equal to one, what can be said about the copolymerization composition equation?

10. When one of the reactivity ratios is much larger compared with the other, what can be said about the copolymer composition?

11. When a terpolymer is formed, what is the significance of the following?

$$r_{12}\, r_{21}\, r_{13}\, r_{31}\, r_{23}\, r_{32} = 0$$

12. When a terpolymer is formed, what is the significance of the following?

$$r_{12}\, r_{21}\, r_{13}\, r_{31}\, r_{23}\, r_{32} = 1$$

13. During multicomponent copolymerization, what ought to be the constraint on the reactivity ratios between $n$ monomers for a state of "ideal copolymerization"?

14. During multicomponent copolymerization, what ought to be the constraint on the reactivity ratios between $n$ monomers for a state of "alternating copolymerization"?

15. When one of the reactivity ratios is much larger compared with the other, what can be said about the terpolymer composition?

16. Can the copolymerization composition equation yield multiple roots? If so, what can be the possible physical significance of a second composition?

17. What is the physical significance of an inflection in the copolymerization composition curve?

18. Can reactivity ratios take on negative values?

19. In the terpolymerization composition equation derived for terpolymers and given by Equations (10.23–10.25), what is the significance of equal reactivity ratios such as $r_{12} = r_{13}$?

20. In the terpolymerization composition equation derived for terpolymers and given by Equation (10.23–10.25), what is the significance of equal reactivity ratios such as $r_{32} = 1$?

21. What is the difference between selectivity and reactivity ratio?

22. What is meant by penultimate effect?

23. Does reactivity ratio depend on reactor choice?

24. What is the dependence of reactivity ratio on temperature and pressure?

25. Does the initiator concentration affect the copolymer composition?

26. Does the termination mechanism affect the copolymer composition? Why not?

27. Can a terpolymer azeotrope exist?

28. Can azeotropy be expected in multicomponent copolymerization?

29. Is there a maxima in the copolymer composition equation?

30. Can diffusion have a role in copolymer composition?

# REFERENCES

1. G. Odian, *Principles of Polymerization,* John Wiley & Sons, New York, 1991.

2. A. Varma and M. Morbidelli, *Mathematical Methods in Chemical Engineering*, Oxford University Press, Oxford, U.K., 1997.

3. K. R. Sharma, Stability Issues in Multicomponent Continuous Mass Copolymerization, Middle Atlantic Regional Meeting of ACS, MARM 02, Fairfax, VA, May 2002.

4. K. R. Sharma, Overview of Continuous Polymerization Process Technology, 233rd ACS National Meeting, Chicago, IL, March 2007.

5. K. R. Sharma, Overview of Continuous Polymerization Process, CHEMCON 2006, Ankaleshwar, December 2006.

6. K. R. Sharma, Continuous Polymerization Process: Some Scale-Up Issues, AIChE Spring National Meeting, Orlando, FL, April 2006.

7. K. R. Sharma, Subcritical Damped Oscillations in Thermal Free Radical Method, 228th ACS National Meeting, Philadelphia, PA, August 2004.

8. K. R. Sharma, Subcritical Damped Oscillatory Concentration in Free Radical Polymerization of Alphamethylstryene, Acrylonitrile and Methacrylonitrile Terpolymer, 229th ACS National Meeting, San Diego, CA, March 2005.

9. K. R. Sharma, Reactions in Circle Representation to Account for Disproportionate Termination and Subcritical Damped Oscillatory Behavior, ANTEC 2004, Annual Technical Conference of the Society of Plastics Engineers, Chicago, IL, May 2004.

10. K. R. Sharma, Subcritical Damped Oscillatory Behavior of Reactions in Circle, 226th ACS National Meeting, New York, September 2003.

11. K. R. Sharma, Damped Oscillatory Graft Length by Free Radical Scheme, 225th ACS National Meeting, New Orleans, LA, March 23rd–March 28th, 2003. Francisco, CA, November 2003.

12. K. R. Sharma, Methacrylonitrile as Termonomer in Alphamethylstyrene and Acrylonitrile Multicomponent Copolymerization, AIChE Spring National Meeting, New Orleans, LA, March 30th–April 3rd, 2003.

13. K. R. Sharma, Graft Composition Stability in Continuous Mass Copolymerization, Middle Atlantic Regional Meeting of ACS, MARM 02, Fairfax, VA, May 2002.

14. K. R. Sharma, Thermal Polymerization Molecular Kinetics Representation by Ramanujan Numbers, 53rd Southeast Regional Meeting of the ACS, SERMACS, Savannah, GA, September 2001.

15. K. R. Sharma, Continuous Copolymerization of Alphamethylstyrene Acrylonitrile, Betamethylstyrene Acrylonitrile Copolymers, Annual Technical Conference for Society of Plastics Engineers, ANTEC 1999, New York, May 1999.

16. K. R. Sharma, Continuous Copolymerization of Betamethylstyrene Acrylonitrile, 90th AIChE Annual Meeting, Miami, FL, November 1998.

17. K. R. Sharma, Modeling Thermal Polymerization of the Polybetamethylstyrene and Grafting in a Twin Screw Extruder, 31st ACS Central Regional Meeting, Columbus, OH, June 1999.

18. K. R. Sharma, Best Reactor Choice for Copolymerization of Betamethylstyrene and Acrylonitrile for Continuous Mass Polymerization, 50th ACS Southeast Regional Meeting, Research Triangle Park, NC, October 1998.

19. K. R. Sharma, Effect of Chain Sequence Distribution on Contamination Formation in Alphamethylstyrene Acrylonitrile Copolymers during its Manufacture in Continuous Mass Polymerization, 5th World Congress of Chemical Engineering, San Diego, CA, July 1996.

# 11 Sequence Distribution of Copolymers

## LEARNING OBJECTIVES

- Dyads and triads in copolymers
- Free radical polymerization
- Geometric distribution
- Dyads and triads in terpolymers
- Nucleotide sequences in DNA
- Amino acid sequences in protein
- Sequence alignment

## 11.1 DYAD AND TRIAD PROBABILITIES IN COPOLYMER

Consider a random copolymer obtained by *free radical copolymerization*. The statistics of copolymerization and dyad probabilities can be calculated as follows. The propagation reactions can be of four types:

$$M_1^* + M_1 \xrightarrow{k_{11}} M_1 M_1^*$$

$$M_1^* + M_2 \xrightarrow{k_{12}} M_1 M_2^*$$

$$M_2^* + M_1 \xrightarrow{k_{21}} M_2 M_1^*$$

$$M_2^* + M_2 \xrightarrow{k_{11}} M_2 M_2^*$$

(11.1)

The rate of irreversible reactions can be written as

$$\frac{d[M_1]}{dt} = -k_{11}[M_1][M_1^*] - k_{21}[M_1][M_2^*]$$

$$\frac{d[M_2]}{dt} = -k_{22}[M_2][M_2^*] - k_{12}[M_2][M_1^*]$$

(11.2)

The free radical species formed may be assumed to be highly reactive and can be assumed to be consumed as rapidly as they are formed. This is referred to as the quasi-steady-state assumption (QSSA). Thus:

$$\frac{d[M_1^*]}{dt} = -k_{11}[M_1][M_1^*] - k_{12}[M_1^*][M_2].$$

$$\frac{d[M_2^*]}{dt} = -k_{22}[M_2][M_2^*] - k_{21}[M_2^*][M_1].$$

(11.3)

$$\frac{d[M_1 M_1^*]}{dt} = k_{11}[M_1][M_1^*].$$

$$\frac{d[M_2 M_2^*]}{dt} = k_{22}[M_2][M_2^*].$$

$$\frac{d[M_1 M_2^*]}{dt} = k_{12}[M_2][M_1^*].$$

(11.4)

$$\frac{d[M_2 M_1^*]}{dt} = k_{21}[M_1][M_2^*].$$

By QSSA:

$$\frac{d[M_1^*]}{dt} + \frac{d[M_1 M_1^*]}{dt} + \frac{d[M_2 M_1^*]}{dt} = 0.$$

$$\frac{d[M_2^*]}{dt} + \frac{d[M_2 M_2^*]}{dt} + \frac{d[M_1 M_2^*]}{dt} = 0.$$

(11.5)

Thus:

$$k_{12}[M_1^*][M_2] = k_{21}[M_1][M_2^*].$$

or

$$\frac{[M_2]}{[M_1]}\frac{k_{12}}{k_{21}} = \frac{[M_2^*]}{[M_1^*]}.$$

(11.6)

The dyad probabilities can be calculated from the selectivity achieved in the parallel reactions. Thus, the probability of the $M_1$-$M_1$ dyad can be found to be

$$P_{11} = \frac{dM_1}{dM_1 + dM_2} = \frac{1}{1 + \dfrac{k_{21}}{k_{11}}\dfrac{M_2^*}{M_1^*}}$$

(11.7)

Combining Equation (8.7) and Equation (8.6):

$$P_{11} = \frac{dM_1}{dM_1 + dM_2} = \frac{1}{1 + \dfrac{1}{r_{12}}\dfrac{[M_2]}{[M_1]}} = \frac{r_{12}[M_1]}{r_{12}[M_1] + [M_2]}$$

(11.8)

where $r_{12}$ is the reactivity ratio of monomer 1 and 2. Then the $M_1$-$M_2$ dyad probability $P_{12}$ can be written as

$$P_{12} = 1 - P_{11} = \frac{[M_2]}{r_{12}[M_1] + 1} \qquad (11.9)$$

In a similar fashion, the $M_2$-$M_2$ dyad probability, $P_{22}$, and $M_2$-$M_1$ dyad probability can be written as

$$P_{22} = \frac{r_{21}[M_2]}{r_{21}[M_2] + [M_1]} \qquad (11.10)$$

and

$$P_{21} = \frac{[M_1]}{r_{21}[M_2] + 1} \qquad (11.11)$$

### Example 11.1

*The styrene and acrylonitrile can be copolymerized by free radical methods using a continuous stirred tank reactor (CSTR). The reactivity ratios $r_{12}$ and $r_{21}$ can be taken as 0.04 and 0.41, respectively. Construct a first-order Markov model using the dyad probabilities derived in Section 11.1.*

Hidden Markov models (HMMs) are constructed by using concepts such as conditional probability. They are used in a variety of applications, and are classified under a useful class of probabilistic models. HMMs are a special case of neural networks, stochastic networks, and Bayesian networks. The dyad probabilities as a function of reactor monomer composition are given in Table 11.1. The probabilities are calculated using Equations (11.8–11.10).

The microstructure of the SAN copolymer with respect to the chain sequence distribution can be found from the first-order hidden Markov model. The HMM architecture is shown in Figures 11.1A and B. The conditional dyad probabilities

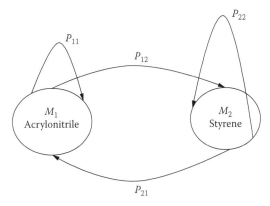

**FIGURE 11.1A**   First-order Markov model to represent chain sequence distribution of SAN copolymer.

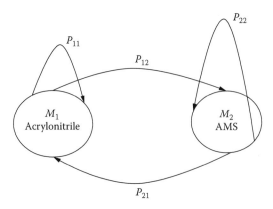

**FIGURE 11.1B** First-order Markov model to represent chain sequence distribution of AMS–AN copolymer.

## TABLE 11.1
## Dyad Probabilities in SAN Copolymer
## Formed in a CSTR

| $M_1$ | $M_2$ | $P_{11}$ | $P_{12}$ | $P_{21}$ | $P_{22}$ |
|---|---|---|---|---|---|
| 0.000 | 1.000 | 0.000 | 1.000 | 0.000 | 1.000 |
| 0.050 | 0.950 | 0.002 | 0.998 | 0.114 | 0.886 |
| 0.100 | 0.900 | 0.004 | 0.996 | 0.213 | 0.787 |
| 0.150 | 0.850 | 0.007 | 0.993 | 0.301 | 0.699 |
| 0.200 | 0.800 | 0.010 | 0.990 | 0.379 | 0.621 |
| 0.250 | 0.750 | 0.013 | 0.987 | 0.448 | 0.552 |
| 0.300 | 0.700 | 0.017 | 0.983 | 0.511 | 0.489 |
| 0.350 | 0.650 | 0.021 | 0.979 | 0.568 | 0.432 |
| 0.400 | 0.600 | 0.026 | 0.974 | 0.619 | 0.381 |
| 0.450 | 0.550 | 0.032 | 0.968 | 0.666 | 0.334 |
| 0.500 | 0.500 | 0.038 | 0.962 | 0.709 | 0.291 |
| 0.550 | 0.450 | 0.047 | 0.953 | 0.749 | 0.251 |
| 0.600 | 0.400 | 0.057 | 0.943 | 0.785 | 0.215 |
| 0.650 | 0.350 | 0.069 | 0.931 | 0.819 | 0.181 |
| 0.700 | 0.300 | 0.085 | 0.915 | 0.851 | 0.149 |
| 0.750 | 0.250 | 0.107 | 0.893 | 0.880 | 0.120 |
| 0.800 | 0.200 | 0.138 | 0.862 | 0.907 | 0.093 |
| 0.850 | 0.150 | 0.185 | 0.815 | 0.933 | 0.067 |
| 0.900 | 0.100 | 0.265 | 0.735 | 0.956 | 0.044 |
| 0.950 | 0.050 | 0.432 | 0.568 | 0.979 | 0.021 |
| 1.000 | 0.000 | 1.000 | 0.000 | 1.000 | 0.000 |

are given in the links of the model. The probability of the next repeat unit depends on the preceding repeat unit probability in the first-order Markov model.

## Example 11.2

*The alphamethyl-styrene and acrylonitrile can be copolymerized by free radical methods using CSTR. The reactivity ratios $r_{12}$ and $r_{21}$ can be taken as 0.03 and 01.4, respectively, from Chapter 10, Table 10.1. Construct a first-order Markov model using the dyad probabilities derived in Section 11.1.*

Hidden Markov models are often discussed as HMMs. They are constructed by using concepts such as conditional probability. They are used in a variety of applications. They are classified under a useful class of probabilistic models. HMMs are a special case of neural networks, stochastic networks, and Bayesian networks. The dyad probabilities as a function of reactor monomer composition are given in Table 11.2. The probabilities are calculated using Equations (11.8–11.10).

### TABLE 11.2
### Dyad Probabilities in AMS–AN Copolymer Formed in a CSTR

| | | $r_{12}$ | 0.03 | $r_{21}$ | 0.14 | | |
|---|---|---|---|---|---|---|---|
| AN | Alphamethyl–Styrene | | $P_{11}$ | $P_{12}$ | $P_{21}$ | $P_{22}$ | |
| $M_1$ | $M_2$ | $F_1$ | $M_1M_1$ | $M_1M_2$ | $M_2M_1$ | $M_2M_2$ |
| 0.000 | 1.000 | 0.000 | 0.000 | 1.000 | 0.000 | 1.000 |
| 0.050 | 0.950 | 0.215 | 0.002 | 0.998 | 0.273 | 0.727 |
| 0.100 | 0.900 | 0.307 | 0.003 | 0.997 | 0.442 | 0.558 |
| 0.150 | 0.850 | 0.359 | 0.005 | 0.995 | 0.558 | 0.442 |
| 0.200 | 0.800 | 0.392 | 0.007 | 0.993 | 0.641 | 0.359 |
| 0.250 | 0.750 | 0.416 | 0.010 | 0.990 | 0.704 | 0.296 |
| 0.300 | 0.700 | 0.433 | 0.013 | 0.987 | 0.754 | 0.246 |
| 0.350 | 0.650 | 0.446 | 0.016 | 0.984 | 0.794 | 0.206 |
| 0.400 | 0.600 | 0.457 | 0.020 | 0.980 | 0.826 | 0.174 |
| 0.450 | 0.550 | 0.467 | 0.024 | 0.976 | 0.854 | 0.146 |
| 0.500 | 0.500 | 0.475 | 0.029 | 0.971 | 0.877 | 0.123 |
| 0.550 | 0.450 | 0.482 | 0.035 | 0.965 | 0.897 | 0.103 |
| 0.600 | 0.400 | 0.489 | 0.043 | 0.957 | 0.915 | 0.085 |
| 0.650 | 0.350 | 0.495 | 0.053 | 0.947 | 0.930 | 0.070 |
| 0.700 | 0.300 | 0.502 | 0.065 | 0.935 | 0.943 | 0.057 |
| 0.750 | 0.250 | 0.510 | 0.083 | 0.917 | 0.955 | 0.045 |
| 0.800 | 0.200 | 0.520 | 0.107 | 0.893 | 0.966 | 0.034 |
| 0.850 | 0.150 | 0.533 | 0.145 | 0.855 | 0.976 | 0.024 |
| 0.900 | 0.100 | 0.556 | 0.213 | 0.787 | 0.985 | 0.015 |
| 0.950 | 0.050 | 0.609 | 0.363 | 0.637 | 0.993 | 0.007 |
| 1.000 | 0.000 | 1.000 | 1.000 | 0.000 | 1.000 | 0.000 |

## Example 11.3

*Calculate the triad concentrations in the SAN copolymer discussed in Example 11.1.*
There can be six different triads formed in the SAN copolymer. These are AAA,
AAS, ASA, SAS, SSA, and SSS, where A represents AN and S represents styrene
monomers.

$$P(AAA) = P_{11}^2 \qquad\qquad (11.12)$$

$$P(AAS) = P_{11}P_{12} \qquad\qquad (11.13)$$

$$P(ASA) = P_{12}P_{21} \qquad\qquad (11.14)$$

$$P(SAS) = P_{21}P_{12} \qquad\qquad (11.15)$$

$$P(SSA) = P_{22}P_{21} \qquad\qquad (11.16)$$

$$P(SSS) = P_{22}^2 \qquad\qquad (11.17)$$

where $P_{11}$, $P_{12}$, $P_{21}$, and $P_{22}$ are given through Equations (11.8–11.18). The triad
concentrations are plotted as ppm (parts per million) versus the AN composition
in the SAN copolymer in Figure 11.2 for the AAA, AAS, and ASA triads. As shown
by Sharma [7], the triad concentrations can play a primary causative role in contamination formation in the product during manufacturing due to the formation of
Grassie chromophores. The AN-AN-AN hexads react on subject to heat, and after
sufficient time, the polymer was found to change in color from colorless to yellow to blue-black. For one product, at manufacturing, the contamination numbers
in PPM from the manufacturing plant matched the hexad and higher sequence
length of AN-AN-AN chain sequence concentrations in the copolymer.

The triad probabilities in SAN, that is, AAA, AAS, ASA (or SAS), SSA, and SSS
were calculated using an MS Excel spreadsheet as given by Equations (11.12–11.17).
It can be seen from Equations (11.14) and (11.15) that the P(SAS) = P(ASA). It can be
seen from Figure 11.3 that the probability of ASA triads reaches a maximum value
at 50% mole fraction of AN in the SAN copolymer. The SSA triads also reach a
maxima at an AN composition in the SAN copolymer of 35% fraction. Concave
curvature was found in the AAS triad concentration versus AN composition curve.
The concave to convex to concave change in curvature was found for SSA triad
concentration change with AN copolymer composition. The AAA triads rise rapidly in a monotonic fashion with increase in AN copolymer composition.

The triad probabilities of alphamethyl-styrene acrylonitrile (AMS–AN)
copolymer, that is, AAA, AAS, ASA (or SAS), SSA and SSS, were calculated using
an MS Excel spreadsheet as given by Equations (11.12–11.17). It can be seen from
Equations (11.14) and (11.15) that P(SAS) = P(ASA).

As discussed in Chapter 12, the AMS-AMS homopolymerization reaction is not
favored at the CSTR reactor temperature due to the ceiling temperature or unzipping effect. Hence, the triad concentrations of P(SSS), P(SSA) can be seen to be 0.

It can be seen from Figure 11.4 that the probability of ASA decreases from a maximum value of 1.0 at smaller compositions of AN in the copolymer. Concave curvature
was found in the AAS triad concentration versus AN composition curve. The AAA triads rise rapidly in a monotonic fashion with increase in AN copolymer composition.

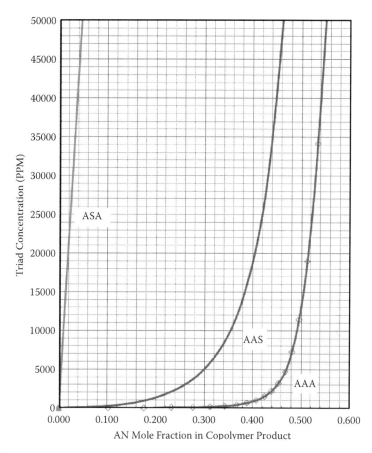

**FIGURE 11.2**    Triad concentrations in PPM in SAN copolymer versus mole fraction AN in copolymer for AAA, AAS, and ASA triads.

## 11.2  DYAD AND TRIAD PROBABILITIES IN TERPOLYMERS

The chain sequence length distribution of DNA can be represented using the geometric distribution. The mean and variance of the geometric distribution would be expected to depend on the mechanism of formation of polynucleotide sequences. For instance, a terpolymer formed by *free radical polymerization* can be modeled with respect to the sequence distribution as follows. Let three termonomers enter a long copolymer chain at $M_1$, $M_2$, and $M_3$ concentrations with reactivity ratios $r_{12}$, $r_{21}$, $r_{23}$, $r_{32}$, $r_{13}$, $r_{31}$. Assume that the bond formation order does not matter in the rate, that is,

$$k_{ij} = [M_j][M_i^*] = k_{ji}[M_i][M_j^*].$$

$$P_{11} = \cfrac{1}{\left[1 + \cfrac{M_2}{r_{12}M_1} + \cfrac{M_3}{r_{13}M_1}\right]} \tag{11.18}$$

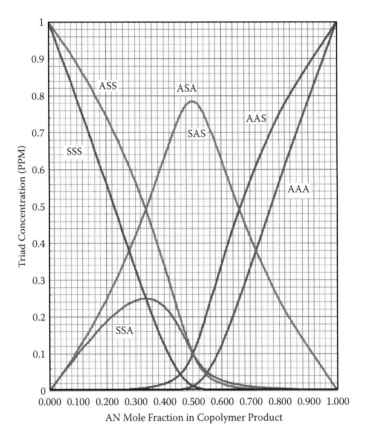

FIGURE 11.3  Triad probabilities in SAN copolymer—AAA, AAS, ASA (or SAS), SSA, ASS, and SSS triads as a function of AN mole fraction in copolymer.

$$\text{Let } \beta = \frac{M_2}{r_{23}M_1}; \quad \gamma = \frac{M_3}{r_{13}M_1}. \tag{11.19}$$

Then the probability of a $M_2M_2$ dyad is

$$P_{11} = \frac{1}{(1+\beta+\gamma)}. \tag{11.20}$$

The sequence length of the repeats of monomer 2 in the chain is given by

$$N_{1x} = \frac{1}{[1+\beta+\gamma]^{x-1}}\left[1 - \frac{1}{(1+\beta+\gamma)}\right]. \tag{11.21}$$

The mean of the distribution is

$$\lambda = \frac{(\beta+\gamma)}{(1+\beta+\gamma)}. \tag{11.22}$$

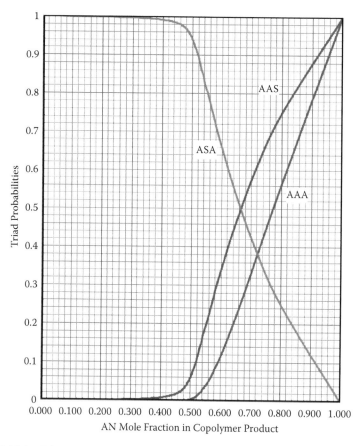

**FIGURE 11.4** Triad probabilities in AMS–AN copolymer—AAA, AAS, ASA (or SAS), and ASS triads as a function of AN mole fraction in copolymer.

The variance, $\sigma^2$, of the distribution can be written as

$$\sigma^2 = \frac{1}{(\beta + \gamma)^2} + \frac{1}{(\beta + \gamma)}. \tag{11.23}$$

For a tetrapolymer with four monomers such as the case for DNA, polynucleotides,

$$P_{11} = \frac{1}{(1 + \beta + \gamma + \delta)}, \tag{11.24}$$

where $\delta = \dfrac{M_4}{r_{24} M_2}$

The sequence length of a single base in the polynucleotide chain can be given by a geometric distribution.

$$N_{1x} = \frac{1}{[1 + \beta + \gamma + \delta]^{x-1}} \left[ 1 - \frac{1}{(1 + \beta + \gamma + \delta)} \right]. \tag{11.25}$$

The mean and variance of the distribution can be written as

$$\lambda = \frac{(\beta + \gamma + \delta)}{(1 + \beta + \gamma + \delta)}; \quad \sigma^2 = \frac{1}{(\beta + \gamma + \delta)^2} + \frac{1}{(\beta + \gamma + \delta)}. \tag{11.26}$$

The polymer compositions can be related to the monomer compositions by simple relations. Thus, the run lengths of A, AA, AAA, and AAAA, etc. for each of the four bases can be calculated. Modifications to $(\beta + \gamma + \delta)$ can be made depending on the mechanism of formation of the polynucleotide chain. The assumption is that the composition of adenine, guanine, cytosine, and thymine in equal proportions in the polymer chain can be used to simplify the terms. The triad and tetrad probabilities such as AGC, AAC, AGG, etc., can be computed to the dyad probabilities.

## Example 11.4

*Discuss the dyad probabilities in SANM, styrene–acrylonitrile–methyl methacrylate terpolymer.*
      Consider the preparation of styrene–acrylonitrile and methyl methacrylate terpolymer in a CSTR. There can be 3×3 = 9 possible dyads formed. These are AA, AS, SA, SS, SM, MS, AM, MA, and MM. The reactivity ratios can be read from Chapter 10, Table 10.1 The terpolymer composition can be calculated using the terpolymerization composition equations for a CSTR given in Equations (10.23–10.25). The dyad probabilities were calculated using an MS Excel spreadsheet and listed in Table 11.3. Let AN be denoted as monomer 1, styrene as monomer 2, and methyl methacrylate as monomer 3. Assuming that the bond formation order does not influence the rate, that is,

$$k_{ij} = [M_j][M_i^*] = k_{ji}[M_i][M_j^*] \tag{11.27}$$

$$P_{11} = \frac{1}{\left[1 + \dfrac{M_2}{r_{12}M_1} + \dfrac{M_3}{r_{13}M_1}\right]} \tag{11.28}$$

$$P_{22} = \frac{1}{\left[1 + \dfrac{M_1}{r_{21}M_2} + \dfrac{M_3}{r_{23}M_2}\right]} \tag{11.29}$$

$$P_{33} = \frac{1}{\left[1 + \dfrac{M_1}{r_{31}M_3} + \dfrac{M_2}{r_{32}M_3}\right]} \tag{11.30}$$

$$P_{12} = \frac{1}{\left[r_{12}\dfrac{M_1}{M_2} + 1 + \dfrac{r_{12}M_3}{r_{13}M_2}\right]} \tag{11.31}$$

## TABLE 11.3
## Dyad Probabilities in SANM, Styrene–Acrylonitrile–Methyl Methacrylate Terpolymer

$(f_3 = 0.1)$

| $r_{12}$ | 0.02 |
|---|---|
| $r_{21}$ | 0.29 |
| $r_{13}$ | 0.4 |
| $r_{31}$ | 1.3 |
| $r_{23}$ | 0.52 |
| $r_{32}$ | 0.46 |

| $f_1$ | $f_2$ | $P_{11}$ | $P_{12}$ | $P_{13}$ | $P_{22}$ | $P_{21}$ | $P_{23}$ | $P_{33}$ | $P_{31}$ | $P_{32}$ |
|---|---|---|---|---|---|---|---|---|---|---|
| 0 | 0.9 | 0.000 | 0.994 | 0.000 | 0.824 | 0.000 | 0.176 | 0.049 | 0.000 | 0.951 |
| 0.05 | 0.85 | 0.158 | 0.993 | 0.024 | 0.700 | 0.142 | 0.158 | 0.050 | 0.266 | 0.797 |
| 0.1 | 0.8 | 0.273 | 0.991 | 0.047 | 0.598 | 0.258 | 0.144 | 0.052 | 0.377 | 0.686 |
| 0.15 | 0.75 | 0.361 | 0.989 | 0.067 | 0.514 | 0.354 | 0.132 | 0.054 | 0.411 | 0.602 |
| 0.2 | 0.7 | 0.431 | 0.987 | 0.085 | 0.442 | 0.436 | 0.122 | 0.056 | 0.407 | 0.536 |
| 0.25 | 0.65 | 0.487 | 0.985 | 0.100 | 0.381 | 0.506 | 0.113 | 0.059 | 0.384 | 0.483 |
| 0.3 | 0.6 | 0.534 | 0.982 | 0.113 | 0.328 | 0.566 | 0.105 | 0.061 | 0.351 | 0.439 |
| 0.35 | 0.55 | 0.573 | 0.979 | 0.123 | 0.282 | 0.619 | 0.099 | 0.064 | 0.315 | 0.403 |
| 0.4 | 0.5 | 0.606 | 0.975 | 0.132 | 0.241 | 0.666 | 0.093 | 0.067 | 0.279 | 0.371 |
| 0.45 | 0.45 | 0.635 | 0.970 | 0.138 | 0.205 | 0.707 | 0.088 | 0.070 | 0.243 | 0.345 |
| 0.5 | 0.4 | 0.660 | 0.964 | 0.143 | 0.173 | 0.744 | 0.083 | 0.074 | 0.209 | 0.321 |
| 0.55 | 0.35 | 0.682 | 0.956 | 0.146 | 0.144 | 0.778 | 0.079 | 0.078 | 0.176 | 0.300 |
| 0.6 | 0.3 | 0.701 | 0.946 | 0.149 | 0.117 | 0.808 | 0.075 | 0.082 | 0.146 | 0.281 |
| 0.65 | 0.25 | 0.718 | 0.933 | 0.150 | 0.093 | 0.835 | 0.072 | 0.087 | 0.117 | 0.264 |
| 0.7 | 0.2 | 0.734 | 0.913 | 0.150 | 0.071 | 0.860 | 0.069 | 0.093 | 0.090 | 0.248 |
| 0.75 | 0.15 | 0.748 | 0.882 | 0.150 | 0.051 | 0.883 | 0.066 | 0.100 | 0.065 | 0.232 |
| 0.8 | 0.1 | 0.760 | 0.826 | 0.149 | 0.033 | 0.904 | 0.063 | 0.107 | 0.042 | 0.215 |
| 0.85 | 0.05 | 0.772 | 0.694 | 0.148 | 0.016 | 0.924 | 0.061 | 0.116 | 0.020 | 0.188 |
| 0.9 | 0 | 0.783 | 0.000 | 0.147 | 0.000 | 0.942 | 0.058 | 0.126 | 0.000 | 0.000 |

$$P_{13} = \cfrac{1}{\left[ r_{13}\cfrac{M_1}{M_3} + \cfrac{r_{13}M_3}{r_{12}M_2} + 1 \right]} \tag{11.32}$$

$$P_{31} = \cfrac{1}{\left[ r_{31}\cfrac{M_3}{M_1} + \cfrac{r_{31}M_1}{r_{32}M_3} + 1 \right]} \tag{11.33}$$

$$P_{21} = \cfrac{1}{\left[ r_{21}\cfrac{M_2}{M_1} + 1 + \cfrac{r_{21}M_3}{r_{23}M_1} \right]} \tag{11.34}$$

$$P_{32} = \cfrac{1}{\left[ r_{32}\cfrac{M_3}{M_2} + 1 + \cfrac{r_{32}M_1}{r_{31}M_2} \right]} \qquad (11.35)$$

$$P_{23} = \cfrac{1}{\left[ r_{23}\cfrac{M_2}{M_3} + 1 + \cfrac{r_{23}M_1}{r_{21}M_2} \right]} \qquad (11.36)$$

## Example 11.5

*Discuss the triad probabilities for the terpolymer discussed in Example 11.4.*
There are 3×3×3 = 27 triads that can be formed in the SANM terpolymer discussed in Example 11.4. These triads and their probabilities are given in Table 11.4. Let AN be denoted by monomer 1, styrene by monomer 2, and methyl methacrylate by monomer 3.

For the monomer composition ratios used in Table 11.3, the triad probabilities with the AN monomer repeat unit as header are given in Table 11.5. The triad probabilities with the styrene monomer repeat unit as header are given in Table 11.6, and the triad probabilities with the methyl methacrylate monomer repeat unit as header are given in Table 11.7.

## Example 11.6

Chaves, P.B., Paes, M.F., Strier, K.B., Mendes, S.L., Louro, I.D. and Fagundes, V. [7] submitted the following DNA sequence in *homo sapiens* to the NCBI with 660 bases. Develop a Markov model of the third order to represent this information. Calculate the transition probabilities and represent the information in the form of a suitable table.

$$\text{Number of triads that need to be studied} = 4^3 = 64 \qquad (11.37)$$

$$\text{Alphabet} = \{A, C, G, T\} \qquad (11.38)$$

Number of transition probabilities that need to be calculated = 4×64 = 256(11.39)

The 256 transition probabilities P(A/AAA), P(G/AAA) ,... etc., are calculated from the information provided in Table 11.8 and presented in the Table 11.9. Columns 3–6 are conditional probability values for the base pair shown at the top of the column, given the preceding triad that occurred in the sequence given in column 2. A triad number is also given to the 64 possible triads for DNA.

## Example 11.7

Develop a first-order Markov model for the DNA sequence given in Example 11.1 in order to represent the first 60 base pairs. Calculate the transition probabilities and represent the information in the form of a suitable diagram

ctatatatcttaatggcacatgcagcgcaagtaggtctacaagacgctacttcccctatc

## TABLE 11.4
## Triad Probabilities in SANM Terpolymer

$P(AAA) = P_{11}^2$      $P(SSS) = P_{22}^2$      $P(MMM) = P_{33}^2$

$P(AAS) = P_{11}P_{12}$      $P(SSA) = P_{22}P_{21}$      $P(MMA) = P_{33}P_{31}$

$P(ASA) = P_{12}P_{21}$      $P(SAS) = P_{21}P_{12}$      $P(MAM) = P_{31}P_{13}$

$P(AAM) = P_{11}P_{13}$      $P(SSM) = P_{22}P_{23}$      $P(MMS) = P_{33}P_{32}$

$P(AMA) = P_{13}P_{31}$      $P(SMS) = P_{23}P_{32}$      $P(MSM) = P_{32}P_{23}$

$P(AMS) = P_{13}P_{32}$      $P(SAM) = P_{21}P_{13}$      $P(MAS) = P_{31}P_{13}$

$P(ASM) = P_{12}P_{23}$      $P(SMA) = P_{23}P_{31}$      $P(MSA) = P_{32}P_{23}$

$P(ASS) = P_{12}P_{22}$      $P(SAA) = P_{21}P_{11}$      $P(MAA) = P_{31}P_{11}$

$P(AMM) = P_{13}P_{33}$      $P(SMM) = P_{23}P_{33}$      $P(MSS) = P_{31}P_{22}$

## TABLE 11.5
## Nine Triad Probabilities, AAA, AAS, ASA, AAM, AMA, ASS, AMM, AMS, and ASM for SANM Terpolymer

| AAA | AAS | ASA | AAM | AMA | AMS | ASM | ASS | AMM |
|---|---|---|---|---|---|---|---|---|
| 0.000 | 0 | 0 | 0 | 0 | 0 | 0.175084 | 0.819391 | 0 |
| 0.025 | 0.156623 | 0.140942 | 0.003828 | 0.006445 | 0.019354 | 0.157205 | 0.694844 | 0.001222 |
| 0.075 | 0.270854 | 0.255648 | 0.012767 | 0.017612 | 0.032061 | 0.142573 | 0.593104 | 0.002439 |
| 0.131 | 0.357631 | 0.350644 | 0.024204 | 0.027536 | 0.040308 | 0.130368 | 0.508434 | 0.003628 |
| 0.186 | 0.425563 | 0.430415 | 0.036528 | 0.034485 | 0.045423 | 0.12002 | 0.436871 | 0.004773 |
| 0.237 | 0.479946 | 0.498134 | 0.048733 | 0.038359 | 0.048291 | 0.111122 | 0.375593 | 0.005864 |
| 0.285 | 0.524198 | 0.556089 | 0.060204 | 0.039623 | 0.049538 | 0.103376 | 0.322532 | 0.006897 |
| 0.328 | 0.5606 | 0.605953 | 0.070595 | 0.038879 | 0.049621 | 0.096553 | 0.276142 | 0.007875 |
| 0.367 | 0.590702 | 0.648941 | 0.079745 | 0.036694 | 0.048878 | 0.090477 | 0.235241 | 0.008803 |
| 0.403 | 0.615546 | 0.685908 | 0.087612 | 0.033545 | 0.047562 | 0.085006 | 0.198913 | 0.009691 |
| 0.435 | 0.635789 | 0.717401 | 0.094233 | 0.029808 | 0.045861 | 0.080018 | 0.166437 | 0.010549 |
| 0.464 | 0.651743 | 0.743644 | 0.099693 | 0.025765 | 0.043908 | 0.075404 | 0.137236 | 0.011393 |
| 0.491 | 0.663345 | 0.764468 | 0.104099 | 0.021622 | 0.041796 | 0.071056 | 0.110848 | 0.012236 |
| 0.516 | 0.669993 | 0.779092 | 0.107569 | 0.017521 | 0.03958 | 0.066845 | 0.086899 | 0.013098 |
| 0.538 | 0.670094 | 0.785566 | 0.11022 | 0.013561 | 0.037274 | 0.062586 | 0.06509 | 0.013996 |
| 0.559 | 0.659785 | 0.779217 | 0.112164 | 0.009802 | 0.03483 | 0.057942 | 0.045195 | 0.014955 |
| 0.578 | 0.628476 | 0.747265 | 0.113501 | 0.006279 | 0.032029 | 0.052093 | 0.027088 | 0.016001 |
| 0.596 | 0.536129 | 0.641419 | 0.114324 | 0.00301 | 0.027835 | 0.042084 | 0.010942 | 0.017168 |
| 0.612 | 0 | 0 | 0.114715 | 0 | 0 | 0 | 0 | 0.0185 |

$$\text{Alphabet} = \{A, C, G, T\} \tag{11.40}$$

Number of transition probabilities that need to be calculated = $4 \times 4 = 16$ (11.41)

The 16 transition probabilities P(A/A), P(G/A),... etc., are calculated from the information provided in Table 11.7. Columns 3–6 are conditional probability values for the base pair shown at the top of the column given the preceding base pair

**TABLE 11.6**

**Nine Triad Probabilities, SSS, SSA, SAS, SSM, SMS, SAM, SMA, SAA, and SMM in SANM Terpolymer**

| SSS | SSA | SAS | SSM | SMS | SAM | SMA | SAA | SMM |
|---|---|---|---|---|---|---|---|---|
| 0.679 | 0 | 0 | 0.145061 | 0.167495 | 0 | 0 | 0 | 0.008561 |
| 0.490 | 0.09932 | 0.140942 | 0.11078 | 0.126236 | 0.003445 | 0.042035 | 0.04188 | 0.00797 |
| 0.358 | 0.154291 | 0.255648 | 0.086047 | 0.098676 | 0.012051 | 0.054205 | 0.102977 | 0.007506 |
| 0.264 | 0.182103 | 0.350644 | 0.067705 | 0.079309 | 0.023731 | 0.05418 | 0.148628 | 0.007138 |
| 0.196 | 0.192902 | 0.430415 | 0.05379 | 0.065156 | 0.036945 | 0.049467 | 0.175399 | 0.006846 |
| 0.145 | 0.192896 | 0.498134 | 0.043031 | 0.054488 | 0.05058 | 0.043281 | 0.186934 | 0.006616 |
| 0.108 | 0.185993 | 0.556089 | 0.034576 | 0.046239 | 0.063867 | 0.036984 | 0.187539 | 0.006438 |
| 0.080 | 0.17471 | 0.605953 | 0.027838 | 0.039724 | 0.076307 | 0.031124 | 0.180713 | 0.006305 |
| 0.058 | 0.160699 | 0.648941 | 0.022405 | 0.034484 | 0.087607 | 0.025888 | 0.169014 | 0.006211 |
| 0.042 | 0.145058 | 0.685908 | 0.017977 | 0.030201 | 0.097626 | 0.0213 | 0.154241 | 0.006153 |
| 0.030 | 0.128525 | 0.717401 | 0.014335 | 0.026651 | 0.106329 | 0.017322 | 0.137635 | 0.006131 |
| 0.021 | 0.111599 | 0.743644 | 0.011316 | 0.023669 | 0.11375 | 0.013889 | 0.120045 | 0.006141 |
| 0.014 | 0.094616 | 0.764468 | 0.008794 | 0.02113 | 0.119969 | 0.010931 | 0.102046 | 0.006186 |
| 0.009 | 0.077802 | 0.779092 | 0.006675 | 0.018937 | 0.125086 | 0.008383 | 0.084025 | 0.006267 |
| 0.005 | 0.061309 | 0.785566 | 0.004885 | 0.017005 | 0.129214 | 0.006187 | 0.066241 | 0.006385 |
| 0.003 | 0.045233 | 0.779217 | 0.003364 | 0.015248 | 0.132467 | 0.004291 | 0.048861 | 0.006547 |
| 0.001 | 0.029637 | 0.747265 | 0.002066 | 0.013526 | 0.134954 | 0.002652 | 0.031991 | 0.006758 |
| 0.000 | 0.014553 | 0.641419 | 0.000955 | 0.011391 | 0.136776 | 0.001232 | 0.015693 | 0.007026 |
| 0.000 | 0 | 0 | 0 | 0 | 0.138027 | 0 | 0 | 0.007365 |

that occurred in the sequence given in column 2. A base pair number is also given to the four possible base pairs: adenine, guanine, cytosine, and thymine.

DNA strings can be generated from a four-letter alphabet {A, C, G, T}. A simple sequence model can be developed by assuming that the sequences have been obtained by independent tosses of a four-sided die. Let the data be represented by D and the model by M. The model M has four parameters, namely, $P_A$, $P_C$, $P_G$, and $P_T$ for the probabilities of the bases adenine, cyotosine, guanine, and thymine.

$$P_A + P_B + P_G + P_T = 1. \tag{11.42}$$

Equation (11.42) is written based on the simple surmise that each sequence member has to be among the four nucleotide base pairs. Further,

$$P(D/M) = P_A{}^{na} P_C{}^{nc} P_G{}^{ng} P_T{}^{nt}, \tag{11.43}$$

where, na, nc, ng, and nt are the number of times the letters A, C, G, and T, respectively, appear in the sequence O.

$$P(D/M) = \prod P_x{}^{nx} \tag{11.44}$$

$$x \, \varepsilon \, A$$

where N: D = {0}, with O = $x^1.....x^N$ where $x_i \, \varepsilon \, A$.

## TABLE 11.7
## Nine Triad Probabilities, MMM, MMA, MAM, MMS, MSM, MAS, MSA, and MSS in SANM Terpolymer

| MMM | MMA | MAM | MMS | MSM | MAS | MSA | MAA | MSS |
|---|---|---|---|---|---|---|---|---|
| 0.002 | 0 | 0 | 0.046261 | 0.167495 | 0 | 0 | 0 | 0.145061 |
| 0.003 | 0.013367 | 0.006445 | 0.040144 | 0.126236 | 0.263656 | 0.042035 | 0.04188 | 0.11078 |
| 0.003 | 0.01967 | 0.017612 | 0.035808 | 0.098676 | 0.373625 | 0.054205 | 0.102977 | 0.086047 |
| 0.003 | 0.022278 | 0.027536 | 0.03261 | 0.079309 | 0.406864 | 0.05418 | 0.148628 | 0.067705 |
| 0.003 | 0.022918 | 0.034485 | 0.030187 | 0.065156 | 0.40176 | 0.049467 | 0.175399 | 0.05379 |
| 0.003 | 0.022493 | 0.038359 | 0.028317 | 0.054488 | 0.377777 | 0.043281 | 0.186934 | 0.043031 |
| 0.004 | 0.021486 | 0.039623 | 0.026863 | 0.046239 | 0.344999 | 0.036984 | 0.187539 | 0.034576 |
| 0.004 | 0.02016 | 0.038879 | 0.025729 | 0.039724 | 0.308737 | 0.031124 | 0.180713 | 0.027838 |
| 0.004 | 0.018658 | 0.036694 | 0.024853 | 0.034484 | 0.271805 | 0.025888 | 0.169014 | 0.022405 |
| 0.005 | 0.017061 | 0.033545 | 0.02419 | 0.030201 | 0.235682 | 0.0213 | 0.154241 | 0.017977 |
| 0.005 | 0.015408 | 0.029808 | 0.023706 | 0.026651 | 0.201113 | 0.017322 | 0.137635 | 0.014335 |
| 0.006 | 0.013719 | 0.025765 | 0.023379 | 0.023669 | 0.168438 | 0.013889 | 0.120045 | 0.011316 |
| 0.007 | 0.011995 | 0.021622 | 0.023187 | 0.02113 | 0.137778 | 0.010931 | 0.102046 | 0.008794 |
| 0.008 | 0.010231 | 0.017521 | 0.023111 | 0.018937 | 0.109131 | 0.008383 | 0.084025 | 0.006675 |
| 0.009 | 0.008412 | 0.013561 | 0.02312 | 0.017005 | 0.082445 | 0.006187 | 0.066241 | 0.004885 |
| 0.010 | 0.006515 | 0.009802 | 0.02315 | 0.015248 | 0.057656 | 0.004291 | 0.048861 | 0.003364 |
| 0.011 | 0.00451 | 0.006279 | 0.023006 | 0.013526 | 0.034767 | 0.002652 | 0.031991 | 0.002066 |
| 0.013 | 0.002357 | 0.00301 | 0.021793 | 0.011391 | 0.014116 | 0.001232 | 0.015693 | 0.000955 |
| 0.016 | 0 | 0 | 0 | 0 | 0 | 0 | 0 | 0 |

## TABLE 11.8
## 660 Base Pairs of DNA in *Homo Sapiens*
## Chaves et al. (in [7])

ctatatatcttaatggcacatgcagcgcaagtaggtctacaagacgctacttcccctatc

atagaagagcttatcacctttcatgatcacgccctcataatcattttccttatctgcttc

ctagtcctgtatgcccttttcctaacactcacaacaaaactaactaatactaacatctca

gacgctcaggaaatagaaaccgtctgaactatcctgcccgccatcatcctagtcctcatc

gccctcccatccctacgcatcctttacataacagacgaggtcaacgatccctcccttacc

atcaaatcaattggccaccaatggtactgaacctacgagtacaccgactacggcggacta

atcttcaactcctacatacttcccccattattcctagaaccaggcgacctgcgactcctt

gacgttgacaatcgagtagtactcccgattgaagcccccattcgtataataattacatca

caagacgtcttgcactcatgagctgtccccacattaggcttaaaaacagatgcaattccc

ggacgtctaaaccaaaccactttcaccgctacacgaccgggggtatactacggtcaatgc

tctgaaatctgtggagcaaaccacagtttcatgcccatcgtcctagaattaattcccta

## TABLE 11.9
## Transition Probabilities in the Third-Order Markov Model to Represent DNA Sequence from *Homo Sapiens*

| Triad No. | Triad | P(A/Triad) | P(G/Triad) | P(C/Triad) | P(T/Triad) |
|---|---|---|---|---|---|
| 1 | AAA | 3/657 | 0 | 6/657 | 3/657 |
| 2 | AAC | 5/657 | 1/657 | 6/657 | 4/657 |
| 3 | AAG | 3/657 | 0 | 1/657 | 1/657 |
| 4 | AAT | 3/657 | 3/657 | 4/657 | 5/657 |
| 5 | ACC | 5/657 | 4/657 | 0 | 3/657 |
| 6 | AGG | 1 | 0 | 2/657 | 2/657 |
| 7 | ACG | 4/657 | 2/657 | 4/657 | 3/657 |
| 8 | ATG | 2/657 | 2/657 | 5/657 | 0 |
| 9 | ATC | 7/657 | 2/657 | 5/657 | 5/657 |
| 10 | AGC | 1/657 | 1/657 | 1/657 | 2/657 |
| 11 | ATT | 4/657 | 2/657 | 4/657 | 1/657 |
| 12 | AGT | 4/657 | 0 | 2/657 | 1/657 |
| 13 | ACT | 7/657 | 1/657 | 5/657 | 3/657 |
| 14 | ATA | 4/657 | 2/657 | 3/657 | 2/657 |
| 15 | AGA | 4/657 | 1/657 | 3/657 | 1/657 |
| 16 | ACA | 4/657 | 3/657 | 3/657 | 6/657 |
| 17 | GGA | 1/657 | 1/657 | 2/657 | 0 |
| 18 | GGC | 1/657 | 2/657 | 1/657 | 1/657 |
| 19 | GGT | 2/657 | 0 | 3/657 | 0 |
| 20 | GCA | 3/657 | 1/657 | 2/657 | 1/657 |
| 21 | GCG | 2/657 | 1/657 | 1/657 | 0 |
| 22 | GCC | 2/657 | 0 | 6/657 | 0 |
| 23 | GCT | 2/657 | 1/657 | 1/657 | 3/657 |
| 24 | GTA | 0 | 2/657 | 3/657 | 3/657 |
| 25 | GTG | 0 | 1/657 | 0 | 0 |
| 26 | GTC | 2/657 | 0 | 4/657 | 4/657 |
| 27 | GTT | 1/657 | 1/657 | 0 | 1/657 |
| 28 | GAA | 3/657 | 2/657 | 3/657 | 1/657 |
| 29 | GAG | 0 | 1/657 | 3/657 | 2/657 |
| 30 | GAC | 1/657 | 6/657 | 2/657 | 3/657 |
| 31 | GAT | 0 | 1/657 | 2/657 | 1/657 |
| 32 | GGG | 0 | 2/657 | 1/657 | 1/657 |
| 33 | CCA | 2/657 | 1/657 | 4/657 | 5/657 |
| 34 | CCG | 2/657 | 1/657 | 2/657 | 1/657 |
| 35 | CCC | 5/657 | 2/657 | 7/657 | 8/657 |
| 36 | CCT | 9/657 | 3/657 | 4/657 | 6/657 |
| 37 | CAA | 4/657 | 3/657 | 3/657 | 5/657 |
| 38 | CAG | 2/657 | 2/657 | 1/657 | 1/657 |
| 39 | CAC | 4/657 | 2/657 | 4/657 | 3/657 |
| 40 | CAT | 3/657 | 4/657 | 8/657 | 4/657 |
| 41 | CGA | 0 | 3/657 | 4/657 | 2/657 |

**TABLE 11.9 (CONTINUED)**

**Transition Probabilities in the Third-Order Markov Model to Represent DNA Sequence from *Homo Sapiens***

| Triad No. | Triad | P(A/Triad) | P(G/Triad) | P(C/Triad) | P(T/Triad) |
|-----------|-------|------------|------------|------------|------------|
| 42 | CGG | 1/657 | 1/657 | 1/657 | 1/657 |
| 43 | CGC | 2/657 | 0 | 2/657 | 3/657 |
| 44 | CGT | 1/657 | 0 | 4/657 | 0 |
| 45 | CTA | 5/657 | 4/657 | 8/657 | 3/657 |
| 46 | CTG | 3/657 | 0 | 3/657 | 3/657 |
| 47 | CTC | 6/657 | 0 | 5/657 | 0 |
| 48 | CTT | 5/657 | 1/657 | 4/657 | 4/657 |
| 49 | TTA | 1/657 | 3/657 | 3/657 | 3/657 |
| 50 | TTG | 2/657 | 1/657 | 1/657 | 0 |
| 51 | TTC | 4/657 | 1/657 | 7/657 | 0 |
| 52 | TTT | 1/657 | 0 | 5/657 | 2/657 |
| 53 | TAA | 2/657 | 0 | 4/657 | 6/657 |
| 54 | TAG | 4/657 | 2/657 | 0 | 3/657 |
| 55 | TAC | 6/657 | 4/657 | 1/657 | 6/657 |
| 56 | TAT | 4/657 | 1/657 | 5/657 | 1/657 |
| 57 | TCA | 5/657 | 1/657 | 4/657 | 7/657 |
| 58 | TCG | 1/657 | 0 | 0 | 2/657 |
| 59 | TCC | 0 | 0 | 10 | 11/657 |
| 60 | TCT | 2/657 | 4/657 | 2/657 | 3/657 |
| 61 | TGA | 4/657 | 1/657 | 1/657 | 1/657 |
| 62 | TGG | 1/657 | 0 | 2/657 | 1/657 |
| 63 | TGC | 3/657 | 1/657 | 3/657 | 1/657 |
| 64 | TGT | 1/657 | 1/657 | 1/657 | 1/657 |

The negative logarithm of Equation (11.44) yields

$$-\log(P(M/D)) = \Sigma\, n_x \log P_x$$

$$x\, \varepsilon A \tag{11.45}$$

Functional regions can be identified from biological sequence data. This includes the problem of how to identify relatively long functional regions such as genes. A site is a short sequence that contains some signal that is often recognized by some enzyme. Examples of nucleotide sequence sites include the origins of replication, where DNA polymerase binds, transcription start and stop sites, ribosome binding in prokaryotes, promoters or transcription factor binding sites, and intron splicing sites.

## 11.3 SEQUENCE ALIGNMENT IN DNA AND PROTEIN SEQUENCES

*Sequence alignment* is a process of lining up two or more sequences to obtain matches among them. Sequence alignment can be used to develop cures for *autoimmune* disorders; *phylogenetic tree construction*; identification of *polypeptide microstruc-ture*, during *shotgun sequencing*, in *gene finding*, in *restriction site mapping*, in *open reading frame (ORF) analysis*; in *genetic engineering*; during *drug design*, and in *protein secondary structure determination*, in *protein folding, clone analysis, pro-tein classification*, etc. An alignment grading function is introduced to keep track of the degree of alignments and pick the optimal alignment.

Optimal global alignment of a pair of sequences can be achieved in $O(n^2)$ time using the Needleman and Wunsch dynamic programming algorithm [1]. A dynamic programming table is filled, and the optimal alignment falls out of the procedure. The different alignments can be identified using traceback procedures. Penalties and rewards are selected such that when the grade of alignment is greater than zero, the number of characters aligned is greater than the number of mismatched characters in the sequence; when the grade of alignment is zero, then the numbers of mismatched characters are equal to each other; and when the grade of alignment is less than zero, then the number of mismatched characters is greater than the number of matched characters. Semiglobal alignment is obtained by awarding no penalty to end gaps or allowing free end gaps. The development of a dynamic programming algorithm con-sists of characterization of the structure of an optimal solution, recursive definition of the grade of an optimal solution, computation of the grade of an optimal solution in a bottom-up fashion, and construction of an optimal solution from computed infor-mation. The space requirement of $O(n^2)$ can be reduced to $O(n)$ using Hirschberg's dynamic array method. Algorithms for finding longest common subsequence in less than quadratic time are presented. The Smith and Waterman algorithm can be used for obtaining optimal local alignment between sequences using dynamic program-ming in $O(n^2)$ and $O(n^2)$ space efficiency. For example, Figure 11.5 shows the local alignment between two sequences S = ACGTAATGT and T = GCGCGCG. The affine gap model can be used to define the penalty for gaps and gap lengths in order to obtain biologically meaningful alignments.

Greedy algorithms can be used for aligning sequences that differ only by a few errors. Myers and Miller [15] have developed a method that can guarantee optimality in $O(en)$ time, where $e$ is much less than $n$ and in $O(m + n)$ space required. These are

| $S_i$ $T_j$ | | A | C | G | T | A | A | T | G | T |
|---|---|---|---|---|---|---|---|---|---|---|
| | 0 | −1 | −2 | −3 | −4 | −5 | −6 | −7 | −8 | −9 |
| G | −1 | −1 | −2 | 0 | −1 | −2 | −3 | −4 | −5 | −6 |
| C | −2 | −2 | +1 | 0 | −1 | −2 | −3 | −4 | −5 | −6 |
| G | −3 | −3 | 0 | +3 | +2 | +1 | 0 | −1 | −2 | −3 |
| C | −4 | −4 | −1 | +2 | +2 | +1 | 0 | −1 | −2 | −3 |
| G | −5 | −5 | −2 | +1 | +1 | +1 | 0 | −1 | +1 | 0 |
| C | −6 | −6 | −3 | 0 | 0 | 0 | 0 | −1 | 0 | 0 |
| G | −7 | −7 | −4 | −1 | −1 | −1 | −1 | −1 | +1 | 0 |

**FIGURE 11.5** Local alignment of sequence S = ACGTAATGT and sequence T = GCGCGCG.

implanted in the Unigene database by NCBI. Other methods for obtaining sequence alignment include the method of significant diagonals, heuristic method, approximate alignments, hamming, etc. PAM (pointed accepted mutation) and BLOSUM (blocks of amino acid substituting matrix) matrices are provided, and the benefits of using them for alignments are outlined. The methods described were applied to sequences with varying microstructures such as alternating, random, and block distributions. The concept of super sequence was introduced.

The x-drop algorithm for global alignment is touched upon. The effect of repeats in a sequence on the dynamic programming procedures is explored in Sharma [7]. The antidiagonal is defined, and banded diagonal methods are explored. What will happen when the dynamic programming table is *sparse*? The implications are explored. The *stability* of global and local alignment is touched upon. Staircase table, inverse dynamic programming, consensus sequence, sequencing errors, and their ramifications are introduced.

### 11.3.1 Global Alignment of a Pair of Sequences

A computer-adaptable method using a dynamic programming algorithm was suggested for *optimal global alignment of two sequences* by Needleman and Wunsch [1]. In global alignment, the two sequences are aligned end to end. Needleman and Wunsch developed a computer-based statistical and general method applicable to the search for similarities in the amino acid sequence of two proteins. A number of researchers have studied the question of how to construct a good grading function for sequence comparison [3–6]. From these findings, it is possible to determine whether significant homology exists between the proteins. Another goal for seeking alignment is to establish a full genetic relationship between the proteins. This information is used to trace their possible evolutionary development. The maximum match can be defined as the largest number of amino acids of one protein that can be matched with those of another protein while allowing for all possible deletions.

Needleman and Wunsch [1] were the first to devise a computationally feasible method for automated sequence alignment. Their application is built on an analogy with the old visual comparison method of alignment. They used a PASCAL array that maps directly to the sequence alignment plot, and their algorithm then operated within this context. They describe the alignment process as "pathways through the array," which are evaluated to find the "maximum match." The algorithm works by progressively building a path through the array, gaining rewards for obtaining matches, and incurring penalties for gaps. This kind of approach constitutes dynamic programming, a common method for optimizing computer algorithms. They obtained alignment of whole myoglobin and human β-hemoglobin and alignment of bovine pancreatic ribonuclease and hen's egg lysozyme.

Given strings S and T with $|S| = n$ and $|T| = m$, an optimal global alignment of S and T can be obtained using dynamic programming. A grade of alignment $G(i,j)$ of string $S(i)$ and $T(j)$ is defined. The grade of optimal alignment of S and T is $G(m,n)$. The dynamic programming method is used to solve for the general problem of computing all grades $G(i,j)$ with $0 \leq i \leq n$ and $0 \leq j \leq m$ in order of increasing $i$ and $j$.

### 11.3.1.1   Algorithm 1 Global Alignment

Basis:

$$G(0,0) = 0 \tag{11.46}$$

$$G(i,0) = G(i-1,0) + \sigma\,(S(i),\,\_),\text{ for } i > 0. \tag{11.47}$$

$$G(0,j) = G(0, j-1) + \sigma\,(\_,\, T(j)),\text{ for } j > 0. \tag{11.48}$$

Recurrence formula:

$$G(i,j) = \max\,(G(i-1,j-1) + \sigma(S(i),T(j)),$$
$$G(i-1,j) + \sigma(S(i), -),\; G(i,j-1) + \sigma(-, T(j)\,). \tag{11.49}$$

The interpretation of the alignment is as follows. Consider the optimal alignment of the first $i$ characters from S and the first $j$ characters from T. In particular, consider the last aligned pair of characters in such an alignment. This last pair must be one of the following:

1. [S($i$), T($j$)], in which case the remaining alignment excluding this pair must be an optimal alignment of S(1), …,S($i$−1) and T($i$)…T($j$−1) (i.e., must have grade G($i$−1,$j$−1) or
2. [S($i$),−], in which case the remaining alignment excluding this pair must have grade G($i$−1,$j$) or
3. [−, T($j$)], in which case the remaining alignment excluding this pair must have grade G($i$, $j$−1). A traceback procedure is used to obtain all the alignments.

### EXAMPLE 11.8 [7]

*Global alignment of two sequences by dynamic programming:*

Demonstrate the dynamic programming method to obtain the global alignment of the two strings with the following sequence distribution:

```
S: a   c   g   t   t   t   g   c   a
T: c   c   a   t   g   c   g   a
```

The grading function used was +2 for a match, −1 for a mismatch, and −1 for an indel.

Recovering the alignments:

```
I. a   c   g   t   t   g   c   -   a
       c   c   a   t   -   g   c   g   a
```

Grade of Alignment: (5×2) − 4 = +6

```
II. a   c   g   t   t   g   c   -   a
        a   c   c   -   a   t   g   c   g   a
```

Grade of Alignment: (5×2) − 4 = +6                                          (11.50)

## EXAMPLE 11.9

When the grade of alignment is 0 and when it is less than 0.

In Example 11.4, what is the meaning when the grade of alignment is (1) 0 and (2) negative or less than zero.

From Figure 11.6, an alignment of two sequences where the grade of alignment is zero can be selected as follows.

A zero grade of alignment can be expected when these two sequences are aligned:

$$\text{S: a} \quad \text{c} \quad \text{g} \quad \text{t} \quad \text{t} \quad \text{g}$$
$$\text{T: c} \quad \text{c} \quad \text{a} \quad \text{t}$$

The tracebacks for alignment of S and T would be

$$\text{I. a} \quad \text{c} \quad \text{g} \quad \text{t} \quad \text{t} \quad \text{g}$$
$$\text{-} \quad \text{c} \quad \text{c} \quad \text{a} \quad \text{t} \quad \text{-}$$

Grade of Alignment = (2×2) – (2×1) – (2×1) = 0

$$\text{II. a} \quad \text{c} \quad \text{g} \quad \text{t} \quad \text{t} \quad \text{g}$$
$$\text{c} \quad \text{c} \quad \text{-} \quad \text{a} \quad \text{t} \quad \text{-}$$

Grade of Alignment = (2×2) – (2×1) – (2×1) = 0

$$\text{III. a} \quad \text{c} \quad \text{g} \quad \text{t} \quad \text{t} \quad \text{g}$$
$$\text{c} \quad \text{c} \quad \text{a} \quad \text{-} \quad \text{t} \quad \text{-}$$

Grade of Alignment = (2×2) – (2×1) – (2×1) = 0

Compared with the alignments in Example 11.8, where the number of characters that were aligned was 5 out of a length of 9 of the mapped string, in Example 11.9 the number of characters that were aligned was 2 out of a length of the mapped string of length 6. A majority of the mapped string in Example 11.9 was matched, and a minority of characters was aligned in Example 11.9. Maybe a positive grade of alignment signifies that more characters are aligned in the string, and a zero grade of alignment denotes that only few characters are aligned.

From Figure 11.7, two strings can be selected where the grade of alignment can be expected to be negative or less than zero. These strings are as follows:

| $T_{(j)}$ | $S_{(i)}$ | a | c | g | t | t | g |
|---|---|---|---|---|---|---|---|
| | 0 —— −1 | −2 | −3 | −4 | −5 | −6 |
| c | −1 | −1 | 1 | 0 | −1 | −2 | −3 |
| c | −2 | −2 | 1 — 0 | −1 | −2 | −3 |
| a | −3 | 0 | 0 | 0 — −1 | −2 | −3 |
| t | −4 | −1 | −1 | −1 | 2 | 1 — 0 |

**FIGURE 11.6** Global alignment of two sequences S = acgttgca and T = ccatgcga by dynamic programming.

|   |   | u | u | c | g | a | u | u | g | u |
|---|---|---|---|---|---|---|---|---|---|---|
|   | 0 | -1 | -2 | -3 | -4 | -5 | -6 | -7 | -8 | -9 |
| c | -1 | -1 | -2 | 0 | -1 | -2 | -3 | -4 | -5 | -6 |
| c | -2 | -2 | -2 | 0 | -1 | -2 | -3 | -4 | -5 | -6 |
| c | -3 | -3 | -3 | 0 | -1 | -2 | -3 | -4 | -5 | -6 |
| g | -4 | -4 | -4 | -1 | 2 | 1 | 0 | -1 | -2 | -3 |
| g | -5 | -5 | -5 | -2 | 1 | 1 | 0 | -1 | 1 | 0 |
| g | -6 | -6 | -6 | -3 | 0 | 0 | 0 | -1 | 1 | 0 |
| u | -7 | -4 | -4 | -4 | -1 | -1 | 2 | 2 | 1 | 3 |
| g | -8 | -5 | -5 | -5 | -2 | -2 | 1 | 1 | 4 | 3 |
| a | -9 | -6 | -6 | -6 | -3 | 0 | 0 | 0 | 3 | 3 |

**FIGURE 11.7**  Global alignment of two sequences, S = acgttg and T = ccat, with grade of alignment = 0.

$$S:\ a \quad c \quad g \quad t$$
$$T:\ c \quad c \quad a$$

The traceback procedure can be used, and the alignments recovered for a grade of alignment −1 would be

$$IV.\ a \quad c \quad g \quad t$$
$$-\quad c \quad c \quad a$$

Grade of Alignment = $(1×2) − (2×1) − (1×1) = −1$

$$V.\ a \quad c \quad g \quad t$$
$$c \quad c \quad - \quad a$$

Grade of Alignment = $(1×2) − (2×1) − (1×1) = −1$

$$VI.\ a \quad c \quad g \quad t$$
$$c \quad c \quad a \quad -$$

Grade of Alignment = $(1×2) − (2×1) − (1×1) = −1$

In the tracebacks of the alignments shown in IV, V, and VI recovered from the alignments shown in Table 11.10, only one character is aligned compared with a length of 4 of the mapped string. The number of misaligned characters is larger than the number of aligned characters.

Thus, the interpretation of the grade of alignment can be made as follows based on the above calculations:

1. When the grade of alignment is greater than zero, the number of characters aligned is greater than the number of mismatched characters in the sequence.
2. When the grade of alignment is zero, then the numbers of matched characters and the number of mismatched characters are equal to each other.

When the grade of alignment is less than zero, then the number of mismatched characters is greater than the number of matched characters.

**TABLE 11.10**

**Transition Probabilities in the First-Order Markov Model to Represent DNA Sequence from *Homo Sapiens***

| # | Given base pair | $P(A/\#)$ | $P(G/\#)$ | $P(C/\#)$ | $P(T/\#)$ |
|---|---|---|---|---|---|
| 1 | A | 3/59 | 4/59 | 4/59 | 6/59 |
| 2 | G | 1/59 | 2/59 | 5/59 | 2/59 |
| 3 | C | 5/59 | 2/59 | 3/59 | 6/59 |
| 4 | T | 8/59 | 2/59 | 4/59 | 2/59 |

## 11.3.2 DYNAMIC PROGRAMMING

Bellman [8] began the systematic study of dynamic programming in 1955. He used a tabular solution method that was called dynamic programming. Prior to Bellman's work, dynamic programming was used in optimization techniques such as in finding the optimal reactor heat exchanger network of the sulfur-trioxide-forming step in the oleum process to manufacture sulfuric acid. Bellman was the first to provide the area with a solid mathematical basis. The time taken for the longest common subsequence problem as suggested by Smith and Waterman [9] using dynamic programming methods is O($mn$). Knuth [10] posed the question of whether subquadratic algorithms for the longest common subsequence (LCS) problem exist. Masek and Paterson [11] answered this question in the affirmative by giving an algorithm that runs in O($mn$)/lg($n$), time where $n \leq m$ and the sequences are drawn from a bounded size. For the special case in which no element appears more than once in an input sequence, Szymanski [12] showed that the problem can be solved in O($n+m$)lg($n+m$) time. In 1970, Don Knuth conjectured that a linear time algorithm for the problem of finding the longest common substring would be impossible. It was shown [7] that the longest common substring of two strings can be found in linear time using a generalized suffix tree.

The solutions to subproblems are combined to solve a given problem in the dynamic programming method. This is similar to the divide-and-conquer principle used a lot in computer algorithms. In the divide-and-conquer strategy, the given problem is split into subproblems. The subproblems are solved recursively and the solutions combined to form the solution of the original problem. When the subproblems are not independent of each other, the dynamic programming method may be applicable. Every subproblem is solved only once, and the results saved in the tabular form in the dynamic programming method compared with the divide-and-conquer method, where more work is done than necessary. Optimization problems can use the dynamic programming method. Many solutions are possible for such problems. Each solution has a grade, and the extremum is of interest as the optimal solution. The development of a dynamic programming algorithm can be broken into four steps [13]:

1. Characterize the structure of an optimal solution.
2. Recursively define the grade of an optimal solution.

3. Compute the grade of an optimal solution in a bottom-up fashion.
4. Construct an optimal solution from computed information.

Two key ingredients to the application of dynamic programming problems are the identification of optimal substructure and overlapping subproblems.

### 11.3.3 Analysis of Time and Space Efficiency

The optimal global alignments can be obtained using dynamic programming in $O(mn)$ time. When $m = n$, the time taken becomes $O(n^2)$.

**Proof:** The $(m+1)(n+1)$ table needs to be filled. Each and every entry is computed with a maximum of 6 table look-ups, 3 additions, and a 3-way maximum in time $c$.

Complexity of the algorithm $= c(n+1)(m+1) = O(mn)$.

Reconstructing a single alignment $= O(n+m)$ time. The space required for retaining the grades of alignment in the table is $(mn)$ also.

### 11.3.4 $O(n)$ Space Solution by Dynamic Array

The space required in dynamic programming during global alignment of two sequences can be reduced from $O(n^2)$ as follows. It can be realized that only the global optimal grade of alignment is needed. So, two dynamic rows at any given time will be sufficient. In order to construct the next row of alignments, the previous row is sufficient. Thus, space required will be $O(2m)$ or $O(m)$. Reconstructing an alignment is somewhat more complicated, but can be achieved in $O(n+m)$ space and $O(nm)$ time with a divide-and-conquer approach [14,15].

### 11.3.5 Subquadratic Algorithms for Longest Common Subsequence

Hunt and Szymanski [16] introduced a fast algorithm for computing longest common subsequences. They provided a running time of $O(r+n)\lg(n)$, where $r$ is the total number of ordered pairs of positions at which the two sequences match. In the worst case, the algorithm has an $O(n^2\lg(n))$ running time. However, for those applications where most positions of one sequence have few matches in the other sequence, a running time of $O(n\lg(n))$ can be expected.

Let S be a finite sequence of elements chosen from some alphabet $\Sigma$. The length of the sequence S is $|S|$. $S[i]$ is the $i$th element of S and $S[i:j]$ is the sequence $S[i]$, $S[i+1]$, $S[i+2]...S[j]$. If U and V are finite sequences, then U is said to be a subsequence of V if there exists a monotonically increasing sequence of integers $r_1$, $r_2,...r_{|U|}$ such that $U[i] = V[r_i]$ for $1 \le i \le |U|$. U is a common subsequence of S and T if U is a subsequence of both S and T. A longest common subsequence is a common subsequence of greatest possible length. Both sequences are assumed to have the same length $n$. The number of elements in set $\{(i,j)$ such that $S[i] = T[j]\}$ is denoted by $r$.

The data structure used in the algorithm of Hunt and Szymanski is $G_{i,k}$. $G_{i,k}$ is an array of threshold grades defined by the smallest $j$ such that $S[1:i]$ and $T[1:j]$ contain a common subsequence of length $k$. Each $G_{i,k}$ may be considered as a pointer that signifies how much of the T sequence is needed to produce a common subsequence of length $k$ with the first $i$ elements of S. Each row of the G array can be seen to be increasing, that is,

Lemma 1:  $G_{i,1} < G_{i,2} < .... < G_{i,p}$ as defined above.

Lemma 2:  $G_{i,k-1} < G_{i+1,k} \le G_{i,k}$

Lemma 3:  $G_{i+1,}k = $ Smallest $j$ such that $S[i+1] = T[i]$
and $G_{j,k}$, if no such $j$ exists $G_{i,k-1} < j < G_{i,k}$

Lemmas 1, 2, and 3 are stated and the proofs are available in [17]. This algorithm can be completed with an $O(n^2 \lg(n))$ time efficiency to determine the length of the common subsequence. This can be refined to improve the running time to $O(r+n)$ $\lg(n)$, and the longest common subsequence can be recovered.

### 11.3.5.1  Algorithm 2: Length of Longest Increasing Subsequence

```
G[0] = 0
Recurrence Formula:
       for i = 1 to n
              G[i] = n+1
  For i = 1 to n
  For j = n to 1 step -1
  If S[i] = T[j] then
  Begin
Find k such that G[k-1] < j ≤ G[k]
G[k] = j
End
Print largest k such that G[k] ≠ n+1
```

A small amount of preprocessing will improve the performance of Algorithm 2 in great measure. The main source of inefficiency in Algorithm 2 is the inner loop $j$ in which the elements that match $S[i]$ are searched for repeatedly in the T sequence. A linked list can be used to eliminate this search step. For each $i$ a list of corresponding $j$ positions is needed such that $S[i] = T[j]$. These lists must be retained in decreasing order of $j$. All positions of the S sequence that contain the same element may be set up to use the same physical list of matching $j$'s.

### 11.3.5.2  Algorithm 3: Find and Print the Longest
### Common Subsequence of S and T

Initialize arrays S[1;n], T[1;n], G[0,n], MATCHLIST [1,n], LINK[1,n], PTR.
Build linked lists

```
        for i = 1 to n
               MATCHLIST[i] = <j₁, j₂,...., jₚ> such
that j₁ > j₂ >.... >jₚ and S[i] = T[jₑ] for 1 ≤ i ≤ i.
```

```
Initialize threshold array:    G[0] = 0
     For i = 1 to n
            G[i] = n+1
            LINK[0] = null
Compute successive threshold grades
     For i = i to n
            For j on MATCHLIST [i]
            Find k such that G[k-1] < j ≤ G[k];
                  If j < G[k] then
                        G[k] = j
                        LINK[K] = NEW NODE (I,J,
LINK[K-1])
                  End:    End
Recover longest common subsequence in reverse order
     k = largest k such that G[k] ≠ n+1
     PTR = LINK[k]
     While PTR ≠ null do
            Print (I,j) pair pointed to by PTR
     Advance PTR;        End
```

The longest common subsequence of the sequences S and T is found and printed by Algorithm 3. The time efficiency of the algorithm is $O((r+n)\lg(n))$, and space required is $O(r+n)$. The key operations in the implementation of Algorithm 4 are the operations of inserting, deleting, and testing membership of elements in a set where all elements are restricted to the first $n$ integers. Such operations can be performed in $O(\lg(\lg(n)))$ time efficiency as shown in [17,18]. Time taken for initializing using this data structure is $O(n\lg(\lg(n)))$.

### 11.3.6 GREEDY ALGORITHMS FOR PAIRWISE ALIGNMENT

When the two sequences that are aligned differ only in sequencing errors, a greedy algorithm [15] can be used that is much faster than traditional dynamic programming approaches and guarantees an optimal alignment. Chao et al. [20] presented greedy algorithms for solving a simple formulation of the alignment problem called the longest common subsequence problem. This problem is equivalent to finding the fewest one-character insertion and deletion operations that will convert one sequence into another. Let S and T be two sequences with sequence lengths $m$ and $n$, and let $e$ denote the minimum number of operations; $e$ is the edit distance between the two sequences when two DNA sequences are considered, for example, where the shorter sequence is very similar to some concatenated region of the longer sequence. A similar region of the longer sequence is determined, and then an optimal set of single-nucleotide changes such as insertions, deletions, or substitutions are computed that will convert the shorter sequence to that region. The grade of alignment scheme is developed to model sequencing errors rather than evolutionary processes.

Greedy alignment algorithms presented by Ukkonen [19] and Myers and Miller [15] are best used when $e$ is a lot smaller compared to $m$ and $n$. The time efficiency of the algorithms is a worst-case of $O(\min(m,n).e)$ and space $O(m+n)$. The space needed is an order of magnitude less compared with that required by the dynamic programming approaches presented by Smith and Waterman [9] and Needleman and Wunsch [1]. The expected-case time efficiency of greedy algorithms can be $O(ne^2)$.

Greedy algorithms for sequence alignment are implemented in the assembly of the Unigene database maintained by the National Center for Biotechnology Information (NCBI). The algorithm suggested by Chao et al. [20] consists of two phases. The interval in the longer sequence that should be aligned with the shorter sequence is located during Phase I. A divide-and-conquer approach is employed to obtain the alignment in Phase II. The end gaps are then added to the alignment.

### 11.3.6.1  Algorithm 4: Tool for Aligning Very Similar DNA Sequences [7]

$$\text{Input: S: } a_0\ a_1\ a_2 \ldots a_{m-1;} |S| = m$$

$$\text{T: } b_0\ b_1\ b_2 \ldots b_{m-1;} |T| = n$$

$$n \geq m$$

Edit graph for sequences S and T is a directed graph with a vertex at each integer grid point $(x,y)$, $0 \leq x \leq m$ and $0 \leq y \leq m$. Let $I(k,c)$ denote the $x$ grade of the farthest point in diagonal $k$ that can be reached from the source (i.e., grid point $(0,0)$) with cost $c$ and that is free to open an insertion gap. The grid point can be (1) reached by a path of cost $c$ that ends with an insertion or (2) reached by path of cost $c-1$, and the gap-open penalty of 1 can be "paid in advance." Let $D(k,c)$ denote the grade of the farthest point in diagonal $k$ that can be reached from the source with cost $c$ and that is free to open a deletion gap. Let $S(k,c)$ denote the $x$ grade of the farthest point in diagonal $k$ that can be reached from the source with cost $c$. With proper initializations, these vectors can be calculated by the following recurrence relation:

$$I\ (k,c) = \text{Max}\{I(k-1,c-1),\ S(k,c-1)\}$$

$$D\ (k,c) = \text{Max}\ \{D(k+1,c-1) + 1,\ S(k,c-1)\}$$

$$S\ (k,c) = \text{Snake}\{k, \max\{S(k,c-1) +1,\ I(k,c),\ D(k,c)\}\}$$

$$\text{Where Snake } (k,x) = \text{Max}\ \{x,\ \max(z: a_x \ldots a_{z-1} = b_{x+k} \ldots b_{z-1+k})\}$$

As the vectors at cost $c$ depend only on those at costs $c$ and $c-1$, a linear-space version of the above relationship can be derived.

## Exact Phase I

Phase I can be accomplished by applying the recurrences for I, D, and S, where all costs in row 0 are initialized to 0. Once row $m$ is reached, the desired interval has been located. Although the worst-case running time for this approach is $O(mn)$, the average running time is $O(n \text{ Dist})$, where Dist is the distance of S and T. The average length of a snaked fragment is a small constant.

## Phase II

Backward vectors $I^*(k,c)$ denotes the $x$ grade of the farthest I node in diagonal $k$ that can reach the sink (i.e., grid point$(m,n)$) with cost $c$. $D^*(k,c)$ is the $x$ grade of the farthest D node in diagonal $k$ that can reach the sink with cost $c$. Let $S^*(k,c)$ denote the $x$ grade of the farthest S node in diagonal $k$ that can reach the sink with cost $c$. After initializations, these vectors can be computed by the following recurrence relation:

$$S^*(k,c) = \text{Snake}^*(k, \text{Min} \{S^*(k,c-1) - 1, D^*(k,c-1), I^*(k,c-1)\}$$

$$D^*(k,c) = \text{Min}\{D^*(k-1,c-1) - 1, S^*(k,c)\}$$

$$I^*(k,c) = \text{Min} \{I^*(k+1,c-1), S^*(k,c)\}$$

where $\text{Snake}^*(k,x) = \text{Min} \{x, \text{Min} \{z; a_z...a_{x-1} = b_{z+k}...b_{x-1+k}\}$

A linear space version of the recurrence relation can be derived. The pseudocode for the linear space algorithm for alignment is as follows:

```
Procedure path(I₁, J₁, Type₁, I₂, J₂, Type₂, Dist)
{       if boundary cases then
               {Output the edit script; return
               Else
               {
               Mid ¬ Dist/2
               Mid ¬ Dist -mid
       A Linear Space Forward Pass Computes S(k,mid),
D(k,mid), and I(k,mid) for J₁ - I₁ - mid ≤ k ≤ J₁ - I₁
+ mid.
       A Linear Space Backward Pass Computes S*(k,mid),
D*(k,mid), and I*(k,mid) for J₂ - I₂ - mid* ≤ k ≤ J₂ - I₂
+ mid*.
       Let K be the diagonal such that that X(K,mid) ≥
X*(K,mid*) where X is S, D or I.
               Path(I₁, J₁, Type₁, X(K,mid), X(K,mid) + K,X,mid)
               Path(X(K,mid),X(K,mid) + K, X, I₂, J₂, Type₂,
mid*)}}
```

## 11.3.7 OTHER METHODS FOR PAIRWISE ALIGNMENT

Pearson and Lipman [21] developed an algorithm where the sequences are searched for diagonals of length $k$ ($k$-tuples) ahead of time. The $k$-tuples are then evaluated, and groups of continuous tuples are labeled *significant diagonals*. A *window space* can be identified in the grid as the region around the most significant diagonals that represent partial matches. In this method, diagonals and regions are used instead of the traceback used in the Needleman and Wunsch's and Smith and Waterman's algorithms. The window size is controllable and hence can speed up the computations. The time taken did not better the $O(n^2)$ needed for the dynamic programming methods. Sharma [7] suggested a heuristic algorithm for approximate global alignment of a pair of sequences with less time efficiency compared with the $O(n^2)$ time needed for dynamic programming. The two sequences are parsed by generating a random index $i$. Depending on the matches and mismatches the indel and gap are introduced. The grade of alignment is calculated, and the next random index is called for. This procedure is repeated until the maximum grade of alignment is reached. The maximum grade of the alignment is reached in $O(en)$ time efficiency, where $e$ is the number of indels/and gaps called for. For some sequences, this may be $O(n)$, but the alignment is approximate. The worst-case time efficiency will revert to $O(n^2)$.

## 11.3.8 GRADING FUNCTIONS DURING GLOBAL ALIGNMENT

The sensitivity of the alignment to the selection of grading function is evaluated by changing the grading function values. Consider two sequences S = uucgauugu and T = cccggguga. A grading function of +1 for match, −1 for mismatch, and indel yields a grade of alignment of −1 (Figure 11.8).

A grading function of +1.5 for match, −1 for mismatch, and indel yields a grade of alignment of +1. A grading function of +2 for match, −1 for mismatch, and indel yields a grade of alignment of +3. (Figure 11.9). The tables were generated using an MS Excel 2007 spreadsheet. The effect of gap penalty can be seen in Figure 11.10.

|   |    | u  | u  | c  | g  | a  | u  | u  | g  | u  |
|---|----|----|----|----|----|----|----|----|----|----|
|   | 0  | -1 | -2 | -3 | -4 | -5 | -6 | -7 | -8 | -9 |
| c | -1 | -1 | -2 | -1 | -2 | -3 | -4 | -5 | -6 | -7 |
| c | -2 | -2 | -2 | -1 | -2 | -3 | -4 | -5 | -6 | -7 |
| c | -3 | -3 | -3 | -1 | -2 | -3 | -4 | -5 | -6 | -7 |
| g | -4 | -4 | -4 | -2 | 0  | -1 | -2 | -3 | -4 | -5 |
| g | -5 | -5 | -5 | -3 | -1 | -1 | -2 | -3 | -2 | -3 |
| g | -6 | -6 | -6 | -4 | -2 | -2 | -2 | -3 | -2 | -3 |
| u | -7 | -5 | -5 | -5 | -3 | -3 | -1 | -1 | -2 | -1 |
| g | -8 | -6 | -6 | -6 | -4 | -4 | -2 | -2 | 0  | -1 |
| a | -9 | -7 | -7 | -7 | -5 | -3 | -3 | -3 | -1 | -1 |

**FIGURE 11.8** Global alignment of sequences S = uucgauugu and T = cccggguga with a + 1 for a match.

|   |   | a | t | t | a | g | a | c | ? | t | a | a | g |
|---|---|---|---|---|---|---|---|---|---|---|---|---|---|
|   | 0 ←—2← 4 ←—6 | 8 | 10 | 12 | 14 | 16 | 18 | 20 | 22 | 24 |   |   |   |
| a | −2 | 2 | 2 | −2 | −1 | −6 | −8 | −10 | −12 | −14 | −16 | −18 | −20 |
| g | 4 | 0 | 1 | 1 | 3 | 2 ←—4 | 6 | 8 | 10 | 12 | 14 | 16 |   |
| c | −6 | −2 | −1 | 0 | −2 | −4 | −3 | −2 ←−1 | −6 | −8 | −10 | −12 |   |
| i | 8 | 4 |   | 1 | 1 | 3 | ?? | 4 | 0 | 2 | 2 | 6 | 8 |
| a | −10 | −6 | −2 | −1 | 3 | 1 | −1 | −3 | −2 | −1 | 0 | −2 | −4 |
| g | 12 | 8 | 4 | 3 | 1 | ?? | ?? | ?? | 1 | 3 | 2 | ?? | 0 |
| g | −14 | −10 | −3 | −5 | −1 | 3 | 4 | 2 | 0 | −2 | −2 | −3 | 1 |

**FIGURE 11.9** Global alignment of sequences S = uucgauugu and T = cccggguga with a + 2 for a match.

|   |   | é | e | h | g | w | a | g | a | e | h |
|---|---|---|---|---|---|---|---|---|---|---|---|
|   | 0 | −16 | −32 | −48 | −64 | −80 | −96 | −112 | −128 | −144 | −160 |
| e | −16 | 18 | 2 | −6 | −14 | −22 | −30 | −38 | −46 | −54 | −62 |
| a | −40 | 2 | 18 | 10 | 6 | −2 | −9 | −17 | −25 | −33 | −41 |
| e | −216 | −6 | 12 | 10 | 13 | 7 | 1 | −6 | −14 | −15 | −23 |
| h | −248 | −14 | 4 | 27 ←—19 | 15 | 9 | 3 | −4 | −7 | 0 |   |
| w | −288 | −22 | −4 | 19 | 27 | 74 | 66 | 58 | 50 | 42 | 34 |
| a | −336 | −30 | −12 | 11 | 31 | 66 | 87 ←—79 | 71 | 58 | 48 |   |
| p | −832 | −38 | −20 | 3 | 23 | 58 | 79 | 92 | 88 ←—78 ←—70 |   |   |

**FIGURE 11.10** Global alignment of S = attagacttaag and T = agctagg with gap penalty.

The global alignment between sequences S = attagacttaag and T = agctagg was obtained with a grading scheme of +2 for an indentity (match), −1 for mismatch and −2 for the gap penalty.

The amino acid sequences S = eehgwagaeh and T = eaehwap were aligned globally using the PAM-250 grading matrix with a gap penalty of −8 (Figure 11.10). From FASTA format: e—glutamic acid; h—histidine; g—glycine; w—tryptophan; a—alanine; p—proline.

Global alignment was obtained between the sequences S = cgccautacgcgaatttta and T = catataaacgct using the following parameters: identity = +4; translation = −2; gap = −8 fixed and transfers = −4 (Figure 11.11).

The *Escherichia coli* promoter sequences were aligned using the grading function of +2 for a match, −1 for a mismatch and −3 for a gap. −10 signal in *E. coli* sequences TATAAT was aligned with the sequence GTTACGTAA (Figures 11.12 and 11.13). The complementary sequence of S was found to match better.

It was found that when the sequences aligned S, T had a *alternating sequence* distribution; the traceback path was along the main diagonal from the bottom-right cell to the top-left cell of the dynamic programming array (Figure 11.14).

The sequences considered were S = ugugugugugugug and T = gugugugugugugu. The grading function for a match was +2 and that for a mismatch was −1.

|   |   | c | g | c | c | g | c | t | a | c | g | c | g | a | a | t | t | t | t | a |
|---|---|---|---|---|---|---|---|---|---|---|---|---|---|---|---|---|---|---|---|---|
|   | 0 | 10 | 20 | 30 | 40 | 50 | 60 | 70 | 80 | 90 | ## | ## | ## | ## | ## | ## | ## | ## | ## | ## |
| c | -10 | 4 | -4 | ?? | -20 | -28 | -36 | -44 | -52 | -60 | -68 | -76 | -84 | -92 | ## | ## | ## | ## | ## | ## |
| a | -28 | -4 | 6 | ?? | -22 | -16 | -24 | -32 | -40 | -48 | -56 | -61 | -72 | -80 | -88 | -96 | ## | ## | ## | ## |
| t | 54 | 12 | 4 | ?? | 24 | 24 | 26 | 20 | 28 | 36 | 44 | 52 | 60 | 68 | 76 | 84 | 92 | ## | ## | ## |
| a | -88 | -20 | -22 | -24 | -26 | ?? | ?? | -28 | -16 | -24 | -32 | -40 | -43 | -53 | -64 | -72 | -80 | -88 | -96 | ## |
| t | ## | -28 | -30 | -32 | -34 | -28 | -30 | ?? | -24 | -26 | -34 | -42 | -50 | -53 | -66 | -60 | -68 | -76 | -84 | -92 |
| a | ## | 36 | 38 | 40 | 42 | 30 | 38 | 32 | 20 | 28 | 36 | 44 | 52 | 46 | 54 | 62 | 70 | 78 | 86 | 80 |
| a | ## | -44 | -46 | -48 | -50 | -38 | -40 | ?? | ?? | -30 | -38 | -46 | -54 | -48 | -42 | -50 | -58 | -66 | -74 | -82 |
| a | ## | -52 | -54 | -56 | -58 | -16 | -18 | -18 | -36 | -38 | -40 | -48 | -56 | -50 | -44 | -52 | -60 | -68 | -76 | -70 |
| c | ## | 60 | 62 | 50 | 52 | 54 | 56 | 56 | 44 | ?? | 40 | 36 | 44 | 52 | 52 | 54 | 62 | 70 | 78 | 78 |
| g | ## | -68 | -56 | -58 | -60 | -67 | -64 | ?? | -52 | -40 | -28 | -36 | -32 | -40 | -48 | -56 | -64 | -72 | -80 | -86 |
| c | ## | -76 | -64 | -52 | -54 | -62 | -70 | -72 | -60 | -48 | -36 | ?? | ?? | ?? | ?? | ?? | ?? | ?? | -80 | -88 |
| t | ## | 84 | 72 | 60 | 62 | 64 | 72 | 56 | 68 | 56 | 44 | 32 | 34 | 42 | 50 | 44 | 52 | 60 | ?? | ?? |

**FIGURE 11.11**    Global alignment using PAM-250 grading matrix.

|   |   | T | A | T | A | A | T |
|---|---|---|---|---|---|---|---|
|   | 0 | -3 | -6 | -9 | -12 | -15 | -18 |
| G | -3 | -1 | -4 | -7 | -10 | -13 | -16 |
| T | -6 | -1 | -2 | -2 | -5 | -8 | -11 |
| T | -9 | -4 | -2 | 0 | -3 | -6 | -6 |
| A | -12 | -7 | -2 | -3 | 2 | -1 | -4 |
| C | -15 | -10 | -5 | -3 | -1 | 1 | -2 |
| G | -18 | -13 | -8 | -6 | -4 | -2 | 0 |
| T | -21 | -16 | -11 | -6 | -7 | -5 | 0 |
| A | -24 | -19 | -14 | -9 | -4 | -5 | -3 |
| A | -27 | -22 | -17 | -12 | -7 | -2 | -5 |

**FIGURE 11.12**    Global alignment with affine gap penalty with translations, gaps, and transfers.

|   |   | A | T | A | T | T | A |
|---|---|---|---|---|---|---|---|
|   | 0 | -3 | -6 | -9 | -12 | -15 | -18 |
| G | -3 | ?? | -4 | -7 | -10 | -13 | -16 |
| T | -6 | ?? | 1 | -2 | -5 | -8 | -11 |
| T | -9 | -7 | -2 | 0 | 0 | -3 | -6 |
| A | -12 | -7 | -5 | 0 | -1 | -1 | -1 |
| C | -15 | -10 | -8 | ?? | -1 | -2 | -2 |
| G | -18 | -13 | -11 | ?? | -4 | -2 | -3 |
| T | -21 | -16 | -11 | -9 | -4 | -2 | -3 |
| A | -24 | -19 | -14 | -9 | -7 | -5 | 0 |
| A | -27 | -22 | -17 | -12 | -10 | -8 | -3 |

**FIGURE 11.13**    Dynamic programming table.

| | | u | g | u | g | u | g | u | g | u | g | u | g | u | g | u | g |
|---|---|---|---|---|---|---|---|---|---|---|---|---|---|---|---|---|---|
| | 0 | -1 | -2 | -3 | -4 | -5 | -6 | -7 | -8 | -9 | -10 | -11 | -12 | -13 | -14 | -15 | -16 |
| g | -1 | -1 | 1 | 0 | -1 | -2 | -3 | -4 | -5 | -6 | -7 | -8 | -9 | -10 | -11 | -12 | -13 |
| u | -2 | 1 | 0 | 3 | 2 | 1 | 0 | -1 | -2 | -3 | -4 | -5 | -6 | -7 | -8 | -9 | -10 |
| g | -3 | 0 | 3 | 2 | 5 | 4 | 3 | 2 | 1 | 0 | -1 | -2 | -3 | -4 | -5 | -6 | -7 |
| u | -4 | -1 | 2 | 5 | 4 | 7 | 6 | 5 | 4 | 3 | 2 | 1 | 0 | -1 | -2 | -3 | -4 |
| g | -5 | -2 | 1 | 4 | 7 | 6 | 9 | 8 | 7 | 6 | 5 | 4 | 3 | 2 | 1 | 0 | -1 |
| u | -6 | -3 | 0 | 3 | 6 | 9 | 8 | 11 | 10 | 9 | 8 | 7 | 6 | 5 | 4 | 3 | 2 |
| g | -7 | -4 | -1 | 2 | 5 | 8 | 11 | 10 | 13 | 12 | 11 | 10 | 9 | 8 | 7 | 6 | 5 |
| u | -8 | -5 | -5 | 1 | 4 | 7 | 10 | 13 | 12 | 15 | 14 | 13 | 12 | 11 | 10 | 9 | 8 |
| g | -9 | -6 | -3 | 0 | 3 | 6 | 9 | 12 | 15 | 14 | 17 | 16 | 15 | 14 | 13 | 12 | 11 |
| u | -10 | -7 | -4 | -1 | 2 | 5 | 8 | 11 | 14 | 17 | 16 | 19 | 18 | 17 | 16 | 15 | 14 |
| g | -11 | -8 | -5 | -2 | 1 | 4 | 7 | 10 | 13 | 16 | 17 | 18 | 21 | 20 | 19 | 18 | 17 |
| u | -12 | -9 | -6 | -3 | 0 | 3 | 6 | 9 | 12 | 15 | 18 | 21 | 20 | 23 | 22 | 21 | 20 |
| g | -13 | -10 | -7 | -4 | -1 | 2 | 5 | 8 | 11 | 14 | 17 | 20 | 23 | 22 | 25 | 24 | 23 |
| u | -14 | -11 | -8 | -5 | -2 | 1 | 4 | 7 | 10 | 13 | 16 | 19 | 22 | 25 | 24 | 27 | 26 |
| g | -15 | -12 | -9 | -6 | -3 | 0 | 3 | 6 | 9 | 12 | 15 | 18 | 21 | 24 | 27 | 26 | 29 |
| u | -16 | -13 | -10 | -7 | -4 | -1 | 2 | 5 | 8 | 11 | 14 | 17 | 20 | 23 | 26 | 29 | 28 |

**FIGURE 11.14**  Dynamic programming table.

| | | a | b | c | n | j | r | q | c | l | c | r | p | m |
|---|---|---|---|---|---|---|---|---|---|---|---|---|---|---|
| | 0 | -1 | -2 | -3 | -4 | -5 | -6 | -7 | -8 | -9 | -10 | -11 | -12 | -13 |
| a | -1 | 2 | -2 | 1 | -3 | 0 | -4 | -1 | -5 | -2 | -6 | -3 | -7 | -4 |
| j | -2 | 1 | 1 | 0 | 0 | -1 | -1 | -2 | -2 | -3 | -3 | -4 | -4 | -5 |
| c | -3 | 0 | 0 | 3 | 2 | 1 | 0 | -1 | 0 | -1 | -1 | -2 | -3 | -4 |
| j | -4 | -1 | -1 | 2 | 2 | 4 | 3 | 2 | 1 | 0 | -1 | -2 | -3 | -4 |
| n | -5 | -2 | -2 | 1 | 4 | 3 | 3 | 2 | 1 | 0 | -1 | -2 | -3 | -4 |
| r | -6 | -3 | -3 | 0 | 3 | 3 | 5 | 4 | 3 | 2 | 1 | 1 | 0 | -1 |
| c | -7 | -4 | -4 | -1 | 2 | 2 | 4 | 4 | 6 | 5 | 4 | 3 | 2 | 1 |
| k | -8 | -5 | -5 | -2 | 1 | 1 | 3 | 3 | 5 | 5 | 4 | 3 | 2 | 1 |
| c | -9 | -6 | -6 | -3 | 0 | 0 | 2 | 2 | 5 | 4 | 7 | 6 | 5 | 4 |
| r | -10 | -7 | -7 | -4 | -1 | -1 | 2 | 1 | 4 | 4 | 6 | 9 | 8 | 4 |
| b | -11 | -8 | -5 | -5 | -2 | -2 | 1 | 1 | 3 | 3 | 5 | 8 | 8 | 7 |
| p | -12 | -9 | -6 | -6 | -3 | -3 | 0 | 0 | 2 | 2 | 4 | 7 | 10 | 9 |

**FIGURE 11.15**  Dynamic programming table for sequences with alternating sequence distribution.

Needleman and Wunsch's [1] seminal article contained the dynamic programming table shown in Figure 11.15.

## 11.4  SUMMARY

The dyad probabilities for copolymers were calculated as a function of reactivity ratios and monomer composition. A first-order Markov model was developed to predict the chain sequence distribution of SAN and AMS–AN copolymers. The 6 triad concentrations for SAN copolymer were calculated. Also, 9 dyads and 27 triads for random terpolymers were calculated and tabulated.

A Markov model of the third order was developed to repeat DNA sequences in *homo sapiens* with 660 bases, and 256 possible triad probabilities were calculated. A dynamic programming method can be used to align two sequences. Global and local alignment can be sought. PAM and BLOSUM matrices with affine gap penalty sequences were discussed. Time taken using dynamic programming methods is $O(n^2)$. Greedy algorithms can be used to obtain alignment in $O(en)$ for certain sequences with fewer errors. that is, $e \ll n$. X-drop alignment for global alignment was touched upon.

## PROBLEMS

1. *Optimal Global Pairwise Alignment*
   Find the optimal global alignment between
   S: aaacccaaggtacg
   and
   T: cacacacacacaca

2. Other than the optimal grade of alignment, how many alignments come to within 1 of the grade of optimal alignment in Problem 1?

3. Can the optimal global alignment in Problem 1 be improved upon by using an affine gap penalty model?

4. *Grading Functions during Optimal Global Alignment*
   Choose the appropriate grading functions to align globally the following strings:
   S: ucucugaugu
   T: ccggccggguga

5. *Reverse of Sequence*
   Consider the string S with the following sequence distribution:
   S: GCUGAAGUAAUAUU
   Construct a string T with sequence architecture that is the reverse of the sequence distribution of S. Using a global optimal alignment and employing dynamic programming, find the optimal grade of the alignment.

6. *Cell Grades during Local Alignment versus Cell Grades during Global Alignment*
   An additional term 0 is used in the maximum term when calculating the grade of alignment when seeking an optimal local alignment compared with the maximum term when calculating the grade of alignment used when seeking a global alignment of two sequences. Why?

7. *Interpretation of Grade of Alignment*
   Can the optimal grade of alignment take on negative numbers?

8. *Gap Penalty*
   Obtain the global optimal alignment between the pair of sequences S and T.
   S: ATTAGACTTAAG
   T: AGCTA
   The suggested grading scheme is 2 for an identity (match), −1 for mismatch, and −2 for the gap penalty; show the initialization, matrix fill, and traceback steps. Recover all the possible alignments.

9. *Longest Common Subsequence*
   Determine the longest common subsequence (LCS) of strings S and T
       with the following sequence distributions:
   S: dcbddb
   T: dcdcdba

10. *Optimal Local Alignment*
    Find the best local alignment between the pair of sequences S and T.
    S: GAGATAGAGACCGD
    T: AGTATGCGAAD.

11. *Optimal Local Alignment of Protein Sequences*
    Show the local alignment between protein sequences between the pair of
        sequences S and T.
    S: KCTNVTQDIGRAD
    T: QHATNDVACD.

12. *Pair of Sequences with no Repetitions of Characters*
    What can be achieved during global alignment of two sequences when
        the sequences have repetitions of characters in them?

13. *BLOSUM and PAM Matrices*
    What are the strengths and weaknesses of BLOSUM and PAM matrices?

14. *Affine Gap Penalty Affixation*
    What are the drawbacks of affine gap penalty affixation?

15. The best substitution matrix for Smith–Waterman comparisons of distant
    homologs is often BLOSUM45. The best matrices for BLAST are differ-
    ent. Why?

16. When will the optimal alignment not be sought by the FASTA and
    BLAST software?

17. *PAM250 Grading Matrix*
    Align the sequences
    S: EEHGWAAEH
    T: EAEHWAP
    using the PAM250 grading matrix and gap penalty of −8. Seek the
        following:
        (a) global alignment, (b) local alignment, (c) global alignment with
        the end gap penalties. For all alignments provide the complete
        dynamic programming matrix. Use SSEARCH, which can be run on
        the Internet from http://workbench.sdsc.edu to align these sequences,
        and compare its alignment with your result.

18. *Affine Gap Penalty with Translation, Gaps, and Transfers*
    Consider strings S and T with the following sequence distribution:
    S: CGCCAUTACGCGAATTTTA
    T: CATATAAACGCT
    Seek a global alignment using the following parameters: Identity = +4;
        Translation = −2; Gap = −8 fixed; and Transfers = −4. Align the two
        sequences, and report their grade of alignment. Revise the algorithm
        to produce a local alignment.

19. Give examples from computational molecular biology when each of the
    following alignment strategies would be appropriate:
    a.  Global alignment with no end gap penalties
    b.  Global alignment
    c.  Local alignment
    d.  Spaces penalty

20. *Award for Matches, Penalty for Mismatch and Gaps*

Use a match grade of alignment of +5, a mismatch penalty of −4, and a gap penalty of 3, and develop a dynamic programming algorithm for aligning the following DNA sequences:

S: acuggagcaucaucgaugcac

T: gagacaucgucgau

21. Escherichia coli *Promoter Sequences*

Align the −10 signal in *E. coli* promoter sequences TATAT with the sequences GTTAGTAA. Use the grading function 2 for a match, −1 for a mismatch, and −3 for a gap. Does the complementary sequence of S match better? What is the time efficiency?

22. *Both Strings with Alternating Sequence Distribution*

Consider two strings S and T with the following alternating sequence distribution:

S: ugugugugugugugug

T: gugugugugugugugu

Using the Needleman and Wunsch global alignment method with a grading function of +2 for matches and −1 for mismatch, indel, and gap, show that the optimal alignment of the two sequences would be in such a fashion that the traceback path would be along the diagonal from the bottom-right cell to the top-left cell of the dynamic programming array.

23. *One String with Block Sequence Distribution Architecture*

Consider the two strings S and T, with one of them, S, having a chain sequence distribution with a block architecture.

S: uuuuccc

T: ucacuuccc

Obtain the optimal local alignment between the two strings S and T. Show that the optimal grade of alignment for the strings is +10. Find the second-best alignment.

24. *Oligonucleotides*

PBMCs are peripheral blood mononuclear cells. ssDNA (single-stranded DNA), can be synthesized with ligands to human PBMC. PBMCs are isolated from whole blood and contain a complex mixture of cell types of B-lymphocytes, T-lymphocytes, and monocytes. Ligands to PBMCs have many uses, including imaging lymph nodes for cancer screening and flow cytometry used for AIDS monitoring. A library of synthetic DNA oligonucleotides containing 40 random nucleotides was created. The sequences of two clones are as follows:

S: aguuggau

T: gugaaaau

Using a grading scheme of +1 for a match, −1/3 for a mismatch, −1 for a gap opening, and −1/3 for a gap extension, obtain a global sequence alignment of strings S and T.

25. *Nonaligned Sequences*

One way to speed up the $O(n^2)$ time efficiency needed for optimal global alignment between two strings S and T is to identify regions of sequences that are misaligned. Suppose there are two sequences S and T with no characters in common or only one or two characters in common. Would it be better to define a distance metric and maximize the distance between the mapped strings? This way the optimal nonalignment can be obtained. What is the biological significance of nonalignment? When would this be preferred to the alignment schemes discussed in the chapter?

26. *DNA Sequence of Simian Varicella Virus*

In nonhuman primates, simian varicella virus (SVV) causes a natural disease that is clinically similar to human varicella-zoster virus (VZV) infections. The SVV and VZV genomes are similar in size and structure and share extensive DNA homology. SVV DNA is 124,138 bp in size, 746 bp shorter than VZV DNA, and 40.4% G + C. The viral genome includes a 104,104-bp unique long component bracketed by 8-bp inverted repeat sequences and a short component composed of a 4904-bp unique short region bracketed by 7557-bp inverted repeat sequences. A total of 69 distinct SVV open reading frames (ORFs) were identified, including three that are duplicated within the inverted repeats of the short component. Each of the SVV ORFs shares extensive homology to a corresponding VZV gene. The only major difference between SVV and VZV DNA occurs at the leftward terminus. SVV lacks a VZV ORF 2 homolog. In addition, SVV encodes an 882-bp ORF A that is absent in VZV, but has homology to the SVV and VZV ORF 4. The results of this study confirm that SVV and VZV are related. This provides further support for simian varicella as a model to investigate VZV pathogenesis and latency. What grading scheme would you suggest for seeking an alignment between SVV and VZV viral genomes?

27. *Greedy Algorithm to Align DNA Sequences*

Obtain the optimal global alignment of two sequences S: ccatacgtggttg-gtt and T: acgg using the greedy method. How is this an improvement over the dynamic programming method of Needleman and Wunsch?

28. *Supersequence for Global Alignment*

Consider the following two sequences S and T:

S: tgttgtccc

T: cttgcctcc

Define a supersequence U, where S and T are subsequences of U. How can U be used in obtaining the optimal global alignment of sequences S and T?

29. *Supersequence for Local Alignment*

How can the supersequence construction shown in Problem 28 be used to obtain the optimal local alignment between sequences S and T?

30. *X-Drop Algorithm for Global Alignment*

The procedure to obtain global alignment between two sequences S and T by the method of dynamic programming calls for creation of a table of size of $m \times n$ at a space efficiency of $O(n^2)$. In order to save space and time, a method is developed in which most of the cells in the array are not filled. As only the traceback from the bottom-right cell to the top-left cell is important, an X-drop procedure can save time and space. The grade of alignment is calculated across the diagonal of the table. When there is a match, the next cell diagonal down can be called for. When there is a mismatch, $x$ cells vertically can be traversed from the diagonal until a match is found. All along, indels/gaps can be called for. Upon finding a match, the procedure can continue either from the diagonal cell branched off from or from the matched cell to the diagonal cell down. This way, the only space required would be $O(k \text{ Max}(m,n))$, where $k$ is the departure from the diagonal in steps and the time taken would be $O(k, \text{Max}(m,n)$. For nearly aligned sequences, $k$ can be small, and the best-case space and time can be linear in time, $O(n)$. Show an example of nearly aligned sequences S and T and the advantages of the X-drop method compared with the recurrence relation discussed.

31. *X-Drop Algorithm for Local Alignment*

How suitable is the X-drop method outlined in Problem 30 to obtain local alignment between sequences S and T? What happened to the guaranteed optimality? Should the entire table be filled to confirm local maxima?

32. *Repeats in a Sequence*

Consider two sequences S and T as follows?

S: acgtacgtacgt

T: ccgatca

It can be seen that acgt repeats three times in the sequence S. When asked to obtain the global alignment between two sequences S and T by the method of dynamic programming as shown, how can you use your knowledge of repeats to cut down the time taken and space efficiency from $O(n^2)$. Which cells can be filled and what cell calculations can be cut down in order to increase the time and space efficiency?

33. *Hirschberg Array for Local Alignment*

Can the dynamic array method suggested by Hirschberg and discussed in the foregoing sections be used to obtain the optimal local alignment using the dynamic programming method of Smith and Waterman. Why not?

34. *Antidiagonal*

Antidiagonal $k$ is the set of all points $(i,j)$ such that $i + j = k$. Thus, antidiagonal $k = 2$ would mean the cells $(2,0)$, $(0,2)$, and $(1,1)$. Antidiagonal wise computation and half-nodes can speed up the sequence alignment process. Given an example of alignment of two sequences, where the antidiagonal computation speeds up the time taken, what is the payoff?

35. *Edit Distance*

The edit distance between two sequences S and T is the minimum cost C of a sequence of editing steps such as insertions, deletions, and changes that convert one sequence into another. A tabulating method was developed to compute C, as well as the corresponding editing sequence in a time efficiency of $O(C\min(m,n))$, and space efficiency of $O(C\min(C,m,n))$ where all editing steps have the same cost independent of the characters involved. If the editing sequence that gives cost C is not required, the algorithm can be implemented in space efficiency of $O(\min(C,m,n))$. Consider two sequences S and T:

S: aacaaagtta

T: attgaaacaa

Convert sequence S to T, and confirm the time and space efficiency of the editing method.

36. *Band Across Diagonal*

In the method of dynamic programming to align two sequences, to obtain the optimal global alignment grade, filling a table of grades with $m$ rows and $n$ columns is called for. It can be seen that the alignments can be recovered using a traceback procedure. The alignments are close to the diagonal of the table. A lot of cells in the table that are far from the diagonal are needed to recover the optimal global alignment. So, a procedure can be developed that calls for computations only across the diagonal in the table from (0,0) to $(m,n)$ and a few cells at the top and bottom of the main diagonal. So, confinement within a band off the main diagonal can reduce the number of computations needed to obtain the optimal grade of alignment and recover the alignments. What is the speed increase expected as a function of the width of the band? What is the space reduction achieved as a function of the width of the band?

37. *Trade-Off between Time Efficiency versus Optimality*

Let us say a trade-off is allowed between time efficiency and optimality. In order to obtain an optimal global alignment between two sequences within twice the optimal grade of alignment, what is the speed increase and space reduction that can be expected?

38. *Global Alignment of Three Sequences*

Show that a dynamic programming method can be used to obtain optimal global alignment of three sequences S, T, and U. The time taken efficiency would be $O(n^3)$, and space required would be also $O(n^3)$. A cube of cells with $m$ rows, $n$ columns, and $o$ floors has to be filled to complete the procedure.

39. *Local Alignment of Three Sequences*

Show that a dynamic programming method can be used to obtain optimal local alignment of three sequences S, T, and U. The time efficiency would be $O(n^3)$, and space efficiency required would be also $O(n^3)$. A cube of cells with $m$ rows, $n$ columns, and $o$ floors have to be filled to complete the procedure.

40. *Affine Gap Penalty*

Consider the local alignment of three sequences S, T, and U. How would you modify the procedure developed in Problem 39 to incorporate the affine gap penalty?

41. *Hirschberg Space Array for Global Alignment of Three Sequences*

Show that the dynamic array concept developed by Hirschberg and discussed in Section 2.7 can be extended to obtaining the optimal grade of alignment during global alignment of three sequences S, T, and U in space efficiency of $O(n^2)$?

42. *Hirschberg Space Array for Local Alignment of Three Sequences*

Can the dynamic array concept developed by Hirschberg and discussed in Section 2.7 be extended to obtaining the optimal grade of alignment during local alignment of three sequences S, T, and U. Why?

43. *Dynamic Programming Table for Global Alignment*

In the method of dynamic programming, to obtain an optimal global alignment of two sequences S and T, a table of grades has to be generated with *mn* cells. Are there two sequences S and T for any set of cell grades in the table? Discuss.

44. *Dynamic Programming Table for Local Alignment*

   In the method of dynamic programming, to obtain optimal local alignment of two sequences S and T, a table of grades has to be generated with *mn* cells. Are there two sequences S and T for each and every set of cell grades in the table? Discuss.

45. *Sparse Table*

   Obtain the optimal local alignment of the following two sequences S and T:

   S: acgtt

   T: acaaa

   Show the dynamic programming table.

   What is the unique feature of the dynamic programming table used to align sequences S and T? Is it a sparse matrix or a sparse table? Once the sparse matrix is recognized, can the time efficiency be increased and space required cut down?

46. Recover the local alignments from the table filled in Problem 45 using the traceback procedure.

47. Note that the grading function had a +2 for matches in table in Problem 45. Should the grade of alignment for a match be −1 Can a table with sparse matrix property such as the one shown in the table in Problem 45 be generated for any two sequences S and T. Why?

48. *Sparse Table*

   During the local alignment procedure of Smith and Waterman using dynamic programming, depending on the nature of the sequences S and T, some tables can be seen to be sparse; that is, many of the cells have zero grade. For example, only half of the table in Problem 45 is filled, and the rest in not filled. How can this knowledge be used to decrease the space required and increase the time efficiency of the method?

49. *Stability of Global Alignment*

   In the dynamic programming method, in order to obtain the global and optimal alignment between two sequences S and T, the traceback procedures originate from the bottom-right cell and end up at the top-left cell. What is it in the procedure that keeps it from taking a detour to the top-right cell or a wavy path?

50. *Stability of Local Alignment*

   In the dynamic programming method, to obtain the local and optimal alignment between two sequences S and T, the traceback procedures originate from the cell with the local maxima and ends up at the cell with the local minima. What is it in the procedure that keeps it from taking a wavy path?

51. *Inverse Dynamic Programming for Global Alignment*

   Define a formal inverse problem of conversion of a filled dynamic programming table (with *m* rows and *n* columns) with grades of alignment in each cell in order to obtain the optimal global alignment into two sequences S and T.

52. *Inverse Dynamic Programming for Local Alignment*

   Define a formal inverse problem of conversion of a filled dynamic programming table (with *m* rows and *n* columns) with grades of alignment in each cell in order to obtain the optimal local alignment into two sequences S and T.

53. *Inverse Dynamic Programming for Local Alignment with Affine Gap Penalty*

Define a formal inverse problem of conversion of a filled dynamic programming table (with *m* rows and *n* columns) with grades of alignment in each cell in order to obtain the optimal local alignment with affine gap penalty into two sequences S and T.

54. *Geometric Distribution*

The dynamic programming methods to align any two sequences S and T requires $O(n^2)$ time and $O(n^2)$ space for guaranteed optimality. When DNA sequences are aligned, a geometric distribution model can be developed to characterize the two sequences S and T. When the sequences are parsed for a match, matched regions are given a positive weight, and when a mismatch is encountered, the model can be called for to check whether the mismatch was on account of some experimental error or because of some biological phenomena. Search across a diagonal is performed, and excursions as in the X-drop method as shown in Problem 30 are allowed. Show that this can result in lower time and less space taken, and that a closer-to-optimal grade of alignment can be achieved.

## REVIEW QUESTIONS

1. What is the maxima in a triad concentration of copolymer mean?
2. Is there a point of inflection in the triad concentration curve for copolymers as a function of monomer concentration?
3. Does the chain sequence distribution of copolymer change with temperature?
4. Why are mutations, conservation, and homology important in sequence alignment?
5. What is the increase in database search cost expected with time and why?
6. What is the connection with sequence alignment during shotgun sequencing?
7. What is the expected role of sequence alignment in *personalized medicine*?
8. How are better drugs designed using sequence alignment methods?
9. What is a trace of alignment?
10. What is the meaning when the optimal grade of alignment is 0?
11. Szymanski showed that for the special case when no element appears more than once in the input sequence, the alignment problem can be solved for in $O(m+n)\lg(n+m)$. This is lower than the $O(n^2)$ time efficiency needed for any two general sequences. Why does the time taken increase when the characters repeat in the sequences?
12. What is the LCS (longest common subsequence) between a pair of sequences S and T?
13. *Sequence Distribution Microstructure*

What would be different about the grade of alignment during pairwise global alignment of two sequences when one of the sequences is (1) randomly distributed, (2) is alternating distributed, or (3) has block architecture?

## REFERENCES

1. S. B. Needleman and C. D. Wunsch, A general method applicable to the search for similarities in the amino acid sequence of two proteins, *J. Mol. Biol.*, Vol. 48, 443–453, 1970.

2. Z. Zhang, S. Schwartz, L. Wagner, and W. Miller, A greedy algorithm for aligning DNA sequences, *J. Comput. Biol.*, Vol. 7, 1, 2, 203–214, 2000.

3. S. F. Altschul, W. Gish, W. Miller, E. W. Myers, and D. J. Lipman, Basic local alignment search tool, *J. Mol. Biol.*, Vol. 215, 3, 403–410, 1990.

4. S. F. Altschul, Amino acid substitution matrices from an information theoretic perspective, *J. Mol. Biol.*, Vol. 219 (3), 555–565, 1991.

5. S. F. Altschul, A protein scoring system sensitive at all Holutiny distances, *J. Mol. Evol.*, Vol. 36, 3, 290–300, 1993.

6. S. F. Altschul and W. Gish, Local alignment statistics, *Methods Enzymol.*, Vol. 266, 460–480, 1996.

7. K. R. Sharma, *Bioinformatics: Sequence Alignment and Markov Models*, McGraw Hill Professional, New York, 2009.

8. R. Bellman, *Dynamic Programming*, Princeton University Press, Princeton, NJ, 1957.

9. T. F. Smith and M. S. Waterman, Identification of common molecular subsequences, *J. Mol. Biol.*, Vol. 147, 195–197, 1981.

10. R. E. Knuth, *The Art of Computer Programming*, Addison-Wesley, Reading, MA, 1997.

11. W. J. Masek and M. S. Paterson, A faster algorithm computing string edit distances, *J. Computer Syst. Sci.*, Vol. 20, 1, 1980, 18–31.

12. T. G. Szymanski, A Special Case of the Maximal Common Subsequence Problem, Technical Report TR-170, Princeton University, Computer Science Laboratory, 1975, Princeton, NJ.

13. T. H. Cormen, C. E. Leiserson, R. L. Rivest, and C. Stein, *Introduction to Algorithms*, MIT Press, Boston, MA, 2001.

14. D. S. Hirschberg, A linear-space algorithm for computing maximal common subsequences, *Commun. ACM*, Vol. 18, 341–343, 1975.

15. E. W. Myers and W. Miller, Optimal alignments in linear space, *Computer Appl. Biosci.*, Vol. 4, 11–17, 1988.

16. J. W. Hunt and T. G. Szymanski, A fast algorithm for computing longest common subsequences, *Commun. ACM*, Vol. 20, 5, 350–353, 1977.

17. P. van Emde Boas, Preserving Order in a Forest in Less than Logarithmic Time, 16th Annual Symp. on Foundations Computer Science, October, 75–84, 1975.

18. Z. Galil and R. Giancarlo, Speeding up dynamic programming with applications to molecular biology, 1989, *Theoretical Computer Sci.*, Vol. 64, 107–118, 1989.

19. E. Ukkonen, Algorithms for approximate string matching, *Information and Control*, Vol. 64, 100–118, 1985.

20. K. M. Chao, J. Zhang, J. Ostell, and W. Miller, A tool for aligning very similar DNA sequences, *CABIOS*, Vol. 13, 1, 75–80, 1997.

21. W. R. Pearson and D. J. Lipman, Improved tools for biological sequence comparison, *Proc. Natl. Acad. Sci. USA*, 85 (8), 2444–2448, 1988.

# 12 Reversible Polymerization

## LEARNING OBJECTIVES

- Derive Clapeyron equation
- Sensible heat integral
- Ceiling temperature
- Entropy, enthalpy, and free energy of reversible polymerization
- Arrhenius relationship for rate constants
- Subcritical damped oscillations during thermal polymerization
- Polyrate of terpolymerization of AMS–AN–Sty
- Enthalpy of random copolymers
- Effect of chain sequence distribution
- Entropy and free energy of copolymerization
- Copolymer composition with and without ceiling temperature effect

During polymerization, in some systems, the reverse step of depolymerization is as favored as the propagation polymerization steps. Some inexpensive monomers, whose polymers properties are desirable, have not achieved the levels of commercial success as some of the other monomers. For example, alphamethyl-styrene, AMS, monomer does not form into polymer above 61°C at atmospheric pressures; whereas, a small beaker filled with AMS monomer left to sit near the window for a few weeks will polymerize *photocatalytically* into polyalphamethyl-styrene. Poly-AMS has interesting properties, such as high glass transition temperature of 171°C. This can be used to design automotive components with increased heat resistance and higher VICAT temperature. VICAT temperature is measured using testing methods as listed in ASTM D 1525 and ISO 306. It is taken as the temperature at which the specimen is penetrated to a depth of 1 mm by a flat-ended needle with a 1 mm$^2$ circular or square cross-section.

The temperature of 61°C is referred to as the *ceiling temperature* of the polymer. This is the temperature above which the monomer to polymer formation and polymer to monomer are equally favored. Systems such as sulfur polymerize only above a certain temperature $T_f$. Such a temperature $T_f$ is called the *floor temperature*. This temperature for sulfur is about 180°C. The pressure changes with the ceiling temperature according to the Clapeyron equation. How does the ceiling temperature change with multicomponent copolymerization? When efforts to form a system of four monomers A, B, C, and D into a multicomponent copolymer are attempted, will a tetrapolymer of A-B-C-D form, or will it form as two copolymers of A-B and C-D? The enthalpy, entropy, and free energy of polymerization of such systems are the subject of this chapter. Prior to this, the thermochemistry of polymerization reactions in general are reviewed.

## 12.1  HEAT EFFECTS DURING POLYMERIZATION

Polymerization reactions in the industry often involve generation of heat or consumption of heat. Heat has to be transferred through the jacket of the kettle to maintain the reactor temperature desired. For exothermic reactions, cooling water is needed to remove the heat generated. For endothermic reactions, steam is supplied to provide the necessary heat needed for the reaction. Polymerization reactions are accompanied by temperature changes during the course of the reaction. The molecular structures of polymers and monomers are different, and they possess different energies. So, on account of the performance of the reaction, heat will be either liberated or consumed. Most vinyl polymerizations are exothermic in nature. Therefore, the design of the reactors that are used to manufacture them in large quantities should include provisions for heat transfer or nonisothermal operation. Nonisothermal operation would require the knowledge of variation of molecular weight with temperature, among other things.

Each of a number of polymerization reactions may be carried out in many different ways. Each reaction carried out in a particular way is accompanied by a particular heat effect. Tabulation of all possible heat effects for all possible polymerization reactions is a herculean task. The heat effects for a desired polymerization reaction can be calculated from data for reactions carried out in a standard way. When heat of polymerization is available at one temperature and pressure, it can be calculated for another temperature and pressure by extrapolation. The Clapeyron equation can be derived as follows to aid this process.

From Appendix A, Equation (A.18) for any process that obeys the first and second law of thermodynamics:

$$dG = VdP - SdT.$$  (12.1)

For polymer,

$$dG_P = V_P dP - S_P dT.$$  (12.2)

For monomer,

$$dG_m = V_m dP - S_m dT.$$  (12.3)

When the monomer and polymer are in equilibrium with each other,

$$dG_p = dG_m$$

or

$$\Delta VdP - \Delta SdT = 0,$$  (12.4)

or

$$\frac{dP}{dT} = \frac{\Delta S}{\Delta V}.$$  (12.5)

For an isobaric and isothermal process,

$$\Delta G = \Delta H - T\Delta S = 0,$$

or
$$\Delta S = \frac{\Delta H}{T}.$$

(12.6)

Most industrial polymerization reactors are operated at constant temperature and constant pressure, and sometimes controlled using computer data acquisition and sensors. So, Equation (12.6) may be well applicable for these systems. Combining Equation (12.6) with Equation (12.5):

$$\frac{dP}{dT} = \frac{\Delta H}{T\Delta V}.$$

(12.7)

Equation (12.7) is the *Clapeyron equation*. It can be used to calculate the enthalpy change of reactions at a desired temperature and pressure when the enthalpy of the reaction at a standard state is known or some temperature and pressure is known. This can be accomplished by integrating Equation (12.7) as

$$\frac{dT}{T} = \frac{\Delta V}{\Delta H}dP.$$

$$\int_{T_1}^{T_2} \frac{dT}{T} = \int_{P_1}^{P_2} \frac{\Delta V}{\Delta H}dP.$$

$$\ln\left(\frac{T_2}{T_1}\right) = \frac{\Delta P\Delta V}{\Delta H}$$

$$\Delta H = \frac{\Delta P\Delta V}{\ln\left(\frac{T_2}{T_1}\right)}$$

(12.8)

## Example 12.1  Effect of Pressure on Ceiling Temperature

The enthalpy change of polymerization of alphamethyl-styrene is −8.4 kcal/mole and change in volume during polymerization is −14.7 mL/mole at 20°C. If the ceiling temperature for polyalphamethyl-styrene at atmospheric pressure is 61°C, what is the ceiling temperature of polyalphamethyl-styrene at 500 atm pressure?

Exponentiation of Equation (12.8) leads to

$$T_2 = T_1 e^{\frac{\Delta P\Delta V}{\Delta H}}$$

(12.9)

The $\Delta H / \Delta V$ can be calculated from the information given at 20°C as 0.571 kcal/mL. Thus:

$$T_2 = 334 \exp\left(\frac{500 * 0.986 * 1.0133E5}{571 * 4.184E6}\right) = 341K. \qquad (12.10)$$

So, the ceiling temperature increases from 61°C to 68°C when the pressure of reaction increases from 1 atm pressure to 500 atm pressure.

The heat associated with a specific polymerization reaction depends on the temperatures of both the monomers and polymer. A *standard* basis that is consistent for treating polymerization heat effects results when the products of polymerization and the monomers are all at the same temperature. Consider a calorimeter method of measurements of heat of polymerization of monomers. The initiator is mixed with the monomer, and the system is a continuous flow CSTR. The polymerization reactions take place in the CSTR. The polymerization products enter a devolatilizer where the monomers are vaporized and removed from the product mix and recycled back to the reactors. The CSTR is water cooled to bring the monomers/polymer to the reactor temperature. There is no shaft work performed by the process. The CSTR is built, so that changes in potential and kinetic energy are negligible. The first law of thermodynamics for open systems can be written for the system as

$$Q = \Delta H. \qquad (12.11)$$

The heat flowing from the calorimeter and absorbed by the water is equal in magnitude to the enthalpy change caused by the polymerization reaction. The enthalpy change of the polymerization $\Delta H$ is called the *heat of polymerization*. Consider a polymerization scheme as follows:

$$nM \rightarrow (M - M - M - M - ... - M)_n = P. \qquad (12.12)$$

The standard heat of polymerization is defined as the enthalpy change when $n$ moles of monomer, $M$ in its standard state at temperature $T$ polymerizes to form 1 mole of polymer $P$ with $n$ repeat units of monomer $M$. For commercial systems, $n$ is 1000 or higher. This is also referred to as *degree of polymerization*. The standard state is a particular state of a species at temperature T and at specified conditions of pressure, composition, and state of matter, that is, gas, liquid, or solid.

A standard state pressure of 1 standard atmosphere (101,325 Pa) has been changed to 1 bar ($10^5$Pa). The standard state for a monomer in liquid form is the pure monomer at 1 bar, liquid or solid. For monomers as gas, the standard state is the pure monomer in the ideal gas state at 1 bar. $C_p^0$ is the standard state heat capacity. The standard heat of polymerization can be written as

$$\Delta H^0 = H_p^0 - nH_m^0. \qquad (12.13)$$

The enthalpy at one temperature and pressure can be calculated from the enthalpy change at another temperature and pressure. Because $H$ is a function of temperature and pressure,

$$H = f(T, P)$$

$$dH = \left(\frac{\partial H}{\partial T}\right)_P dT + \left(\frac{\partial H}{\partial P}\right)_T dP. \tag{12.14}$$

Because

$$\left(\frac{\partial H}{\partial T}\right)_P = C_p,$$

$$dH = C_p dT + \left(\frac{\partial H}{\partial P}\right)_T dP. \tag{12.15}$$

Standard state polymerization is conducted at constant pressure. So, the term with $dP$ in Equation (12.15) would drop to zero, and

$$dH_i^0 = C_{pi}^0 dT. \tag{12.16}$$

The subscript is added to the standard enthalpy to allow for representation of different species such as monomer and polymer. The standard heat-capacity change of polymerization may be defined as

$$\Delta C_p^0 = C_{PP}^0 - n C_{pm}^0. \tag{12.17}$$

Combining Equations (12.17), (12.13), and (12.16):

$$\Delta H^0 = \Delta H_{ref}^0 + \int_{T_{ref}}^{T} \Delta C_p^0 dT, \tag{12.18}$$

where $\Delta H^0$ and $\Delta H^0{}_{ref}$ are heats of polymerization at temperature $T$ and reference temperature $T_f$, respectively. Evaluation of the integral in Equation (12.18) requires the knowledge of the temperature dependence of heat capacity. Although polymers behave differently from small molecules in the way the heat capacity changes with temperature, especially near the glass transition temperature, the heat capacity is usually expressed in terms of the first few terms of the power series in temperature. Thus:

$$\frac{C_p}{R} = a + bT + cT^2 + \frac{d}{T^2}. \tag{12.19}$$

Applying Equation (12.19) in Equation (12.18):

$$\int_{T_{ref}}^{T} \frac{\Delta C_p^0}{R} dT = \int \Delta a T_{ref} (\theta - 1) + \int \frac{\Delta b T_{ref} (\theta + 1)}{2}$$

$$+ \int \frac{\Delta c (\theta^2 + \theta + 1) T_{ref}^2}{3} + \int \frac{\Delta d}{\theta T_{ref}^2}.$$

(12.20)

Equation (12.20) is the sensible heat integral [1].

## 12.2 CEILING TEMPERATURE DURING REVERSIBLE POLYMERIZATION

The total Gibbs free energy of a closed system at constant $T$ and $P$ must decrease during a spontaneous irreversible process. During a reversible process, the Gibbs free energy change must be zero:

$$dG_{total} = 0.$$

(12.21)

If a mixture of monomers is not in equilibrium, any polymerization that occurs at constant temperature, $T$, and constant pressure, $P$, must lead to a decrease in the total Gibbs free energy of the system. A reaction variable $\varepsilon$ may be defined. The fundamental property relation for a single phase system for the total differential of the Gibbs free energy can be written as a function of temperature, pressure, and chemical potential as follows:

$$d(nG) = (nV)dP - (nS)dT + \sum_i \mu_i dn_i.$$

(12.22)

Due to polymerization in a closed system, each $dn_i$ may be replaced by the product $v_i d\varepsilon$. Equation (12.22) then becomes

$$d(nG) = (nV)dP - (nS)dT + \sum_i v_i d\varepsilon_i.$$

(12.23)

Due to the fact that $nG$ is a state function, the right-hand side of Equation (12.23) is an exact differential:

$$\sum_i v_i \mu_i = \left( \frac{\partial (nG)}{\partial \varepsilon} \right)_{T,P} = \left( \frac{\partial G_{total}}{\partial \varepsilon} \right)_{T,P}.$$

(12.24)

At equilibrium, Equation (12.24) becomes equal to zero. The *fugacity* of a species can be defined as

$$\mu_i = C_i(T) + RT\ln(\hat{f}_i),\tag{12.25}$$

where $C_i(T)$ is an integration constant, a function of temperature only for a given species. The fugacity of a species captures the real fluid behavior as different from the ideal gas behavior. In terms of standard state,

$$G_i^0 = C_i(T) + RT\ln(f_i^0).\tag{12.26}$$

The difference between Equation (12.25) and Equation (12.16) can be seen to be

$$\mu_i - G_i^0 = RT\ln(\frac{\hat{f}_i}{f_i^0}).\tag{12.27}$$

Combining Equation (12.27) with Equation (12.24) at equilibrium

$$\sum_i v_i(\mu_i - G_i^0 = RT\ln(\frac{\hat{f}_i}{f_i^0})) = 0,\tag{12.28}$$

or

$$\ln\left(\prod_i\left(\frac{\hat{f}_i}{f_i^0}\right)^{v_i}\right) = \frac{-\sum_i v_i G_i^0}{RT}.\tag{12.29}$$

The product over all species $i$ is denoted by the symbol $\prod$. In exponential form, Equation (12.29) becomes

$$\left(\prod_i\left(\frac{\hat{f}_i}{f_i^0}\right)^{v_i}\right) = K,\tag{12.30}$$

where

$$K = e^{-\frac{\Delta G^0}{RT}}$$

$$\ln(K) = \frac{-\Delta G^0}{RT}.\tag{12.31}$$

Consider the free radical reactions during formation of homopolymer $P$ from monomer $M$. Let the initiator be denoted by $I$:

$$I \xrightarrow{k_D} 2I^* \quad \text{(Initiation reaction)} \tag{12.32}$$

$$M + I^* \rightarrow M^*$$

$$M^* + M \underset{k_p}{\Longleftrightarrow} MM^* \quad \text{(Propagation and depropagation)} \tag{12.33}$$

$$MM^* \xrightarrow{k_{dp}} M + M^*$$

$$M_m^* + M_n^* \xrightarrow{k_t} M_{n+m}$$

$$\tag{12.34}$$

$$M_m^* + M_n^* \xrightarrow{k_{tp}} M_l + M_k \quad \text{(Termination reactions)}$$

The rate of propagation and the rate of depropagation reactions are assumed to obey the first-order kinetic rate. The rate constants, $k_p$ and $k_{dp}$ obey the *Arrhenius* relationship for temperature dependence of rate constant:

$$k_p = A_p e^{-\frac{E_p}{RT}} \tag{12.35}$$

$$k_{dp} = A_{dp} e^{-\frac{E_{dp}}{RT}}. \tag{12.36}$$

From the *transition state theory,* the frequency factor is given by

$$A_p = \frac{k_B T}{h} e^{-\frac{\Delta S_p}{R}} \tag{12.37}$$

$$A_{dp} = \frac{k_B T}{h} e^{-\frac{\Delta S_{dp}}{R}}. \tag{12.38}$$

Therefore,

$$\Delta S^0 = \Delta S_p - \Delta S_{dp} = -R \ln \left( \frac{k_p}{k_{dp}} \right). \tag{12.39}$$

At equilibrium, the rate of propagation is equal to the rate of depropagation. Thus:

$$k_p[M][M^*] = k_{dp}[M^*]$$

or

$$K = \frac{k_p}{k_{dp}} = \frac{1}{[M_e]}$$ (12.40)

where $M_e$ is the equilibrium monomer concentration. From Equation (12.31)

$$K = \frac{1}{[M_e]} = e^{-\frac{\Delta H^0}{RT}} e^{\frac{\Delta S^0}{R}}$$ (12.41)

$$\ln(K) = -\ln(M_e) = -\frac{\Delta H^0}{RT} + \frac{\Delta S^0}{R}$$ (12.42)

At equilibrium, Equation (12.42) can be rearranged to obtain an expression for the ceiling temperature of the monomer–polymer system, $T_c$, as

$$T_c = \frac{\Delta H_p}{R\ln(M_e) + \Delta S_p}$$ (12.43)

where $\Delta H^0 = \Delta H_p$ is the heat of polymerization, and $\Delta S^0 = \Delta S_p$ is the entropy of polymerization. Further, it can be realized that $\Delta H_p = E_p - E_{dp}$.

The ceiling temperature of some monomers is given in Table 12.1 [2]. Further, the van't Hoff reaction isobar can be calculated as

$$\frac{\partial \ln K}{\partial T} = \frac{\Delta H_p}{RT^2}$$ (12.43a)

The entropy of a system is a measure of the statistical probability or degree of disorder of that system. The depropagation of one polymer molecule into many oligomer fractions is accompanied by an increase in *translational entropy* as this is a dissociative process. Forward propagation or polymerization results in decrease in translational entropy. Entropy values for lot of monomers are similar to each other. Even though the monomers have a wide range of ceiling temperatures, they have a narrow range of entropy of polymerization. Thus, $\Delta S_p$ $\Delta H_p / T_c = 25 - 30$ cal. K$^{-1}$mole$^{-1}$. This equation is analogous to Trouton's rule that has been found in common for vaporization of several liquids. Steric hindrance affects enthalpy of polymerization but does not affect the entropy of polymerization. The rotational entropy of the macromolecular chain will increase by increase in rigidity. For example, the entropy of polymerization of styrene and alphamethyl-styrene are 25 and 26 eu (entropy units), respectively, differing only by 1 eu. The enthalpy of polymerization of styrene and alphamethyl-styrene vary by 10 kcal/mole. Side-group steric

**TABLE 12.1**
**Ceiling Temperature of Some Monomers**

| Monomer | Standard State | Ceiling Temperature $T_c$ (°C) | Equilibrium Monomer Fraction $[M_e]$ |
|---|---|---|---|
| Ethylene | Gas | 407 | 760 mmHg |
| Methacrylonitrile | | 177 | |
| Propylene | Gas | 300 | 760 mmHg |
| Styrene | Gas | 235 | 760 mmHg |
| Alphamethyl-styrene | Liquid | 61 | 760 mmHg |
| Sulfur | Liquid | 159 (floor temperature) | 1 |
| Methacrylic Acid | Gas | 173 | 760 mmHg |

**TABLE 12.2**
**Entropy of Polymerization for Some Polymers**

| Polymer | $-\Delta S$ (CalK$^{-1}$Mole$^{-1}$) |
|---|---|
| Polyethylene, PE | 41.5 |
| Polypropylene, PP | 49.0 |
| Polystyrene, PS | 24.9 |
| Polybutadiene, PBd | 21.2 |
| Polymethyl methacrylate, PMMA | 9.6 |
| Polytetrafluorethylene, PTFE | 26.8 |
| Polyvinyl chloride, PVC | 21.2 |

hindrance results in lowering of ceiling temperature on account of steric repulsion of the side groups. Monomer resonance and higher bond energies implies large vibrational frequencies of the bond within the monomer. A third of the entropy of polymerization of polystyrene can be attributed to vibrational energy.

The entropy of polymerization can be calculated from the measurement of equilibrium monomer concentration, entropies of monomer and polymer, and from the ratio of the kinetic frequency factors $A_p$ and $A_{dp}$. The entropy of polymerization of some common polymers is given in Table 12.2.

## 12.3 SUBCRITICAL OSCILLATIONS DURING THERMAL POLYMERIZATION

In the continuous mass polymerization of acrylonitrile butadiene styrene (ABS) and high impact polystyrene (HIPS), the thermal polymerization of styrene and alpham-ethyl-styrene is of increasing concern [3]. Scale-up issues remain, especially for the higher heat product lines of Monsanto, Dow, and General Electric. The free radical polymerization reactions have been modeled using the three broad steps of initiation, propagation, and termination.

### 12.3.1 THERMAL INITIATION BY DIELS–ALDER DIMERIZATION

$$2S \rightarrow D \tag{12.44}$$

$$S + D \rightarrow D^* \tag{12.45}$$

$$D^* + S \rightarrow S^* \tag{12.46}$$

$$S^* + S \rightarrow S^* \tag{12.47}$$

The net initiation of grafting can be written as

$$4S \xrightarrow{k_{IG}} S^*. \tag{12.48}$$

The initiation rate can be written as

$$R_I = k_{IG} \, [S]^4 \tag{12.49}$$

In a similar fashion, the graft propagation and matrix termination reactions can be written as

$$R_P = k_p \, [S] \, [Sn^*] \tag{12.50}$$

$$R_T = k_t \, [Sn^*]^2 \tag{12.51}$$

The concentration of the radical chains $Sn^*$ can be derived from the quasi-steady state assumption of radicals and is written as

$$[S_n^*] = \sqrt{\frac{k_{IG}}{k_t}} [S]^2 \tag{12.52}$$

The rate of propagation of the thermal polymerization can then be written as

$$R_p = k_p \sqrt{\frac{k_{IG}}{k_t}} [S]^3 \tag{12.53}$$

Thus, a third-order dependence on the monomer concentration is expected. The number averaged molecular weight can be given by the ratio of the propagation reaction to the termination reaction and is

$$M_n(graft)) = \frac{k_p}{\sqrt{k_{IG} k_t}} [S] \tag{12.54}$$

The number average molecular weight by thermal polymerization is inversely proportional to the monomer concentration.

Subcritical damped oscillations can be detected upon representing the propagation reactions in the form of a circle and the conditions where it occurs can be identified [4].

The copolymerization of alphamethyl-styrene and acrylonitrile is another example of reactions that can be represented by use of reaction in circle scheme. The oligomer (A), becoming a decamer (B), can become a 1000 mer or polymer (C) and depolymerize or unzip to a dimer (A). The unzipping reaction is favored because of the low ceiling temperature of the alphamethyl-styrene monomer. With the advent of biotechnological routes to preparing the substances of interest, the design of these systems in bioreactors is of interest.

Oscillations in concentration that were initially reported by Bray in 1921 and Belousov in 1959 were discounted for the violation of the second law of thermodynamics. Oscillations of concentration about a nonequlibrium value of the extent of reaction do not violate the second law [4]. When it was realized that systems far from thermodynamic equilibrium could exhibit oscillations, interest in these and other oscillating reactions rose and gave rise to the study of dissipative structures. Kondepudi and Prigogine [5] developed a simple model that demonstrated clearly how a nonequilibrium system could become unstable and make a transition to an oscillatory state. Koros and Noyes [in 5] performed a thorough study of the reaction mechanism, and identified the key steps in the rather complex Belousov–Zhabotinsky reaction and showed how the oscillations arise from only three variables. The BZ reaction is basically the catalytic oxidation of an organic compound such as malaonic acid, $CH_2(COOH)_2$, with bromate. After an induction period, the oscillatory behavior can be seen in the variation of the concentration of the $Ce^{4+}$ ion due to which there is a change in color from colorless to yellow.

Let a system of reactions be represented by a circle. Although any number of species can be present in this circle, let us consider (a) the system of three species in a circle, (b) the system of four species in a circle, (c) the system of eight species in a circle, and (d) general case. The kinetics of a system of three reactants in a circle as shown in the schematic in a circular fashion can be written as

$$\frac{dC_A}{dt} = -k_1 C_A + k_3 C_c \qquad (12.55)$$

$$\frac{dC_B}{dt} = -k_2 C_B + k_1 C_A \qquad (12.56)$$

$$\frac{dC_c}{dt} = -k_3 C_c + k_2 C_B \qquad (12.57)$$

where $C_A$, $C_B$, and $C_C$ are the concentrations of reactants A, B, and C at a given instant t. If the initial concentration of A is given by $C_{A0}$ and that of B and C is zero, and obtaining the Laplace transforms of Equations (12.55–12.57)

$$(s + k_1)\bar{C}_A = C_{A0} + k_3 \bar{C}_C \qquad (12.58)$$

$$(s + k_2)\bar{C}_B = k_2 \bar{C}_A \qquad (12.59)$$

$$(s + k_3)\bar{C}_C = k_2 \bar{C}_B \qquad (12.60)$$

Eliminating $C_B$ and $C_C$ between Equations (12.58–12.60), the transformed expression for the instantaneous concentration of reactant A can be written as

$$(s + k_1)\bar{C}_A = C_{A0} + \frac{k_3 k_2 k_1 \bar{C}_A}{(s + k_3)(s + k_2)} \qquad (12.61)$$

or
$$\bar{C}_A = \frac{C_{A0}(s + k_3)(s + k_2)}{s(s^2 + s(k_1 + k_2 + k_3) + k_1 k_2 + k_1 k_3 + k_2 k_3)} \qquad (12.62)$$

The inversion of Equation (12.62) can be obtained by use of the residue theorem. The three simple poles can be recognized in Equation (12.62). Further, it can be realized that when the poles are complex, *subcritical damped oscillations* can be expected in the concentration of the reactant. This is when the quadratic $b^2 - 4ac < 0$. This can happen when

$$(k_3 - k_2 - k_1)^2 - 4k_2 k_1 < 0 \qquad (12.63)$$

or

$$(k_3 - k_2 - k_1) < 2\sqrt{k_1 k_2} \qquad (12.64)$$

or

$$k_3 < \left(\sqrt{k_2} + \sqrt{k_1}\right)^2 \qquad (12.65)$$

This expression is *symmetrical* with respect to reactants A, B, and C. When the relation holds, that is, when one reaction rate constant is less than the square of the sum of the square root of the rate constants of the other two reactions, the subcritical damped oscillations can be expected in the reactant concentration.

In order to obtain the concentration profiles as a function of time of the reactants A, B, and C, the Equations (12.55–12.57) are decoupled by differentiating it again with respect to time. Thus:

$$\frac{d^2C_A}{dt^2} + \alpha\frac{dC_A}{dt} + \gamma C_A = C_{A0}k_2k_3 \tag{12.66}$$

where

$$\alpha = k_1 + k_2 + k_3$$
$$\gamma = k_1k_2 + k_3k_1 + k_2k_3$$

In a similar manner, the equations for B and C can be written. The auxiliary quadratic equation with constant coefficients for the homogeneous part of the solution is solved for the particular integral combined to give the general solution for the three reactants with time as

$$C_A = C_1 \exp(-\alpha t/2) \exp(t\sqrt{(\alpha^2 - 4\gamma)}) + C_2 \exp(-\alpha t/2) \exp(t\sqrt{(\alpha^2 - 4\gamma)}) + C' \tag{12.67}$$

$$C_B = C_3 \exp(-\alpha t/2) \exp(t\sqrt{(\alpha^2 - 4\gamma)}) + C_4 \exp(-\alpha t/2) \exp(t\sqrt{(\alpha^2 - 4\gamma)}) + C'' \tag{12.68}$$

$$C_C = C_5 \exp(-\alpha t/2) \exp(t\sqrt{(\alpha^2 - 4\gamma)}) + C_6 \exp(-\alpha t/2) \exp(t\sqrt{(\alpha^2 - 4\gamma)}) + C''' \tag{12.69}$$

Differentiating Equations (12.67–12.69) with respect to time and multiplying both sides by $\exp(\alpha t/2)$, it can be seen that at infinite time the LHS becomes zero as the reaction rate drops to zero. In order for RHS also to be zero, $C_1$, $C_3$, and $C_5$ need be set to zero. Then, at infinite time, the concentrations of A, B, and C add to $A_0$. As the reactions are symmetrical and simple first order, a good approximation is that

$$C' = C'' = C''' = A_0/3 \tag{12.70}$$

The remaining three constants are solved from the initial concentrations for A, B, and C as $A_0$, 0, and 0. Thus, for $\alpha^2 > 4\gamma$,

$$C_A = C_{A0}/3\,(1 + 2\exp(-\alpha t/2)\exp(-t/2\sqrt{(\alpha^2 - 4\gamma)}\,) \tag{12.71}$$

$$C_B = C_{A0}/3\,(1 - \exp(-\alpha t/2)\exp(-t/2\sqrt{(\alpha^2 - 4\gamma)}\,) = C_C \tag{12.72}$$

The solution is bifurcated and, for values of $\alpha^2 < 4\gamma$, the argument within the square root sign in Equations (12.71) and (12.72) become imaginary and, using De Moivre's theorem and equating the real parts,

$$C_A = C_{A0}/3\,(1 + 2\exp(-\alpha t/2)\cos(t/2\sqrt{(4\gamma - \alpha^2)}\,) \tag{12.73}$$

$$C_B = C_{A0}/3 \, (1 - \exp(-\alpha t/2) \, \cos(t/2\sqrt{(4\gamma - \alpha^2)})) = C_C \qquad (12.74)$$

Thus for certain values of the rate constants, the concentration becomes subcritical damped oscillatory.

## 12.3.2 Four Reactions in a Circle

The equivalent Laplace transformed expression for concentration of reactant A for a system of four reactions in circle, assuming that all the reactions in the cycle obey simple, first-order kinetics, can be derived as

$$\bar{C}_A = \frac{C_{A0}(s+k_3)(s+k_2)(s+k_1)}{s(s^3 + s^2(k_1+k_2+k_3+k_4) + s(k_1k_2 + k_1k_3 + k_2k_3 + k_1k_4 + k_2k_4 + k_3k_4) + k_1k_2k_3 + k_1k_2k_4 + k_1k_3k_4 + k_2k_3k_4)}$$

$$(12.75)$$

The conditions where the concentration can be expected to exhibit subcritical damped oscillations can be found when the roots of the following equation becomes complex:

$$s^3 + \alpha s^2 + \beta s + \chi = 0, \qquad (12.76)$$

where

$$\alpha = k_1 + k_2 + k_3 + k_4 \qquad (12.77)$$

$$\beta = k_2k_3 + k_1k_3 + k_1k_2 + k_1k_4 + k_4k_3 + k_1k_4 \qquad (12.78)$$

$$\chi = k_1k_2k_3 + k_1k_2k_4 + k_1k_3k_4 + k_2k_3k_4 \qquad (12.79)$$

It can be seen that $\alpha$ is the sum of all four reaction rate constants, $\beta$ is the sum of products of all possible pairs of the reaction rate constants, and $\chi$ is the sum of products of all possible triple products of rate constants in the system of reactions in circle. Equation (12.76) can be converted to the depressed cubic equation by use of the substitution:

$$x = s - \frac{\alpha}{3} \qquad (12.80)$$

This method was developed in the Renaissance period [6]. The depressed cubic without the quadratic term will then be

$$x^3 + \left(\beta - \frac{\alpha^2}{3}\right)x + \chi + \frac{2\alpha^3}{27} - \frac{\alpha\beta}{3} = 0 \qquad (12.81)$$

Let
$$\left(\beta - \frac{\alpha^2}{3}\right) = B; \left(\chi + \frac{2\alpha^3}{27} - \frac{\alpha\beta}{3}\right) = C.$$
(12.82)

Then, Equation (12.81) becomes

$$x^3 + Bx + C = 0.$$
(12.83)

The complex roots to Equation (12.83) shall occur when $D > 0$, where

$$D = \frac{B^3}{27} + \frac{C^2}{4}.$$
(12.84)

Thus the conditions when subcritical damped oscillations can be expected for a system of four reactions in circle are derived.

### 12.3.3  GENERAL CASE OF $n$ REACTIONS IN CIRCLE

The general case, of which the Krebs cycle with eight reactions in circle is a particular case can be obtained by extension of the expressions derived for three reactions in circle and four reactions in circle. Aliter to this would be the method of eigenvalues and eigenvectors. The cases when $\lambda$, eigenvalues are imaginary represent situations in the system, when the concentration of the species will exhibit suberitical damped oscillations. These are given by the characteristic equation:

$$Det|K - \lambda I| = 0$$
(12.85)

The size of the K matrix depends on the number of reactions in circle. For $n$ reactions in circle K would be an $n \times n$ matrix. For the case of the Krebs cycle, it would be an $8 \times 8$ matrix.

Upon expansion, an 8th-order polynomial equation in $\lambda$ arises. Eight roots of the polynomial exist. Even if all the values in the characteristic matrix are real, some roots may be complex. When complex roots occur, they appear in pairs. The roots of the polynomial are called *eigenvalues* of the characteristic matrix. The polynomial equation is called the *eigenvalue equation*.

## 12.4  THERMAL TERPOLYMERIZATION OF ALPHAMETHYL-STYRENE, ACRYLONITRILE, AND STYRENE

Noting from the product technical brochures, there is significant commercial interest in manufacturing multicomponent copolymers as indicated by BASF, Bayer, etc. Alphamethyl-styrene (AMS) is used as the primary monomer choice. Products that offer higher heat resistance compared with styrene–acrylonitrile (SAN) copolymer can be made by copolymerization of AMS with AN, thereby circumventing the ceiling temperature effect found during homopolymerization of AMS. The ceiling temperature as indicated by Sawada [2] at atmospheric pressure for homopolymerization of AMS is 61°C. Ceiling temperature as seen in Section 12.2 is the temperature and pressure above which the depolymerization is as preferred as the forward polymerization reactions. Recently, Sharma [7] presented the chain sequence distribution of the AMS–AN copolymer as geometric distribution and as a function of the monomer compositions and reactivity ratios, neglecting effects due to Markov second-order statistics and used it to predict the potential for the formation of Grassie chromophores [8] in the copolymer. Here, the need for control of the chain sequence distribution to improve product performance and ease of manufacturing was highlighted.

One other method for the control of the chain sequence distribution is the addition of a termonomer such as styrene. It would be desirable from product performance considerations such as heat resistance, color, etc., to add a termonomer to the AMS–AN copolymer. As pointed out by Shwier [12], $f_{an}$, the weight fraction of AN in the monomers has a dominant effect on the copolymerization kinetics of AMS–AN using standard free radical initiators. The effect of adding styrene as a termonomer on the polyrates and molecular weight of copolymer formed is not well understood with scarce data available in the literature. Further, a thermal initiation method of free-radical chain polymerization might further improve the product color and might offer commercial possibilities provided other process considerations such as polyrates, molecular weight of copolymer formed, sensitivity of molecular weight to temperature in the reactor, and sensitivity of molecular weight distribution to reactor operating temperature are known. In this study, terpolymerization kinetics of AMS–AN–STY using thermal initiation is provided. The range of weight fractions of monomers varied in the study are $f_{sty} = 0{:}1{-}0{:}45$; $f_{an} = 0{:}1{-}0{:}4$; $f_{ams} = 0{:}2{-}0{:}75$; for the weight fraction in the monomers of styrene, acrylonitrile, alpha–methylstyrene, respectively. Empirical correlations are provided for the polyrate, weight average molecular weight, $M_w$, in the temperature range of 398–418 K.

### 12.4.1 Experimental

Tube polymerization free radical techniques were utilized to effect polymerization to small conversions. Sixteen different monomer compositions were used at three different temperatures, that is, 398, 408, and 418 K. An oil bath was used to provide the constant temperature. The polymer, when formed after 60 min at 398 K, 45 min at 408 K, and 30 min at 418 K, was isolated by using precipitation methods, followed by drying and weighing. Details of the technique are provided in Shwier [12]. The part of the copolymer after drying was used to measure the number average molecular weight ($M_n$), and weight average molecular weight ($M_w$), using gel permeation chromatography (GPC) at the PASC at Monsanto, Indian Orchard, Massachusetts. The composition of the monomers used in the study along with the composition of the product formed, as calculated from the terpolymer monomer–polymer composition dependence equation provided by Odian [13], is provided in Table 12.3. The reactivity ratios used were the same as for those used for respective copolymerizations—AMS, 1; AN, 2; STY, 3; $r_{12}$, 0.1; $r_{21}$, 0.06; $r_{13}$, 0.3; $r_{31}$, 1.3; $r_{23}$, 0.04; $r_{32}$, 0.4. The polyrate

**TABLE 12.3**

**Monomer and Polymer Compositions of Samples Used in Tube Polymerization Study AMS (Wt Fraction) AN (Wt Fraction) Sty (Wt Fraction) AMS (Wt Fraction) AN (Wt Fraction) STY (Wt Fraction)**

| AMS(mon) | AN(mon) | Sty(mon) | AMS(poly) | AN(poly) | STY(poly) |
|----------|---------|----------|-----------|----------|-----------|
| 0.45 | 0.1 | 0.45 | 0.295 | 0.251 | 0.454 |
| 0.75 | 0.1 | 0.15 | 0.511 | 0.297 | 0.192 |
| 0.4 | 0.2 | 0.4 | 0.266 | 0.361 | 0.373 |
| 0.7 | 0.2 | 0.1 | 0.489 | 0.397 | 0.114 |
| 0.3 | 0.3 | 0.4 | 0.212 | 0.421 | 0.367 |
| 0.6 | 0.3 | 0.1 | 0.451 | 0.44 | 0.108 |
| 0.2 | 0.4 | 0.4 | 0.156 | 0.464 | 0.374 |
| 0.5 | 0.4 | 0.1 | 0.42 | 0.468 | 0.112 |

was calculated by dividing the conversion, as measured by weight percent of polymer, made by the time allowed for polymerization.

## 12.4.2 RESULTS

A pseudo-first-order rate expression was used to express the polyrate:

$$R_p = k_0 e^{-\frac{E}{RT}}[M],$$ (12.86)

where $M$ is the monomer concentration expressed in mol/L.

$$K_0 e^{-E/RT} = K_p$$ (12.87)

This approach is suggested here in order to circumvent the need to quantify the mechanism of thermal initiation of styrene and/or alphamethyl-styrene such as the reaction order of three as suggested in the literature [11–13]. The $K$ value at a given temperature was calculated by regression with two independent variables, that is, $f_{an}$ and $f_{sty}$, the weight fraction in the monomers of acrylonitrile and styrene, respectively. Thus, at 398 K, the following expression for $K$ is obtained, with $r^2 = 0.976$, and standard error of estimate = 0.065 for 16 data points:

$$k_p = 0.479 f_{an}^{0.706} f_{sty}^{0.292}.$$ (12.88)

In a similar fashion, expressions for $K$ were obtained at 408 and 418 K:

$$k_p = 0.836 f_{an}^{0.694} f_{sty}^{0.274}$$ (12.89)

TABLE 12.4

**$C_1$ and $C_2$ Values in the Rate Expression for Terpolymerization of AMS–AN–Sty using Thermal Initiation Methods**

| Temperature (K) | $C_1$ | $C_2$ |
|---|---|---|
| 398 | 0.706 | 0.292 |
| 408 | 0.691 | 0.274 |
| 418 | 0.668 | 0.241 |

$$k_p = 1.384 f_{an}^{0.668} f_{sty}^{0.241} \qquad (12.90)$$

with coefficient of determination $r^2 = 0.981$; standard error of estimate $= 0.056$ for $N = 16$ data points at 408 K with $r^2 = 0.96$; and standard error of estimate $= 0.079$ for $N = 16$ data points at 418 K. The generalized expression for the polyrate was obtained by using the Arrhenius dependence of the first-order pseudo rate constant and can be written as

$$k_p = 2.1 * 10^9 e^{-\frac{8328}{T}} f_{an}^{c_1} f_{sty}^{c_2} \qquad (12.91)$$

The values for $C_1$ and $C_2$ obtained from the regression are given in Table 12.4. In order to illustrate the goodness of fit using average values for $C_1 = 0.688$ and $C2 = 0.262$, the experimental values are shown in Figure 12.1 [7] along with the regression line.

In a similar fashion, an empirical relation for the weight average molecular weight of the polymer made was obtained:

$$M_w = 2.4 * 10^{-7} e^{\frac{7760}{T}} R_p^{C_3} f_{ams}^{C_4}. \qquad (12.92)$$

The values of $C_3$ and $C_4$ obtained from the regression along with the statistics of the regression are provided in Table 12.5.

A pseudo-first-order rate expression was used to represent polyrate and molecular weight data for terpolymerization of AMS–AN–STY. The weight fraction AN in the monomers was the primary dominating variable on the polyrate and the weight fraction of styrene the secondary variable. The dependence of the polyrate on the weight fraction of AMS and AN changed little with temperature in the range of 398 K to 418 K. The polyrate dependence on temperature was Arrhenius in nature. The weight-averaged molecular weight ($M_w$), increased with the polyrate with roughly a square root dependence and was inversely proportional to the weight fraction AMS. The dependence changed little with the temperature range between 398 K and 418 K. The molecular weight decreased with temperature with other things constant and the

$$\begin{pmatrix}
(-k_1 - \lambda) & 0 & 0 & 0 & 0 & 0 & 0 & k_8 \\
k_1 & (-k_1 - \lambda) & 0 & 0 & 0 & 0 & 0 & 0 \\
0 & k_2 & (-k_3 - \lambda) & 0 & 0 & 0 & 0 & 0 \\
0 & 0 & k_3 & (-k_4 - \lambda) & 0 & 0 & 0 & 0 \\
0 & 0 & 0 & k_4 & (-k_5 - \lambda) & 0 & 0 & 0 \\
0 & 0 & 0 & 0 & k_5 & (-k_6 - \lambda) & 0 & 0 \\
0 & 0 & 0 & 0 & 0 & k_6 & (-k_7 - \lambda) & 0 \\
0 & 0 & 0 & 0 & 0 & 0 & k_7 & (-k_8 - \lambda)
\end{pmatrix}$$

**FIGURE 12.1**  Characteristic matrix for system of eight reactions in circle.

**TABLE 12.5**
**$C_3$ and $C_4$ Values in Empirical Relation for $M_w$ in Terpolymerization of AMS–AN–Sty Using Thermal Initiation Methods**

| $T$ (K) | $C_3$ | $C_4$ | $r^2$ | Standard Error | N (Number of Data Points) |
|---|---|---|---|---|---|
| 398 | 0.424 | −0.392 | 0.991 | 0.033 | 16 |
| 408 | 0.434 | −0.382 | 0.975 | 0.051 | 16 |
| 418 | 0.486 | −0.281 | 0.999 | 0.091 | 16 |

dependence on temperature was exponential. Further modeling from first principles may be needed to fully understand the nature of the interdependence in the terpolymerization kinetics. The polydispersity, that is, a measure of the molecular weight distribution, $Mw = Mn$, was 1.843 averaged over the 48 data points with a standard deviation of 0.157.

## 12.5   REVERSIBLE COPOLYMERIZATION

### 12.5.1   Copolymer Composition

Three types of copolymerization can be expected:

1. Both monomers have high ceiling temperatures ($T_{c1}$, $T_{c2} > T_r$)
2. One monomer low ceiling temperature ($T_{c1} < T_r$)
3. Both monomers have low ceiling temperatures ($T_{c1}$, $T_{c2} < T_r$),

where $T_r$ is the reactor temperature, and $T_{c1}$ and $T_{c2}$ are the ceiling temperatures of monomer 1 and 2, respectively. Consider the propagation reactions for Type I as discussed in Section 12.5.1:

$$M_1^* + M_1 \xrightarrow{k_{11}} M_1 M_1^*$$

$$M_1^* + M_2 \xrightarrow{k_{12}} M_1 M_2^*$$

$$M_2^* + M_1 \xrightarrow{k_{21}} M_2 M_1^* \tag{12.93}$$

$$M_2^* + M_2 \xrightarrow{k_{11}} M_2 M_2^*$$

In a CSTR, the exit composition and the reactor compositions are the same. For a CSTR, the polymer composition can be calculated as a function of the monomer composition as follows:

$$F_1 = \frac{d[M_1]}{d[M_1] + d[M_2]}$$

$$= \frac{k_{11}[M_1][M_1^*] + k_{21}[M_1][M_2^*]}{k_{11}[M_1][M_1^*] + k_{21}[M_1][M_2^*] + k_{22}[M_2][M_2^*] + k_{12}[M_2][M_1^*]} \tag{12.93a}$$

Let the reactivity ratios be

$$r_{12} = \frac{k_{11}}{k_{12}}; \quad r_{21} = \frac{k_{22}}{k_{21}}.$$

The rate of irreversible reactions can be written as

$$\frac{d[M_1]}{dt} = -k_{11}[M_1][M_1^*] - k_{21}[M_1][M_2^*]$$

$$\frac{d[M_2]}{dt} = -k_{22}[M_2][M_2^*] - k_{12}[M_2][M_1^*]. \tag{12.94}$$

The free-radical species formed may be assumed to be highly reactive and can be assumed to be consumed as rapidly as formed. This is referred to as the quasi-steady-state assumption (QSSA). Thus:

$$\frac{d[M_1^*]}{dt} = -k_{11}[M_1][M_1^*] - k_{12}[M_1^*][M_2]$$

$$\frac{d[M_2^*]}{dt} = -k_{22}[M_2][M_2^*] - k_{21}[M_2^*][M_1]. \tag{12.95}$$

$$\frac{d[M_1 M_1^*]}{dt} = k_{11}[M_1][M_1^*].$$

$$\frac{d[M_2 M_2^*]}{dt} = k_{22}[M_2][M_2^*].$$

$$\frac{d[M_1 M_2^*]}{dt} = k_{12}[M_2][M_1^*]. \tag{12.96}$$

$$\frac{d[M_2 M_1^*]}{dt} = k_{21}[M_1][M_2^*].$$

By QSSA,

$$\frac{d[M_1^*]}{dt} + \frac{d[M_1 M_1^*]}{dt} + \frac{d[M_2 M_1^*]}{dt} = 0$$

$$\frac{d[M_2^*]}{dt} + \frac{d[M_2 M_2^*]}{dt} + \frac{d[M_1 M_2^*]}{dt} = 0 \tag{12.97}$$

Thus:

$$k_{12}[M_1^*][M_2] = k_{21}[M_1][M_2^*].$$

or

$$\frac{[M_2]}{[M_1]}\frac{k_{12}}{k_{21}} = \frac{[M_2^*]}{[M_1^*]} \tag{12.98}$$

The copolymer composition can be calculated from the selectivity achieved in the parallel reactions. Thus, the composition of monomer 1 in the polymer is found to be

$$F_1 = \frac{dM_1}{dM_1 + dM_2} = \frac{1}{1 + \dfrac{k_{21}}{k_{11}}\dfrac{M_2^*}{M_1^*}}. \tag{12.99}$$

Combining Equation (12.99) and Equation (12.98):

$$F_1 = \frac{dM_1}{dM_1 + dM_2} = \frac{1}{1 + \dfrac{1}{r_{12}}\dfrac{[M_2]}{[M_1]}} = \frac{r_{12}[M_1]}{r_{12}[M_1] + [M_2]}. \tag{12.100}$$

The composition of repeat unit 2 can be calculated as $F_2 = 1 - F_1$:

$$F_2 = \frac{dM_2}{dM_1 + dM_2} = 1 - F_1 = \frac{r_{21}[M_2]}{r_{21}[M_2] + [M_1]} \tag{12.101}$$

The copolymer composition with the monomer composition can be seen to vary with the reactivity ratios as shown in Figure 12.2. The copolymer composition as a

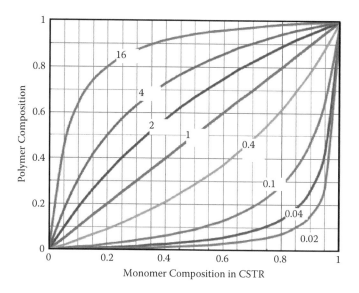

**FIGURE 12.2** Copolymer composition versus monomer composition for different reactivity ratios.

function of monomer composition in CSTR as a function of reactivity ratios is shown for eight different systems in Figure 12.2. It can be seen that as the reactivity ratio $r_{12}$ was varied from 0.02 to 16 with the reactivity ratio $r_{21}$, remaining the same at 0.41, the polymer composition increases with the increase in reactivity ratio.

For case (2) when one of the monomers has a low ceiling temperature, Equation (12.93) may be modified as

$$M_1^* + M_1 \xrightarrow{k_{11}} M_1 M_1^*$$

$$M_1^* + M_2 \xrightarrow{k_{12}} M_1 M_2^* \qquad (12.102)$$

$$M_2^* + M_1 \xrightarrow{k_{21}} M_2 M_1^*$$

By the QSSA,

$$\frac{d[M_1]}{d[M_2]} = \frac{1-F_2}{F_2} = r_{12}\frac{f_1}{f_2} + 1. \qquad (12.103)$$

where

$$f_1/f_2 = M_1/M_2.$$

or

$$F_2 = \frac{f_2}{2f_2 + r_{12}f_1} \qquad (12.104)$$

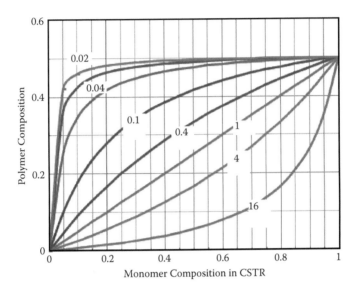

**FIGURE 12.3**  The copolymer composition as a function of monomer composition with one monomer with lower ceiling temperature.

If there was no ceiling temperature, this would have been

$$F_2 = \frac{r_{21}f_2^2 + f_1 f_2}{r_{21}f_2^2 + 2f_1 f_2 + r_{12}f_1^2}.$$

$$(12.105)$$

The composition of the monomer that has the lower ceiling temperature and polymer composition is shown in Figure 12.3 for eight different reactivity ratios.

### 12.5.2  HEAT OF COPOLYMERIZATION

For each of the three or four propagation reactions during copolymerization, when ceiling temperature effects are pronounced as indicated in Equation (12.102), there can be associated a heat of polymerization $H_{11}$, $H_{12}$, and $H_{21}$, respectively. The net heat of copolymerization can then be written as

$$\Delta H = n(P_{11}H_{11} + P_{12}H_{12})$$

$$(12.106)$$

where $n$ is the degree of polymerization and $P_{11}$, $P_{12}$, and $P_{21}$ are dyad probabilities. The dyad probabilities calculated in Chapter 11 can be used in Equation (12.106), and

$$\Delta H = n \left( \frac{r_{12}[M_1]H_{11}}{r_{12}[M_1]+[M_2]} + \frac{[M_2]H_{12}}{r_{12}[M_1]+1} \right)$$

$$(12.107)$$

where $[M_1]$, $[M_2]$ are the comonomer concentrations in the CSTR or instantaneous concentrations in a PFR, and $r_{12}$ and $r_{21}$ are the reactivity ratios. The conversion in the CSTR reactor in terms of the comonomer concentrations can be written as follows:

$$X_1 = \frac{[M_1]}{[M_1]+[M_2]}$$

$$X_2 = 1 - X_1 = \frac{[M_2]}{[M_1]+[M_2]}$$

(12.108)

The heat of copolymerization for random copolymers with pronounced ceiling temperature effect of monomer 2, given by Equation (12.107), can be written as

$$\Delta H = n\left(\frac{r_{12}X_1 H_{11}}{(r_{12}-1)X_1 +1} + \frac{(1-X_1)H_{12}[M]}{[M]r_{12}X_1 +1}\right)$$

(12.109)

where $[M]$ is the total concentration of the comonomers, $[M] = [M_1] + [M_2]$.

Four different heats of copolymerization dependence on conversion of comonomer 1 in CSTR can be seen, as shown in Figure 12.4. From bottom to top, the heats of copolymerization $H_{11}$ and $H_{12}$ are (1) both negative, (2) one negative and the other positive, (3) one zero and the other positive, or (iv) both positive. Further, it can be seen that there is a maxima in the heat of copolymerization at high conversions when both enthalpy of reactions are negative.

Similar equations can be written for entropy of copolymerization. The dyad probabilities can be used to estimate the probability of the type of bonds that form the random copolymer. The free energy of the copolymerization can be given by

$$\Delta G = \Delta H - T\Delta S.$$

(12.110)

The ceiling temperature of the copolymer can be derived in the same manner as for homopolymers as given by Equation (12.43). The conversions of the comonomer when the free energy expression becomes negative are the conditions when the reactions become favorable for occurrence. The stability of the reactions can be calculated from the spinodal condition that

$$\frac{\partial^2(\Delta G)}{\partial X_1^2} > 0.$$

(12.111)

The same treatment outlined earlier can be extended to multicomponent copolymers for $n$ monomers.

## 12.6 SUMMARY

During polymerization in some systems, the reverse step of depolymerization is favored as the propagation steps. The temperature above which the reverse depropagation reactions are favored over the forward propagation reactions is called the

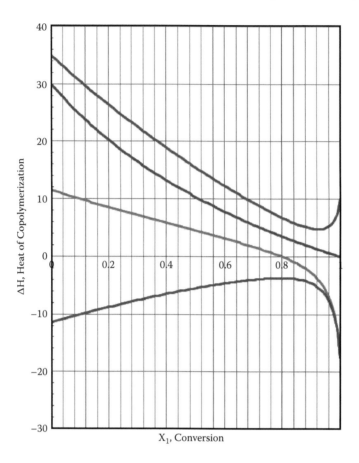

**FIGURE 12.4**   Heat of copolymerization as a function of conversion in CSTR.

*ceiling temperature,* and the temperature below which the forward propagation reactions are favored over the reverse depropagation reactions is referred to as *floor temperature.* The Clapeyron equation is derived for polymerization reactions from the combined first and second laws of thermodynamics. It relates the rate of change of equilibrium pressure and equilibrium temperature with the enthalpy change of reaction, volume change of reaction, and temperature of the reaction. The effect of pressure on the ceiling temperature was illustrated using a worked example. Enthalpy change was expressed as a function of heat capacity, which varies with temperature. The ceiling temperature was represented theoretically using Gibbs free energy and chemical potential. The equilibrium rate constant $K$ can be expressed as

$$\Delta G^0 = -RT \ln(K)$$

$$K = \frac{k_p}{k_{dp}} = \frac{1}{M_e}.$$

Free energy at standard state can be expressed in terms of enthalpy and entropy. The Gibbs free energy change of the reaction at any arbitrary temperature and pressure can be calculated from the Gibbs free energy change at standard state by extrapolation. Ceiling temperature, $T_c$, and entropy of polymerization for some monomers were tabulated.

Thermal polymerization kinetics that obey the Diels–Alder mechanism were studied. Expressions for rate of propagation and molecular weight were derived. The rate of propagation varies as the cube of monomer concentration. Subcritical damped oscillations can be expected under certain conditions. Reaction in a circle representation of free radical reactions was used to analyze the stability of reactions. Oscillations in concentration were discussed for systems of three and four reactions in a circle and the results generalized to $n$ reactions in a circle.

The ceiling temperature constraint in the homopolymerization of AMS can be circumvented by copolymerization with AN to prepare multicomponent random microstructures that offer higher heat resistance than SAN. The feasibility of a thermal initiation of free radical chain polymerization is evaluated by an experimental study of the terpolymerization kinetics of AMS–AN–Sty. Process considerations such as polyrates, molecular weight of polymer formed, sensitivity of molecular weight, molecular weight distribution, and kinetics to temperature were measured. The range of weight fractions of monomers of styrene, AN, and AMS varied in the study were $f_{sty} = 0.1 - 0.45$; $f_{AN} = 0.1 - 0.4$; $f_{AMS} = 0.2 - 0.75$, respectively. Empirical correlations are provided for the polyrate, and weight average molecular weight, $M_w$, in the temperature range 398 K–418 K.

Three types of copolymerization can be expected: (1) both monomers have high ceiling temperatures; (2) one monomer has a low ceiling temperature; (3) both monomers have low ceiling temperatures. The copolymer composition equation was modified for a system where reversible reactions are pronounced.

A mathematical model for heat of copolymerization as a function of chain sequence distribution for random copolymers was developed. Four different types of enthalpy of copolymerization, $\Delta H_p$ versus $X_1$, conversion can be expected when $\Delta H_{11}$ and $\Delta H_{12}$ are (1) both negative; (2) one negative and the other positive; (3) one zero and the other positive; and (4) both positive.

## PROBLEMS

1. Calculate the floor temperature of sulfur at 100 atm pressure.
2. Use Equation (12.43) and Tables 12.1 and 12.2 and calculate the enthalpy of polymerization at the ceiling temperature of styrene.
3. Extrapolate the results in Problem 2 to another pressure at 100 atm.
4. Discuss the effect of chain sequence distribution on the enthalpy of copolymerization of methylmethacrylate and styrene copolymer.
5. Discuss the effect of chain sequence distribution on the enthalpy of copolymerization of styrene acrylonitrile copolymer.
6. Discuss the effect of chain sequence distribution on enthalpy of terpolymerization of alphamethyl-styrene acrylonitrile and styrene.
7. Discuss the effect of chain sequence distribution on enthalpy of terpolymerization of alphamethyl-styrene methyl methacrylate and acrylonitrile.

8. Discuss the copolymer composition curve for sulfur and acrylonitrile at 100°C.

9. Discuss the effect of chain sequence distribution on enthalpy of copolymerization for sulfur and acrylonitrile at 100°C.

10. Discuss the copolymer composition curve for sulfur and alphamethylstyrene at 100°C.

11. Discuss the effect of chain sequence distribution on enthalpy of copolymerization for alphamethyl-styrene and sulfur at 100°C.

12. Derive the terpolymer composition equation when the reaction temperature is below the ceiling temperature of all three monomers.

13. Derive the terpolymer composition equation when the reaction temperature is below the ceiling temperature of two of the three monomers.

14. Derive the terpolymer composition equation when the reaction temperature is below the ceiling temperature of one of the three monomers.

15. Derive the terpolymer composition equation when the reaction temperature is below the ceiling temperature of none of the three monomers.

16. Express the heat of terpolymerization in terms of the chain sequence distribution of the terpolymer when the reaction temperature is below the ceiling temperature of all three monomers.

17. Express the heat of terpolymerization in terms of the chain sequence distribution of the terpolymer when the reaction temperature is below the ceiling temperature of two of the three monomers.

18. Express the heat of terpolymerization in terms of the chain sequence distribution of the terpolymer when the reaction temperature is below the ceiling temperature of one of the three monomers.

19. Express the heat of terpolymerization in terms of the chain sequence distribution of the terpolymer when the reaction temperature is below the ceiling temperature of none of the three monomers.

20. Derive an expression for ceiling temperature of copolymer in terms of the equilibrium monomer concentration, entropy of copolymerization, and enthalpy of copolymerization.

21. Derive the Clapeyron equation for copolymers with one monomer with a high ceiling temperature and one monomer with a low ceiling temperature.

22. Derive the Clapeyron equation for copolymers with both monomers with low ceiling temperatures.

23. Derive the Clapeyron equation for terpolymers with two monomers with low ceiling temperatures.

24. Derive the Clapeyron equation for terpolymers with three monomers with low ceiling temperatures.

25. Derive the Clapeyron equation for terpolymers with one monomer with low ceiling temperatures.

26. Discuss ideal reversible copolymerization when the product of the reactivity ratios is nearly one.

27. Discuss alternating reversible copolymerization when the product of the reactivity ratios is zero.

28. Discuss the enthalpy of copolymerization when one of the reactivity ratios is much higher compared with another reactivity ratio.

29. Discuss the enthalpy of copolymerization when one of the reactivity ratios is zero.

## REVIEW QUESTIONS

1. What does it mean when a polymer has both a floor temperature and a ceiling temperature?
2. What happens to the Clapeyron equation given by Equation (12.7) when the volume change of reactions is small and can be taken as zero?
3. What happens to the molecular weight distribution because of reversible polymerization?
4. Can the enthalpy change of copolymerization be expressed as joule/ mole of monomer or is it better to represent this information as joule/ mole of polymer?
5. What is the difference between reversible polymerization and attainment of a certain molecular weight and irreversible polymerization and attainment of the same molecular weight?
6. What does the relation, $\Delta G = 0$ mean for polymer systems?
7. What is the effect of pressure on the equilibrium rate constant?
8. What happens to the heat capacity of the polymer formed using reversible polymerization?
9. Can copolymerization circumvent the floor temperature effect?
10. What does the spinodal condition mean for polymerization reactions?
11. What are the three types of random copolymerization?
12. How can the theory of heat of copolymerization for random copolymers be extended to alternating copolymerization?
13. How is the heat of polymerization for condensation polymers different from random free radical copolymerization?
14. What is the meaning of entropy of copolymerization becoming positive?
15. Can the mathematical model developed for random copolymers for heat of copolymerization be extended to multicomponent copolymers with $n$ monomers?
16. What are the four different types of enthalpy of copolymerization behavior with conversion of copolymerization?

## REFERENCES

1. J. M. Smith, H. C. Van Ness, and M. M. Abbott, *Introduction to Chemical Engineering Thermodynamics,* 7th edition, McGraw Hill Professional, New York, 2005.
2. H. Sawada, *Thermodynamics of Polymerization*, Marcel Dekker, New York, 1976.
3. K. R. Sharma, A Statistical Design to Define the Process Capability of Continuous Mass Polymerization of ABS at the Pilot Plant, 214th ACS National Meeting, Las Vegas, NV, 1997.
4. K. R. Sharma, Analysis of Damped Oscillatory Kinetics in Simple Reactions in Circle, 35th Great Lakes Regional Meeting of the ACS, GLRM 03, Chicago, IL, June, 2003.
5. D. Kondepudi and I. Prigogine, *Modern Thermodynamics, From Heat Engines to Dissipative Structures*, John Wiley & Sons, New York, 1998.
6. Ans Magna (1501–1576), Solution to the Cubic Equation, Renaissance Mathematics, Internet Website.
7. K. R. Sharma, Thermal terpolymerization of alphamethyl styrene, acrylonitrile and styrene, *Polymer*, Vol. 41, 1305–1308, 2000.

8. K. R. Sharma and R. Chawla, Evaluation of Experimental Data with Theoretical Model for Thermal Polymerization of the Alphamethylstyrene Acrylonitrile Styrene, 31st ACS Central Regional Meeting, Columbus, OH, June 1999.

9. K. R. Sharma, Comparison of Experimental Data with Theoretical Model for the Terpolymerization of Alphamethylstyrene Acrylonitrile and Styrene, 32nd ACS Middle Atlantic Regional Meeting, Madison, NJ, May 1999.

10. K. Renganathan, Chain sequence distribution on the formation of contamination in alphamethylstyrene acrylonitrile copolymer, *Fifth World Congress of Chemical Engineering, Proceedings*, San Diego, CA, 1996.

11. N. Grassie and L. C. McNeil, *Polym. Sci.,* 30, 37, 1958.

12. C. E. Schweir and W. C. Wu, Kinetics of off-azeotrope copolymerization of α methylstyrene with acrylonitrile, *Polym. Chem.*, 32, 1, 553, 1991.

13. G. Odian, *Principles of Polymerization*, John Wiley & Sons, New York, 1981.

14. A. Husain and A. E. Hamielec, Thermal polymerization of styrene, *Appl. Polym. Sci.,* 22, 1207, 1978.

15. W. A. Pyror and L. D. Lasswell, in G. H. Williams (Eds). *Advances in Free Radical Chemistry*, Elsevier Science, London, 1975.

16. A. W. Hui and A. E. Hamielec, Thermal polymerization of styrene at high conversions and temperatures: An experimental study, *Appl. Polym. Sci.*, 16, 749, 1972.

# Appendix A: Maxwell's Relations

Five important measures of energy are introduced here. These are

1. Internal energy, $U$ (J/mole)
2. Enthalpy, $H$ (J/mole)
3. Gibbs free energy, $G$ (J/mole)
4. Helmholtz free energy, $A$, (J/mole)
5. Entropy, $S$ (J/K/mole)

These are also called *state functions*. Some important relationships among the state functions, $U$, $H$, $G$, $A$, and $S$, are as follows:

$$H = U + PV \qquad (A.1)$$

$$G = H - TS \qquad (A.2)$$

$$A = U - TS \qquad (A.3)$$

Therefore, $G$ may also be written as $A + PV$ or $U - TS + PV$. $A$ may also be written as $G - PV$.

The free energy, $G$, of a system, is the amount of energy that can be converted to work at constant temperature and pressure. It is named after the thermodynamicist Gibbs. Helmholtz free energy, $A$, of a system, is the amount of energy that can be converted to work at constant temperature. Enthalpy was first introduced by Clapeyron and Clausius in 1827, and represented the useful work done by a system. Entropy of a system, $S$, represents the unavailability of the system energy to do work. It is a measure of randomness of the molecules in the system. It is central to the quantitative description of the second law of thermodynamics. Internal energy, $U$, is the sum of the kinetic energy, potential energy, and vibrational energy of all the molecules in the system.

From the first law of thermodynamics, which will be formally introduced in Appendix B, it can be seen that

$$dQ + dW = dU, \qquad (A.4)$$

where $dQ$ is the heat supplied from the surroundings to the system, $dW$ is the work done on the system, and $dU$ is the internal energy change. When work is done by the system, $dW = -PdV$.

or

$$dQ - PdV = dU. \qquad (A.5)$$

For a reversible process from the mathematical statement of the second law of thermodynamics, it can be seen that $dQ = TdS$. Hence,

$$TdS - PdV = dU \qquad (A.6)$$

It may be deduced from Equation (A.6) that

$$\left(\frac{\partial U}{\partial S}\right)_V = T \qquad (A.7)$$

$$\left(\frac{\partial U}{\partial V}\right)_S = -P \qquad (A.8)$$

The reciprocity relation can be used to obtain the corresponding Maxwell relation. The reciprocity relation states that the order of differentiation does not matter. Thus,

$$\frac{\partial^2 U}{\partial S \partial V} = \frac{\partial^2 U}{\partial V \partial S}. \qquad (A.9)$$

Combining Equations (A.7) and (A.8) with Equation (A.9):

$$\left(\frac{\partial T}{\partial V}\right)_S = -\left(\frac{\partial P}{\partial S}\right)_V. \qquad (A.10)$$

In a similar fashion, expressions can be derived from $dH$ as follows:

$$dH = d(U + PV) = dU + PdV + VdP \qquad (A.11)$$

Equation (A.5) is the first law of thermodynamics.

$$dH = dQ - PdV + PdV + VdP = dQ + VdP = TdS + VdP. \qquad (A.12)$$

It may be deduced from Equation (A.12) that

$$\left(\frac{\partial H}{\partial S}\right)_P = T \qquad (A.13)$$

$$\left(\frac{\partial H}{\partial P}\right)_S = V. \qquad (A.14)$$

The second derivative of the four measures of energy, i.e., internal energy, $U$, enthalpy, $H$, Gibbs free energy, $G$, and Helmholz free energy, $A$, can be obtained with respect to two independent variables from among temperature, pressure, volume, and entropy. The order of differentiation does not matter as long as the function is "analytic." This property is used to derive the Maxwell relations as follows:

$$\frac{\partial^2 H}{\partial S \partial P} = \frac{\partial^2 H}{\partial P \partial S}. \qquad (A.15)$$

Combining Equations (A.13) and (A.14) with Equation (A.15):

$$\left(\frac{\partial T}{\partial P}\right)_S = \left(\frac{\partial V}{\partial S}\right)_P.$$

(A.16)

In a similar fashion, the corresponding Maxwell relation can be derived from $dG$:

$$dG = d(H - TS) = dH - TdS - SdT$$

(A.17)

Combining Equation (A.17) with the first law of thermodynamics given by Equation (A.12):

$$dG = TdS + VdP - TdS - SdT$$

$$= VdP - SdT.$$

(A.18)

It may be deduced from Equation (A.18) that

$$\left(\frac{\partial G}{\partial P}\right)_T = V$$

(A.19)

$$\left(\frac{\partial G}{\partial T}\right)_P = -S.$$

(A.20)

The order of differentiation does not matter when the property is an analytic function of the two variables under consideration. Thus:

$$\frac{\partial^2 G}{\partial P \partial T} = \frac{\partial^2 G}{\partial T \partial P}.$$

(A.21)

Combining Equations (A.19) and (A.20) with Equation (A.21):

$$\left(\frac{\partial V}{\partial T}\right)_P = -\left(\frac{\partial S}{\partial P}\right)_T.$$

(A.22)

In a similar fashion, the corresponding Maxwell relation can be derived from $dA$

$$dA = d(U - TS) = dU - TdS - SdT.$$

(A.23)

Combining Equation (A.23) with the first law of thermodynamics given by Equation (A.6):

$$dA = TdS - PdV - TdS - SdT$$

$$= -PdV - SdT.$$

(A.24)

It may be deduced from Equation (A.24) that

$$\left(\frac{\partial A}{\partial V}\right)_T = -P \tag{A.25}$$

$$\left(\frac{\partial A}{\partial T}\right)_V = -S. \tag{A.26}$$

The order of differentiation does not matter when the thermodynamic property is an analytic function of the independent variables. Thus:

$$\frac{\partial^2 A}{\partial V \partial T} = \frac{\partial^2 A}{\partial T \partial V}. \tag{A.27}$$

Combining Equations (A.25) and (A.26) with Equation (A.27):

$$\left(\frac{\partial P}{\partial T}\right)_V = \left(\frac{\partial S}{\partial V}\right)_T. \tag{A.28}$$

# Appendix B: Five Laws of Thermodynamics

## B.1 PIONEERS IN THERMODYNAMICS

Jack Daniels set his air conditioner at 63°F and poured himself a glass of freshly squeezed orange juice from his refrigerator and, upon shutting off the incandescent lamps powered by the electric supply, drove his Jeep Cherokee Laredo with an 8 cylinder engine to work. The air-conditioner, refrigerator, power plant, and the internal combustion engine that is central to the automobile are all examples of heat engines. *Heat* is converted to *work* or work to heat at an *efficiency* less than one. Chemical engineering thermodynamics is the study of the laws, methods, and issues involved in the design and operations of such machines. The first, second, and third laws of thermodynamics will be developed and used to design these machines. The refrigerator is an example of a *Carnot* refrigerator, the electrical supply may come from a steam power plant operated in a *Rankine cycle,* the automobile is an example of an *Otto* cycle, and the air conditioner an example of a cooling device in thermal management.

The onset of the industrial revolution was marked by the invention of *steam engines.* Thermodynamics was developed in the nineteenth century based upon the need to describe the operation of steam engines and to set forth the limits of what the steam engines could accomplish. The railroad was an insignia of the British Empire. The laws that govern the development of *power* from heat and the applications of heat engines were discussed in this new discipline. The first and second laws of thermodynamics deals with internal energy, $U$ (J/mole), heat, $Q$ (J/mole), work done, $W$ (J/mole), and entropy, $S$ (J/K/mole). These are all *macroscopic* properties. They do not reveal microscopic mechanisms. System and surroundings are defined prior to application of the laws of thermodynamics.

The fundamental dimensions that would be used are as follows:

1. Length, $L$ (m)
2. Time, $t$ (s)
3. Mass, $M$ (kg or mole)
4. Temperature, $T$ (K)

The International System of Units, SI, is preferred in textbooks. A meter is defined as the distance traveled by light in vacuum during $1/299,792,458$ th of a second. A kg (kilogram) is set as the mass of a platinum/iridium cylinder kept at the International Bureau of Weights and Measures at Sev'res, France. Kelvin is a unit of temperature and is given as $1/273.16$ th of the thermodynamic temperature of the triple point of water. The amount of substance with as many molecules as there are atoms in 0.012 kg of $C_{12}$ is 1 gram mole of the substance. One gram mole of any substance consists of the Avogadro number of molecules (6.023 E 23 molecules/mole).

The word *thermodynamics* is coined from the Greek; *theme* means heat, and *dynamis* represents power. *Heat* means energy in transit and *dynamis* relates to movement. Thermodynamics is a branch of physics where the effects of changes in temperature, pressure, and volume on physical systems are studied at the macroscopic scale by analyzing the collective motion of their particles by use of statistics. The essence of thermodynamics is the study of movement of energy and how energy instills movement. The study includes the discussion of the three laws of thermodynamics, efficiency of engine and refrigerator, entropy, equation of state, thermodynamic potential, internal energy, system, and surroundings. The term *thermodynamics* was coined by James Joule in 1858 to designate the science of relations between heat and power. The first book on thermodynamics was written in 1859 by William Rankine, who was originally a trained physicist. He taught at University of Glasgow as a civil and mechanical engineering professor. Thermodynamics may be classified as classical thermodynamics and statistical thermodynamics.

Otto von Guericke designed the world's first vacuum pump in 1650. Robert Boyle and Robert Hooke built an air pump in 1656. Pressure exerted by a fluid was found to be inversely proportional to volume, according to the Boyle's law. Denis Pipin, an associate of Boyle, built a bone digester that was used to raise high pressure steam. The idea of a piston and cylinder emanated from Pipin, although Tom Savery built the first engine in 1697.

The father of thermodynamics is Sadi Carnot. He wrote Reflections on the Motive Power of Fire in 1824. This was a discourse on heat, power, and engine efficiency. The Carnot engine, Carnot cycle, and Carnot equations are named after him.

Credit is given to Rankine, Clausius, Thompson, and Kelvin for the three laws of thermodynamics. Chemical engineering thermodynamics is the study of interrelation of heat with chemical reactions or with a physical change of state within the laws of thermodynamics. J. W. Gibbs, between 1873 and 1876, authored a series of papers on the equilibrium of heterogeneous substances. He developed the criteria for whether a process would occur spontaneously. Graphic analyses and the study of energy, entropy, volume, temperature, and pressure were introduced. The early twentieth century chemists G. N. Lewis, M. Randall, and E. A. Guggenheim began to apply the mathematical methods of Gibbs to the analysis of chemical processes. Classical thermodynamics originated in the 1600s. The laws of thermodynamics were developed into the form we use today in the late 1800s. The preclassical period is the 250 years between 1600 and 1850. Thermometry originated first, and this was followed by the hypotheses of an *adiabatic wall* and led to calorimetry.

The preclassical period was filled with discussions that were confused and controversial. Galileo may be credited with the discovery of thermometry. Galileo attempted to quantitate the subjective experiences of hot and cold. In the Hellinistic era, air was known to expand upon supply of heat. Galileo used this in his bulb and stem device; we to this day use this device, which we call a *thermometer*. It used to be called a baro-thermoscope. Toricelli, a student of Galileo's, developed the barometer. He showed that the time taken to drain an open tank using an orifice at the bottom is proportional to the square root of the height of the fluid in the tank. Liquids used in the thermometer evolved from water, alcohol, and gas to mercury in the modern era. The thermometry needed two reference temperatures. These were chosen to be the freezing point of water and boiling point of water, both at atmospheric pressure. The temperature of a mixture of two liquids at two different temperatures may be obtained by calculating a weighted average of the two. Joe Black, in 1760, suggested a modification to the mixing rule by the use of the specific heat. He pointed out that *heat* and not temperature was conserved during the mixing process. These discussions form the subject of *metaphysics*. Twenty years later, Count Runford showed by experimentation that mechanical work was an infinite source of a calorie. He called for the revival of a mechanical concept of heat.

Only a century later did Maxwell, Boltzmann, and Gibbs connect the microscale energy to the macroscale calorimetry. In 1824, S. Carnot's ideas led to the replacement of caloric theory by the first and second laws of thermodynamics. The concepts of *heat reservoirs, reversibility,* and the requirement of a temperature difference to *generate work from heat* were introduced. Carnot cycle was analogous to a waterfall in a dam. In 1847, Helmhotlz came up with the principle of conservation of energy. Joule established the equivalence of mechanical, electrical, and chemical energy to heat. Caloric was later split into energy and entropy. Heat and work were forms of energy and were asymmetric. Entropy is conserved in a reversible process, and energy is conserved during a Carnot cycle. These developments occurred in 1850 when Clausius, Kelvin, Maxwell, Planck, Duhem, Poincare, and Gibbs presented their works.

## B.2  HEAT

Energy transfers from a hot body to a cold body in a spontaneous manner when brought in contact with each other. The degree of hotness or coldness is defined by a quantity called *temperature*. The units of temperature, $T$, of a system are °C, Celsius, or F, Fahrenheit. The conversion of Fahrenheit to Celsius can be given by

$$T(°C) = \frac{5(T(°F) - 32)}{9} \tag{B.1}$$

Thermometers are used to measure temperature. They are made of liquid-in-glass constructs. A uniform tube filled with a liquid, such as mercury and alcohol, is allowed to expand depending on the degree of hotness/coldness of the system under scrutiny and the length of the column is measured. The length of the column is calibrated against standard reference points such as the freezing point of water at atmospheric pressure at 0°C, and boiling point of water at atmospheric pressure 100°C. These two points are divided into 100 equal spaces called *degrees*.

The thermodynamic temperature scale is defined by the Kelvin scale.

The conversion of °C, degrees Celsius, to K, degree Kelvin, can be given by

$$T(°K) = T(°C) + 273.15 \tag{B.2}$$

The lower limit of the Kelvin scale is 0° K or −273.15°C. The International Temperature Scale of 1990, ITS-90, is used for calibration of thermometers. Fixed points used are the triple point of Hydrogen at $-259.35°C$ and the freezing point of silver at $961.78°C$. The Rankine temperature scale can be directly related to the Kelvin scale:

$$T(R) = 1.8 \, T \, (K). \tag{B.3}$$

$Q$ is the amount of heat in Joules that is transferred from surroundings into the system. Although the temperature difference is the driving force, the energy transfer is $Q$ in Joules of energy. The heat transfer is transient in nature. The study of heat transfer is a separate subject in itself and is discussed in detail elsewhere [7] and in Chapter 11. The modes of heat transfer, conduction, convection, and radiation and of late microscale mechanisms such as wave heat conduction is discussed in Chapter 9.

Calorie is defined as the quantity of heat when 1 g of water is heated or cooled by one unit of temperature. BTU, British thermal unit, is the quantity of heat that when transferred can effect a one degree Fahrenheit, $F$, change in one pound of water. The SI unit of energy is the joule. One joule equals 1 newton-meter, N.M.

The modern notion of heat stemmed from the experiments conducted by James P. Joule in 1850. He placed known quantities of water, oil, and mercury in an insulated container and agitated the fluid with a rotating stirrer. The amount of work done on the fluid by the stirrer and the temperature changes of the fluid were accurately recorded. He observed that a fixed amount of work was required per unit mass for every degree of temperature raised on account of stirring. A quantitative relationship was established between heat and work. Heat was recognized as a form of energy.

The concepts of *adiabatic wall* and *diathermal wall* are used in discussions about heat engines and heat and work interactions. Consider an object A at a temperature $T_A$ immersed in a fluid at a different temperature $T_B$. The temperature of an object A will attain the temperature of fluid B after a certain time. This is the transient response of a step change in temperature at the interfaces of object A. Should the temperature of object A be relatively unchanged after a certain time after the step change in temperature, the wall of object A separating it from fluid B is said to be an *adiabatic wall*. Should the temperature of object A reach the temperature of fluid B instantaneously, the wall separating object A from fluid B is said to be a *diathermal wall*. Depending on the thermal response characteristics of object A, the transient response of temperature $T_A$ to the fluid temperature $T_B$ for all other materials would lie somewhere in between the adiabatic wall and diathermal wall. This adiabatic wall and diathermal wall are idealizations that get used in thermodynamic discussions.

## B.3  SYSTEM, SURROUNDINGS, AND STATES OF A SYSTEM

A *closed system* is defined as a set of components under study, whose boundaries are impervious to mass flow. Surroundings are the rest of the universe other than the closed system. An open system is defined as a set of components under study whose boundaries permit mass flow across the interfaces. If the closed system is bounded by an adiabatic wall, then it is said to be an *isolated system*. Composite systems consist of two or more systems. Restraints are barriers in a system that do not permit certain changes. In a simple system, there are no adiabatic walls, impermeable walls, or external forces. The phase of a system is the state of matter it is in. The phase rule can be written as

$$F = C - P + 2, \tag{B.4}$$

where $F$ is the degrees of freedom, $C$ is the number of components, and $P$ is the number of phases in the system.

A thermodynamic state is defined as a condition of a system characterized by properties of the system that can be reproduced. States can be at stable, unstable, or metastable equilibrium. The states can be in nonequilibrium as well. Equilibrium states are those where the macroscale changes are invariant with time. These will figure in the discussions on fugacity and vapor–liquid equilibrium.

*For closed systems with prescribed internal restraints, there exist stable equilibrium states that are characterized by two independent variable properties completely in addition to the masses of the chemical species initially introduced.*

A change of state is characterized by a change of one property at the minimum. *Path taken* refers to the description of changes in the system during a change of state. When the intermediate values during a path are at equilibrium states, the path is said to be quasi-static.

*All systems with prescribed internal restraints will change in a fashion as to approach one and only one stable equilibrium state for each of the subsystems during processes with no net effect on the environment. The entire system is said to be in equilibrium.*

Properties of the system may be classified as primitive and derived. Experimental measurements define the primitive property of a system. Properties that can only be defined by changes in the state are derived properties. These can be derived from the primitive properties.

## B.4  PERPETUAL MOTION MACHINE

Some machines are devised and some processes are designed in a fashion that they are infeasible. They claim perpetual motion. They either violate the conservation of energy principle, or they disobey the Clausius inequality. The types of designs with sustained, undamped motion that violate the conservation of energy principle are referred to as Perpetual Motion Machines of the first kind (PMM1). The designs with sustained, undamped motion that violate the second law of thermodynamics are referred to as the Perpetual Motion Machines of the second kind (PMM2). Although the second law of thermodynamics will be formally introduced later, simply stated, no machine can be devised or no process can be designed whose sole effect is to convert all heat to work. Some heat will have to be discarded to the surroundings. Heat cannot flow from a low temperature to a higher temperature in a spontaneous fashion. Heat can only travel from a hot temperature to a cold temperature in a spontaneous manner.

### Example B.1    Water Screw Perpetual Motion Machine

Water from a tray falls and spins a water wheel. This powers a set of gears and pumps and gets the water back to the tray. Can this last forever?

No. Frictional effects will results in reduced water at the water wheel in subsequent cycles. Any design otherwise is a violation of PMM1. The law of conservation of energy is violated.

### Example B.2    Maxwell's Demon

Consider two containers filled with gas at the same temperature, $T$. When a molecule with a higher than average velocity in one container moves toward the wall separating the two containers, a gate-keeper demon opens the partition, grabs the molecule, and allows the molecule to reach the second container. On account of this the average velocity of the remaining molecules in container, would be lower, and hence the first container's temperature would have lowered from $T$. The molecules in the second container will have an average velocity higher than the initial on account of which the temperature of the second container is expected to rise. Heat has transferred from container A to container B. Is this a violation of the second law of thermodynamics?

### Example B.3    Brownian Ratchet

A gear referred to as a ratchet allows for rotation in one direction, and a pawl prevents rotation in the other direction. The ratchet is connected to a paddle wheel immersed in a bath at temperature $T_A$. The molecules undergo Brownian motion. The molecular collisions with the paddle wheel result in a torque on the ratchet. Continuous motion of the ratchet may be expected. Work can be extracted. With no heat gradient, work is extracted. Is this a PMM1 or PMM1?

### Example B.4    Bhaskara's Wheel

Bhaskara (1114–1185) was a twelfth century AD mathematician and astronomer. He headed up the astronomical observatory at Ujjain, India.

Several moving weights are attached to a wheel. The wheels fall to a position further from the center of the wheel after half a rotation (Figure B.1).

Since weights further from the center apply a greater torque, the wheel may be expected to rotate forever. Moving weights may be hammers on pivoted arms, rolling balls, mercury in tubes, etc. Is this a PMM1 or a PMM2?

### Example B.5    Self-Flowing Flask

Robert Boyle suggested that siphon action may be used to fill the flask by itself. Is this possible? Why or why not?

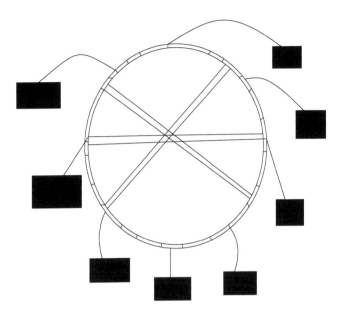

**FIGURE B.1** Bhaskara's unbalanced wheel.

## Example B.6    Orffyreus Wheel

In 1712, Bessler demonstrated a self-moving wheel that was later capable of lifting weights once set in motion. In 1717, he constructed a wheel 3.7 m in diameter and 14 in. thick. The wheel was found to be moving at 2 rpm after 2 weeks by officials. Where does the energy for the motion come from?

## Example B.7    Heat Capacity Relations for Real Gas

It can be shown that for any gas,

$$C_p - C_v = \left(\frac{\partial V}{\partial T}\right)_P \left(P + \left(\frac{\partial U}{\partial V}\right)_T\right). \tag{B.5}$$

Hint: Use the relation in change of variable in obtaining partial derivates:

$$\left(\frac{\partial f}{\partial x}\right)_z = \left(\frac{\partial f}{\partial x}\right)_y + \left(\frac{\partial f}{\partial y}\right)_x \left(\frac{\partial y}{\partial x}\right)_z.$$

## Example B.8

It can be shown that for any gas,

$$C_p - C_v = -\left(\frac{\partial P}{\partial T}\right)_V \left(\left(\frac{\partial H}{\partial P}\right)_T - V\right). \tag{B.6}$$

## Example B.9

Derive the Gibbs–Helmholtz equation given in the form

$$\Delta G = \Delta H + T \left(\partial(\Delta G)/\partial T\right)_{p1,p2,n1,n2}. \tag{B.7}$$

## Example B.10

It has been shown that

$$\Delta H = \Delta U + P \left(\partial(\Delta H)/\partial P\right)_s. \tag{B.8}$$

## Example B.11

It has been shown that

$$\Delta U = \Delta A + S \left(\partial(\Delta U)/\partial S\right)_v. \tag{B.9}$$

## Example B.12

It has been shown that

$$\Delta A = \Delta U + T \left(\partial(\Delta A)/\partial T\right)_v. \tag{B.10}$$

## Example B.13

Given that $\kappa$ is the compressibility factor $-1/V(\partial V/\partial P)$ and $\alpha$ the coefficient of thermal expansion, show that

$$\left(\frac{\partial S}{\partial V}\right)_T = \frac{\alpha}{\kappa}. \tag{B.11}$$

## B.5   FIVE LAWS OF THERMODYNAMICS

The commonly referred to and used five laws of thermodynamics are as follows;

**Zeroth Law of Thermodynamics:** *If two systems are in thermal equilibrium with a third system, then the two systems are in thermal equilibrium with each other.*

**First Law of Thermodynamics:** *The total quantity of energy is a constant and when energy is consumed in one form, it appears concurrently in* *another form. Energy exists in many forms.* The mathematical statement of the first law of thermodynamics for closed systems can be written as

$$Q + W = \Delta U. \tag{B.12}$$

$Q$ is the heat energy needed for the system supplied from the surroundings. $W$ is the work done on the system. When the work is done by the system, then a negative sign should precede the work contribution, Equation (B.12) $\Delta U$ is the internal energy change of the system. The sign convention used in Equation (B.12) is from the recommendations of IUPAC (the International Union of Pure and Applied Chemistry). In differential form, Equation (B.12) may be written as

$$\Delta Q + \Delta W = \Delta U. \tag{B.13}$$

For an open system, the first law of thermodynamics may be written as

$$Q + W - \Delta m(U + u^2/2 + gz) = \Delta\,(mU)/\Delta t, \tag{B.14}$$

where $u$ is the fluid flow rate across the control volume, and $z$ is the height of the fluid. The mass flow rate is $m$ in mole/s, heat and work rates are $Q$ and $W$.

In terms of enthalpy, $H = U + PV$, the first law for the open systems can be written as

$$d(mU)/dt + \Delta m(H + u^2/2 + zg) = Q + W. \tag{B.15}$$

**Second Law of Thermodynamics:** *No process can be affected and no machine can be devised whose sole effect is the complete conversion of heat absorbed to work done by the system. Some heat has to be discarded to the surroundings.*

or

*It is impossible to construct a process whose sole effect is the transfer of heat from a low temperature to a higher temperature. Heat flows spontaneously from higher temperature to a lower temperature and not from a lower temperature to a higher temperature.*

The first statement is the Kelvin–Planck statement of the second law of thermodynamics. As a corollary, it is not possible to affect a cyclic process that can convert heat absorbed by a system completely into work done by the system. Mathematically stated, the second law of thermodynamics can be written as

$$\Delta S_{tot} \geq 0. \tag{B.16}$$

Thus, each and every process proceeds in such a direction that the total entropy change associated with it is positive. In the limiting case of reversible operation, the entropy change would be zero. It is impossible to affect a process whose entropic change is negative. The inequality given by Equation (B.16) is also referred to as the Clausius inequality.

**Third Law of Thermodynamics:** The third law of thermodynamics was developed by Nernst and is referred to as the *Nernst postulate* or *Nernst theorem*. It states that entropy of pure substances approach zero when the temperature of the substance is brought to 0 Kelvin.

*If the entropy of each element in crystalline state with perfect structure be taken as zero at the absolute zero of temperature, every substance has a finite positive entropy; but at the absolute zero of temperature the entropy may become zero, and does so become in the case of crystalline substances with perfect structure.*

**Fourth Law of Thermodynamics:** The Onsager reciprocal relations are referred to as the fourth law of thermodynamics. The relation between forces and flows for systems not in equilibrium but in a state of local equilibrium is provided in the Onsager relations.

$$J_i = \sum_j L_{ij} F_j \tag{B.17}$$

$J$ represent the flows, $F$ the forces, and $L$ the phenomenological coefficients. $i$ and $j$ denote the different flows and forces. Thus, concentration difference may be one force, temperature difference another force, momentum difference another force, etc., and the flows can be heat transfer, mass transfer, and momentum transfer. Onsager showed that from analysis of a positive definite matrix, the cross-coefficients in Equation (B.17) have to be equal. Thus:

$$L_{ij} = L_{ji} \tag{B.18}$$

**Example B.14**

Evaluate the difference $\left(\dfrac{\partial U}{\partial T}\right)_P - \left(\dfrac{\partial U}{\partial T}\right)_V$

Assumption: Fluid is an ideal gas.

$$C_v = \left(\frac{\partial U}{\partial T}\right)_V \tag{B.19}$$

$$\left(\frac{\partial U}{\partial T}\right)_P = \left(\frac{\partial (H - PV)}{\partial T}\right)_P = C_P - P\left(\frac{\partial V}{\partial T}\right)_P \tag{B.20}$$

$$\left(\frac{\partial U}{\partial T}\right)_p - \left(\frac{\partial U}{\partial T}\right)_V = C_p - C_v - PV\beta = R(1 - \beta T), \qquad \text{(B.21)}$$

where $\beta$ is the compressibility factor.

## B.6   REVERSIBILITY AND EQUILIBRIUM

When two systems are closed by adiabatic walls except for one through which they come in contact with each other, the states of the two systems change for some time and cease after a while. This condition is referred to as the state of thermal equilibrium. When two systems are in thermal equilibrium each with a third system, they should also be in thermal equilibrium with each other. This shall be stated formally as the zeroth order of thermodynamics as Guggenheim introduced it.

Spontaneous transfer of heat, such as in the example stated earlier, is generally *irreversible* in nature. In addition to the weightless pulleys and frictionless planes, a reversible process is one where the changes in a series of states that are at equilibrium with each other. Change in a continuous succession of equilibrium states is said to be *reversible*. It is quasi-static. In a piston–cylinder assembly, the work done during a reversible process is more than during the irreversible one. When the weight in gauge is removed suddenly, the process would be irreversible. A reversible process would be more gradual.

Entropy can be defined during a reversible process as follows:

$$\Delta S = Q_{rev}/T$$

$$\text{(B.22)}$$

$$T dS = dQ.$$

For a irreversible process,

$$T dS > dQ. \qquad \text{(B.23)}$$

For a reversible process,

$$T dS = dQ. \qquad \text{(B.24)}$$

# Appendix C: Glass Transition Temperature

One of the properties that is unique to polymers is the glass transition phenomena and existence of the *glass transition temperature*. Often, polymers are characterized by three thermal transition temperatures:

1. Glass transition temperature, $T_g$
2. Crystalline melting temperature, $T_m$
3. Degradation temperature, $T_d$

Upon gradual heating at some temperature, the crystalline domains in the polymer will begin to melt. This temperature is called the *crystalline melting temperature* [1]. At the glass transition temperature, the amorphous domains of the polymer assume the glassy state properties. Prior to this temperature, the polymer was in a rubbery state. The glassy state is characterized by brittleness, stiffness, and rigidity. Upon cooling of the hot polymer, the translational, rotational, and vibrational energies of the polymer molecules tend to decrease. When the energy state of the polymer reaches a point where the translational and rotational energies are for all practical purposes zero, crystallization is possible.

Depending on the symmetry, the molecules tend to form into an organized lattice arrangement at the onset of crystallization. Crystallinity is a salient property in polymers. It affects the polymer's mechanical and optical properties. Crystallinity evolves during the processing of polymers as a result of temperature changes and applied stress. One example here is the formation of PET (polyethylene terepthalate) bottles.

When crystalline regions in the polymer become large enough, they begin to scatter light and make the polymer translucent. In some polymers, delocalized regions crystallize in response to an applied electric field. This phenomenon is responsible for liquid crystalling polymers (LCP) display. Increase in crystallization increases density, resistance to chemical attack, and mechanical properties even at high temperatures. The stronger bonds between the chains contribute to the improved mechanical properties at high temperature. Crystallization phenomena can be used to explain the occurrence of biaxial texture in PET bottles and texture strengthening in nylon fibers [2].

When temperature is changed, the properties of thermoplastics change. Critical temperatures include the melting temperature, glass transition temperature, etc. Chains that were mobile readily stiffen upon melting. In crystalline solids, the chains have difficulty with movement, and, in amorphous solids, the chains move under stress. Beyond the glassy temperature, only local movement of chain segments is allowed. Below the melting temperature, the thermoplastics can be either amorphous or crystalline. Often, engineered thermoplastics contain regions that are amorphous and crystalline. Crystallinity is introduced by slow cooling or by application of stress that can unentangle the chains by a process referred to as *stress-induced crystalliza-tion*. Formation of crystalline regions in the otherwise amorphous solid results in strengthening of thermoplastics. The bonds within the polymer chains are by cova-lent bonds, and bonds between the chains are made by weak van der Waals forces. Upon application of tensile stress, the weak bonds between the chains are overcome and the chains can rotate and slide relative to one another.

A semi-logarithmic plot of modulus of elasticity and temperature reveals rigid, leathery, rubbery, and viscous regions. At very high temperatures, the polymer may begin to *degrade* or decompose. At temperatures above the degradation tem-perature $T_d$, the covalent bonds in the linear chain may be destroyed and the poly-mer may begin to burn or char. Heat stabilizers may be added to delay burning plastics. Polymer degradation may be by other means, such as exposure to oxygen and ultraviolet rays. Carbon black is added to improve the resistance of plastics to UV degradation.

The melting temperature for thermoplastics is not one sharp value but is a gradual range of temperatures. Below the melting temperature, the polymer chains are in a twisted and intertwined state. Below the melting temperature, the polymer behaves in a rubbery manner. When stress is applied, both elastic and plastic deformation of the polymer occurs. At lower temperatures, the bonding between the chains is stron-ger, and the polymer becomes stiffer and stronger, and a leathery state is observed.

Below the glass transition temperature, the polymer becomes brittle, hard, and is glasslike. The glass transition temperature, $T_g$, is 0.5–0.75 times the $T_m$, the melting temperature value. Thermoplastics become brittle at lower temperatures. The brittle-ness of the polymer used for some of the O-rings ultimately may have caused the 1986 Challenger disaster [3]. The lower weather temperature at the time of launch may have caused the embrittlement of the rubber used as O-rings for the booster rockets. The DBTT, ductile-to-brittle transition temperature, has become an impor-tant consideration in design of O-rings since then.

## C.2   GLASS TRANSITION TEMPERATURE OF COPOLYMERS

In industrial practice, in order to obtain a product with a dial-in glass transition tem-perature, $T_g$, copolymerization of two different monomers $A$ and $B$ are undertaken such that $T_{gA} \le T_g \le T_{gB}$.

The glass transition temperature, $T_g$, of the random copolymer, $T_g$ lies between that of the glass transition temperature of the homopolymers of $A$ and $B$, $T_{gA}$ and $T_{gB}$, respectively. Mathematical models that can predict the glass transition temperature of the copolymer as a function of the copolymerization parameters have been devel-oped in the literature. Some of the models are presented in the following text.

The Fox equation [4] for glass transition temperature of copolymers is as follows:

$$\frac{1}{T_g} = \frac{w_A}{T_{gA}} + \frac{w_B}{T_{gB}}, \tag{C.1}$$

where $w_A$ and $w_B$ are the weight fractions of the repeat units A and B in the random copolymer, respectively. The Fox equation is a special case of the Gordon and Taylor [5] equation given below:

$$T_g = \frac{w_A T_{gA} + K w_B T_{gB}}{w_A + K w_B}.$$

(C.2)

The parameter $K$ is depends on the model. Thus,

$$K_{G-T} = \frac{\rho_1}{\rho_2} \frac{\Delta\alpha_2}{\Delta\alpha_1}.$$

(C.3)

Equation (C.3) is applicable for the Gordon–Taylor volume additivity model, and

$$K_{DM-G} = \frac{\mu_1 \gamma_2}{\mu_2 \gamma_1}.$$

(C.4)

Equation (C.4) is applicable for the Di Marzio–Gibbs [6] flexible bond additivity model, where $\rho$ refers to the component density, $\mu$ and $\gamma$ refers to the masses and the numbers of flexible bonds of the monomers, and $\Delta\alpha = \alpha_{\text{melt}} - \alpha_{\text{glass}}$ the increments of thermal expansion coefficient at the glass transition temperature. Equation (C.2) reverts to Equation (C.1) when the densities of the two commoners are equal and the constant $K_{G-T}$ for volume additivity can be approximated as $K_F = T_{gA}/T_{gB}$. Equations (C.1) and (C.2) have also been used to predict glass transition temperatures of binary polymer blends as well.

Johnston [7] took into account the chain sequence distribution of copolymers in the prediction of the glass transition temperatures. Sharma [8], and Sharms et al. [9] found that this model compared well with experimental data even for systems obtained from reversible copolymerization. Johnston's dyad model for prediction of copolymer glass transition temperature can be written as follows:

$$\frac{1}{T_g} = \frac{w_A P_{AA}}{T_{gA}} + \frac{w_B P_{BB}}{T_{gB}} + \frac{w_{AB} P_{AB} + w_{BA} P_{BA}}{T_{gAB}}.$$

(C.5)

The dyad probabilities can be calculated as explained in Chapter 11. The glass transition temperature of polymer blends is discussed in Chapter 6.

## C.3 THERMODYNAMICS OF GLASS TRANSITION

van Krevelen and te Nijenhius [10] gave an expression for $(C_p - C_v)$, the difference in heat capacity and te Nijenhius measured as constant pressure and constant volume for polymers as follows:

$$C_p - C_v = \frac{\beta^2 V T}{\kappa},$$

(C.6)

where $\beta$ and $\kappa$ are the volume expansivity and compressibility, respectively, as defined in Chapter 2. Should the experimental data for the evaluation of Equation (C.6) not be available, a universal expression can be used [10]:

$$C_p - C_v = 0.00511 \frac{C_p^2 T}{T_m}. \tag{C.7}$$

Although the glass transition temperature is not a thermodynamic transition point, it bears resemblance to second-order Ehrenfest transition:

$$\frac{\partial T_g}{\partial p} = \frac{T_g \Delta\beta}{\Delta C_p} = \frac{\Delta\kappa}{\Delta\beta}. \tag{C.8}$$

Other investigators discussed at length the thermodynamics of glass transition and concluded that this phenomenon is not a real second-order transition because the glassy state is not completely characterized by $P$, $V$, and $T$. Experimental data showed large deviations from Equation (C.8).

Glassy state attainment is widespread and has been found in organic liquids, biomaterials, inorganic melts, metals, and alloys. The vitrification process has been studied theoretically since 1930. Specific heats were measured, as well as entropies of glycerol in the liquid, crystalline, and glassy state. The entropy of supercooled liquid below $T_g$ can only be estimated. Linear extrapolation leads to negative entropy at zero temperature. This is a violation of the second law of thermodynamics. This is referred to as the *Kauzmann paradox*.

Di Marzio and Gibbs [6] made enormous strides in the theoretical basis for glass transition temperature. They came within 50 K of experimental observations when making theoretical predictions.

## REFERENCES

1. G. Odian, *Principles of Polymerization,* 3rd edition, Wiley Interscience, New York, 1991.
2. D. R. Askeland and P. P. Phule, *The Science and Engineering of Materials,* Thompson, Toronto, Canada, 2006.
3. S. D. Sagan *The Challenger Launch Decision: Risky Technology, Culture and Deviance at NASA*, University of Chicago Press, Chicago 1996.
4. T. G. Fox and P. J. Flory, *J. Appl. Phys.* 21, 581, 1950.
5. M. Gordon and J. S. Taylor, *J. Appl. Chem, USSR*, 2, 493, 1952.
6. E. A. Di Marzio and J. H. Gibbs, *Polym. Sci.*, 40, 121, 1959.
7. N. W. Johnston, *J. Macromol. Sci. Rev. Macromol. Chem.*, C14, 215, 1976.
8. K. R. Sharma, Glass Transition Temperature of Copolymers and Blends: A Survey of Mathematical Models, 214th ACS National Meeting, Las Vegas, NV, September 1997.
9. K. R. Sharma, V. K. Sharma, and M. Ibrahim, Comparison of Experimental Data on Glass Transition Temperature with the Entropic Difference Model for Copolymers, 2nd Annual Green Chemistry and Engineering Conference, National Academy of Sciences, Washington, DC, July 1998.
10. D. W. van Krevelen and K. te Nijenhius, *Properties of Polymers: Their Correlation with Chemical Structure; Their Numerical Estimation and Prediction from Additive Group Contributions*, 4th edition, Elsevier, Amsterdam, Netherlands, 2009.

# Appendix D: Statistical Distributions

## D.1  PROBABILITY AND STATISTICS

A gambler's dispute in 1654 led to the creation of the *mathematical theory of probability.* This was accomplished by two pioneers, B. Pascal and P. De Fermat. A French noble-man with an interest in gambling called Pascal's attention to an apparent contradiction concerning a popular dice game. The game consisted of throwing a pair of dice one at a time. The problem was to decide whether or not to bet even money on the occurrence of at least one "double six" during 24 throws. A well-established gambling rule led the nobleman to believe that betting on a double six in 24 throws would be profitable, but his own calculations indicated just the opposite. This problem, among others led to an exchange of famous letters between Pascal and Fermat in which the fundamental prin-ciples of probability theory were formulated for the first time. The probability of a double six on two throws can be calculated for a fair six-sided pair of dices as

$$P(X = (6,6)) = 1/6*1/6 = 1/36 \tag{D.1}$$

Thus, in 36 throws, the chances of occurrence of a double six are once. Equation (D.1) takes into account the occurrence of two independent events and that all values will occur with equal likelihood. Bernoulli and de Moivre were big contributors as the subject developed rapidly in the eighteenth century. In 1912, P. de Laplace pub-lished his book. *Theorie Analytique des Probabilities.* This widened the scope of probability to many scientific and practical problems. The theory of errors, actuarial mathematics, and statistical mechanics are some examples of applications developed in the nineteenth century. Mathematical statistics is one important branch of applied probability with applications in a wide variety of fields such as genetics, psychology, economics, and engineering. Important contributors to probability since Laplace were Chebychev, Bell, Markov, Von Mises, and Kolmogorov. The search for a widely acceptable definition of probability took nearly three centuries. This was resolved finally by the axiomatic approach suggested by Kolmogorov.

## D.2  THREE DEFINITIONS OF PROBABILITY

The *classical definition of probability* states that the probability $P(A)$ of an event A is determined a priori without actual experimentation. It is given by

$$P(A) = \frac{N_A}{N}, \tag{D.2}$$

where $N$ is the number of possible outcomes and $N_A$ is the number of outcomes that are favorable to the event $A$. In the die experiment, $A$ is the double 6 and $N$ is 36.

The *axiomatic definition of probability* uses the set theory. A certain event $X$ is the event that occurs in every trial. The union $A + B$ of two events $A$ and $B$ is the event that occurs when $A$ or $B$ both occur. The intersection $AB$ of the events $A$ and $B$ is the event that occurs when both events $A$ and $B$ occur. The events $A$ and $B$ are mutually exclusive if the occurrence of one of them excludes the occurrence of the other. Three postulates are given. The probability $P(A)$ of an event $A$ is

$$P(A) \geq 0. \tag{D.3}$$

The probability of the certain event equals 1. If the events $A$ and $B$ are mutually exclusive, then

$$P(A + B) = P(A) + P(B). \tag{D.4}$$

The axiomatic approach is credited to Kolmogorov. The *relative frequency approach to the definition of probability* states that the probability $P(A)$ of an event $A$ is the limit:

$$P(A) = Lim_{n \to \infty} \frac{n_A}{n}, \tag{D.5}$$

where $n_A$ is the number of occurrences of $A$ and $n$ is the number of trials. Probabilities are used to define frequencies and are defined as limits of such frequencies. Both $n_A$ and $n$ must be large. This approach was suggested by von Mises.

## D.3 CONDITIONAL PROBABILITY AND BAYES' THEOREM

The conditional probability of an event $A$ given the event $G$, denoted by $P(A/G)$, is defined by

$$P(A/G) = P(AG)/P(G) \text{ assume that } P(G) \text{ is not zero.} \tag{D.6}$$

If $G$ is the subset of $A$, then $P(A/G) = 1$. If $A$ is a subset of $M$, then

$$P(A/G) = P(A)/P(G) \geq P(A). \tag{D.7}$$

Bayes proposed a theorem in 1763 that was later named after him. Laplace gave its final forms years later and can be stated as follows:

$$P(A_i/B) = \frac{P(B/A_i)P(A_i)}{P(B/A_1)P(A_1) + \ldots\ldots + P(B/A_n)P(A_n)}. \tag{D.8}$$

The conditional probability can also be written in terms of intersection of sets as

$$P(A/G) = \frac{P(A \cap G)}{P(G)} \tag{D.9}$$

$$P(A \cap G) = P(A/G)P(G) = P(G/A)P(A). \tag{D.10}$$

The Bayes theorems can be stated then as

$$P(A/G) = \frac{P(G/A)P(A)}{P(G)}. \tag{D.11}$$

## D.4  INDEPENDENT EVENTS AND BERNOULLI'S THEOREM

Two events A and B are said to be independent if

$$P(A\cap B) = P(A)P(B) \tag{D.12}$$

Suppose one repeatedly runs independent trials of an experiment in $1 - p$. Then, the probability that there are exactly $k$ successes in these $n$ trials is given by $^nC_k\, pkqn{-}k$. Let $A$ and $B$ be small positive numbers, then there is a value of $n$ large enough, so that the probability that the ratio of the successes in $n$ trials is not within $A$ and $p$ is less than $B$. In other words, if the experiment is run long enough, the fraction of successes is likely to be close to the correct probability.

## D.5  DISCRETE PROBABILITY DISTRIBUTIONS

### D.5.1  BINOMIAL AND MULTINOMIAL DISTRIBUTIONS

The binomial distribution gives the discrete probability distribution of obtaining $n$ successes out of $N$ Bernoulli trials. The result of each Bernoulli trial is true with probability $p$ and false with probability $q = 1 - p$ (Figure D.1).

$$f(x) = P(X = x) = {}^nC_x\, p^x\, q^{n-x} \tag{D.13}$$

$$= n!/x!(n - x)!\, p^x\, q^{n-x} \tag{D.14}$$

$$= x = 0,1,2,\ldots, n$$

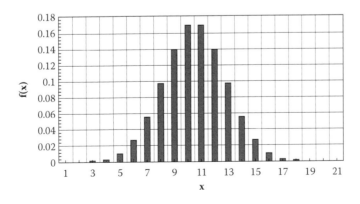

**FIGURE D.1**  Binomial distribution with $n = 21$; $p = q = 0.5$.

The mean, variance, skewness, and kurtosis of the binomial distribution are given in the following equations:

$$\text{Mean, } \mu = np$$

$$\text{Variance, } \sigma^2 = npq \tag{D.15}$$

$$\text{Skewness, } \alpha_3 = (q - p)/(npq)^{1/2}$$

$$\text{Kurtosis, } \alpha_4 = 3 + (1-6pq)/(npq)1/2$$

$A1, A2, \ldots, Ak$ events can occur with probabilities $p1, p2,\ldots, pk$ where, $p1 + p2 +\ldots+ pk = 1$. $X1, X2,\ldots, Xk$ are random variables, respectively, giving the number of times that $A1, A2.,\ldots, Ak$ can occur in a total of $n$ trials so that $X1 + X2 +\ldots\ldots+ Xk = n$. Then, the multinomial distribution can be given by

$$P(X1 = n1, X2 = n2,\ldots, Xk = nk) = n!/n1!n2! ..nk! \, p1n1p2n2...pknk \tag{D.16}$$

where $n1 + n2 +,\ldots,+ nk = n$. The joint probability function for random variables $X1 + X2 +,\ldots, +Xk = n$. This is the general form of binomial distribution and the general term in the multinomial expression of $(p1 + p2 +,\ldots,+pk)n$.

## D.5.2  POISSON DISTRIBUTION

$$f(x) = P(X = x) = \lambda^x \exp\left(\frac{-\lambda}{x!}\right), \, x = 0,1,2,\ldots,n \tag{D.17}$$

$$\text{Mean, } \mu = \lambda$$

$$\text{Variance, } \sigma^2 = \lambda$$

$$\text{Skewness, } \alpha_3 = 1/\lambda^{1/2}$$

$$\text{Kurtosis, } \alpha_4 = 3 + 1/\lambda$$

A good example of Poisson distribution is the number of typos generated by a good typist. The probability of 1 or 2 typos per page is high, and the probability of generating 10 typos per page is slim. In a similar fashion, the time taken at the teller counter at the bank can also be fit to a Poisson distribution (Figure D.2). The probability of the event of the transaction taking 5 minutes or 8 minutes may be high, and the probability of it taking 1 hour will be low. Yet another example is the arrival of students late to class. The probability of the students arriving on time or 5 minutes before the hour is high, and the chances that they will arrive ½ hour late will be on the lower side.

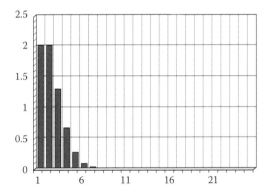

**FIGURE D.2**   Poisson distribution with $\lambda = 2.0$.

### D.5.3   HYPERGEOMETRIC DISTRIBUTION

This is an example of sampling with replacement. Suppose a box contains $b$ blue marbles and $r$ red marbles. Let us perform $n$ trials of an experiment in which a marble is chosen at random, its color is observed, and the marble is put back in the box.

$$f(x) = P(X = x) = {}^bCx \; {}^rC_n - x/{}^{b+r}C_n \tag{D.18}$$

$$\text{Mean, } \mu = nb/(b+r)$$

$$\text{Variance, } \sigma^2 = nbr \; ( b + r - n)/(b+ r)2/(b+ r+1).$$

### D.5.4   GEOMETRIC DISTRIBUTION

$$f(x) = P(X = x) = p \; q^{x-1}, x = 1,2,... \tag{D.20}$$

$$\text{Mean, } \mu = 1/p$$

$$\text{Variance, } \sigma^2 = q/p2$$

The chain sequence distribution in copolymers was shown to be modeled using the geometric distribution.

## D.6   CONTINUOUS PROBABILITY DISTRIBUTIONS

### D.6.1   UNIFORM DISTRIBUTION AND CAUCHY DISTRIBUTION

$$f(x) = \frac{1}{(b-a)} \quad a \le x \le b$$

$$= 0, \text{ otherwise} \tag{D.21}$$

$$\text{Mean, } \mu = \tfrac{1}{2}(a + b)$$

$$\text{Variance, } \sigma^2 = 1/12 \; (b - a)2$$

A good example of uniform distribution is the contact time of solid particles at the heat exchanger surfaces in a circulating fluidized bed boiler (CFB). The Cauchy distribution is given by

$$f(x) = \frac{a}{\pi(x^2 + a^2)}, \quad a > 0, \ -\infty < x < \infty \tag{D.22}$$

Mean, $\mu = 0$

Variance and higher moments do not exist.

This distribution is also called *Lorentz distribution* by physicists. It forms the solution to the differential equation that can be used to describe forced resonance.

### D.6.2   GAMMA AND CHI-SQUARED DISTRIBUTIONS

$$f(x) = x^{\alpha-1} \frac{\exp\left(\dfrac{-x}{\beta}\right)}{\beta^{\alpha}\Gamma(\alpha)} \quad x > 0 \tag{D.23}$$

$$= 0, \ x \le 0$$

where $\Gamma(\alpha)$ is the gamma function.

Mean, $\mu = \alpha$

Variance, $\sigma^2 = \alpha\beta^2$

$\Gamma(n) = \int_0^\infty t^{n-1} \exp(-t)dt, \ n > 0$

$\Gamma(n + 1) = n\Gamma(n)$                       (recurrence formula)

$\Gamma(1) = 1$

when $n$ is a positive integer, then

$\Gamma(n + 1) = n!$

$X1, X2,...,$ and $X\gamma$ are $\gamma$ independently normally distributed random variables with mean zero and variance 1. Consider the random variable

$$\chi2 = X12 + X22 + ...........+ X\gamma2 \tag{D.25}$$

where $\gamma$ is the number of degrees of freedom. A special case of gamma distribution with $\alpha = \gamma/2, \ \beta = 2$.

Mean, $\mu = \gamma$

$$\text{Variance, } \sigma^2 = 2\gamma$$

$$f(x) = x(\gamma/2) - 1\exp(-x/2)/[2\gamma/2 \ \Gamma \ (\gamma/2), \ x > 0 \tag{D.26}$$

$$= 0, \ x \le 0$$

This is the chi-squared distribution.

## D.6.3  STUDENT T DISTRIBUTION

$$f(t) = \frac{\Gamma\dfrac{(\gamma+1)}{2}(1+\dfrac{t^2}{\gamma})^{-(\gamma+1)/2}}{\sqrt{\gamma\pi}\,(\Gamma(\dfrac{\gamma}{2})}, \quad -\infty < t < \infty \tag{D.27}$$

$$\text{Mean, } \mu = 0$$

$$\text{Variance, } \sigma^2 = \gamma/(\gamma - 2), \ (\gamma > 2)$$

where $\gamma$ is the number of degrees of freedom.

## D.6.4  NORMAL DISTRIBUTION

$$f(x) = \left(\frac{1}{\sigma\sqrt{2\pi}}\right)\exp{-\frac{1}{2}\left(\frac{x-\mu}{\sigma}\right)^2} \tag{D.35}$$

The normal distribution was first introduced by de Moivre, who approximated binomial distributions for large $n$ (Figure D.3). His work was extended by Laplace, who used the normal distribution in the error analysis in experiments conducted. Legendre came up with the method of least squares. Gauss, by 1809, justified the normal distribution for experimental errors. The name *bell curve* was coined by Galton and Lexis.

## D.6.5  GENERALIZED NORMAL DISTRIBUTION

The generalized normal distribution was introduced by Sharma [21] to capture the periodicity in pressure fluctuations in addition to the random component. In addition to the mean and standard deviation, the number of saddle points can also be used to characterize the periodicity of pressure fluctuations.

$$f(x) = A \ \exp(-Bx - Cx2 - Dx4) \tag{D.36}$$

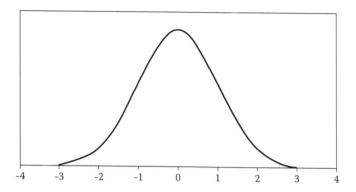

**FIGURE D.3**  Normal distribution with zero mean and unit variance.

$A, B, C,$ and $D$ are parameters that can be obtained by a least squares fit of the experimental data. This can also be referred to as *Sharma distribution*. There is periodicity found in DNA sequence. This periodicity can be represented using Sharma distribution.

## D.7  STATISTICAL INFERENCE AND HYPOTHESIS TESTING

Hypothesis testing is the use of statistics to determine the probability that a given hypothesis is true. The usual process of hypothesis testing consists of four steps.

1. Formulate the null hypotheses $H_0$ (commonly, that the observations are the result of pure chance) and the alternative hypothesis $H_1$ (commonly, that the observations show a real effect combined with a component of chance variation).
2. Identify a test statistic that can be used to assess the truth of the null hypothesis.
3. Compute the $P$-value, which is the probability that a test statistic at least as significant as the one observed would be obtained, assuming that the null hypothesis were true. The smaller the $P$-value, the stronger the evidence against the null hypothesis.
4. Compare the $P$-value to an acceptable significance value $\alpha$ (sometimes called an *alpha value*). If $p \leq \alpha$, then the observed effect is statistically significant, the null hypothesis is ruled out, and the alternative hypothesis is valid.

Type I error is an error in a statistical test that occurs when a true hypothesis is rejected (a false negative in terms of the null hypothesis). Type II error is an error in a statistical test that occurs when a false hypothesis is accepted (a false positive in terms of the null of hypothesis).

# Subject Index